VICTORIAN NIGHTSHADES

Victorian Literature and Culture Series

HERBERT F. TUCKER, EDITOR

WILLIAM R. MCKELVY, JILL RAPPOPORT, AND
ANDREW M. STAUFFER, ASSOCIATE EDITORS

Victorian Nightshades

How the Solanaceae Shaped the Modern World

Elizabeth A. Campbell

UNIVERSITY OF VIRGINIA PRESS
CHARLOTTESVILLE AND LONDON

The University of Virginia Press is situated on the traditional lands of the Monacan Nation, and the Commonwealth of Virginia was and is home to many other Indigenous people. We pay our respect to all of them, past and present. We also honor the enslaved African and African American people who built the University of Virginia, and we recognize their descendants. We commit to fostering voices from these communities through our publications and to deepening our collective understanding of their histories and contributions.

University of Virginia Press
© 2025 by the Rector and Visitors of the University of Virginia
All rights reserved
Printed in the United States of America on acid-free paper

First published 2025

1 3 5 7 9 8 6 4 2

LIBRARY OF CONGRESS CATALOGING-IN-PUBLICATION DATA
Names: Campbell, Elizabeth A., author.
Title: Victorian nightshades : how the solanaceae shaped the modern world / Elizabeth A. Campbell.
Description: Charlottesville : University of Virginia Press, [2025] | Series: Victorian literature and culture series | Includes bibliographical references and index.
Identifiers: LCCN 2024026118 (print) | LCCN 2024026119 (ebook) | ISBN 9780813952536 (hardcover) | ISBN 9780813952543 (paperback) | ISBN 9780813952550 (ebook)
Subjects: LCSH: Solanaceae—Social aspects—History—19th century. | Solanaceae—Development—History—19th century.
Classification: LCC QK495.S7 C36 2025 (print) | LCC QK495.S7 (ebook) | DDC 583/.959309034—dc23/eng/20240716
LC record available at https://lccn.loc.gov/2024026118
LC ebook record available at https://lccn.loc.gov/2024026119

Cover art: Atropa Belladonna from *Medicinal Plants of the Russian,* 1892, scanned by Ivan Burmistrov. (iStock.com/ivan-96)
Cover design: Cecilia Sorochin

For Ted

CONTENTS

	List of Illustrations	xi
	Acknowledgments	ix
1.	A Family Plot	1
2.	Bittersweet: The Climbing Nightshade	18
3.	Dulcamara: Affairs and Elixirs of Love	42
4.	Belladonna: The Deadly Nightshade	73
5.	Victoria's Secrets: Sex, Drugs, and Belladonna	106
6.	The Triumph of the Potato	148
7.	Sublime Tobacco: Now Let Us Praise the Deadliest Nightshade	193
8.	Back to the Garden: Petunias, Peppers, Eggplants, and Tomatoes	244
	Notes	283
	Bibliography	313
	Index	335

ILLUSTRATIONS

1. Petunia and bittersweet, John Lindley, *The Vegetable Kingdom* 2
2. *Solanum dulcamara*, C. A. Johns, *Flowers of the Field* 22
3. Woody nightshade, Rebecca Hey, *The Moral of Flowers* 28
4. Illustration of flower writing, Frederic Shoberl, *The Language of Flowers* 45
5. Black nightshade, John Lindley, *Ladies' Botany* 58
6. Nightshade tribe, Anne Pratt, *Flowering Plants, Grasses, Sedges, and Ferns of Great Britain* 59
7. Sigr. Lablache as Dr. Dulcamara in *L'elisir d'amore*, British Museum 68
8. "Dr. Dulcamara in Dublin," John Tenniel, *Punch* 70
9. "Dr. Dulcamara Up to Date; or Wanted, a Quack-Quelcher," Linley Sambourne, *Punch* 71
10. *Atropa belladonna*, James Sowerby, from John T. Boswell Syme, *English Botany* 74
11. "Atropa Belladonna, or Deadly Nightshade," *Pharmaceutical Journal* 80
12. *Atropa belladonna*, John Lindley, *Medical and Oeconomical Botany* 81
13. Belladonna, Walter Crane, *A Floral Fantasy in an Old English Garden* 123
14. Walter Crane, *A Flower Wedding* 124
15. "The Song of the Passée Belle," Linley Sambourne, *Punch* 127
16. Frederick Sandys, *Medea* 145
17. *Solanum tuberosum*, *Transactions of the Horticultural Society of London* 152
18. "Darwin Potato," *Gardeners' Magazine* 160
19. "The Real Potato Blight of Ireland," William Newman, *Punch* 172
20. "A Sketch of the Great Agi-Tater," William Heath 173
21. "'Save Me from My Friends!,'" John Tenniel, *Punch* 174
22. "The Irish Cinderella and Her Haughty Sisters, Britannia and Caledonia," John Leech, *Punch* 175

ILLUSTRATIONS

23. "Consolation for the Million.—The Loaf and the Potato," John Leech, *Punch* — 178
24. "Grand Vegetable Banquet to the Potato on His Late Recovery," *Punch* — 179
25. "The Baked Potato Man," H. G. Hine and W. G. Mason, from Henry Mayhew, *London Labour and the London Poor* — 189
26. "Alfred Crowquill," frontispiece, Andrew Steinmetz, *Tobacco: Its History, Cultivation, Manufacture, and Adulterations* — 207
27. Frontispiece, F. W. Fairholt, *Tobacco: Its History and Associations* — 210
28. Tobacconist's shop sign, F. W. Fairholt, *Tobacco: Its History and Associations* — 213
29. Duke of Wellington pipe, F. W. Fairholt, *Tobacco: Its History and Associations* — 215
30. "Packing Cigarettes," John Wallace, from Joseph Hatton, "A Day in a Tobacco Factory" — 227
31. "Making Cigars," John Wallace, from Joseph Hatton, "A Day in a Tobacco Factory" — 228
32. "Stripping the Leaves," John Wallace, from Joseph Hatton, "A Day in a Tobacco Factory" — 229
33. "Woman's Emancipation," John Tenniel, *Punch* — 235
34. "Bloomerism—an American Custom," John Leech, *Punch* — 236
35. "Something More Apropos of Bloomerism," John Leech, *Punch* — 237
36. "Punch's Fancy Portraits.—No. 45. Ouida," Linley Sambourne, *Punch* — 238
37. Lady in smoke, M. B. Prendergast, from J. M. Barrie, *My Lady Nicotine* — 240
38. John Wallace, title page, *The Smoker's Garland* — 241
39. John Wallace, frontispiece, *The Smoker's Garland* — 242
40. *Petunia nyctaginiflora violacea*, Joseph Paxton, *Paxton's Magazine of Botany* — 248
41. *Capsicum annum*, George Spratt, *Flora Medica* — 260
42. Sutton's Royal Dwarf Cluster Tomato, *Gardeners' Chronicle* — 271
43. Nisbet's Victoria Tomato, *Gardeners' Chronicle* — 272
44. Advertisement, "Hothouses for the Million," *Gardeners' Chronicle* — 275
45. "Mr. and Mrs. Vinegar at Home," Arthur Rackham, frontispiece, Flora Annie Steel, *English Fairy Tales* — 281

ACKNOWLEDGMENTS

This book would not have been written without the help of the many people over the past seventeen years who encouraged me to turn what began as a sport (in both the recreational and botanical senses of the term) into a full-blown scholarly project. A Fellowship from the Center for the Humanities at Oregon State University provided funding for research, while the personal interest taken by David Robinson and Wendy Madar gave that support even greater impetus. Tara Williams, my friend-in-Chaucer and teaching partner in an honors course, "Magic and Witchcraft in Medieval and Victorian Literature," not only entertained my ideas in class and out, but also spared time from her own research to guide and direct me in all things technical. The many other colleagues at Oregon State who have kindly offered their assistance in a variety of ways include Kerry Ahearn, Rebecca Olson, Bob Burton, Vicki Tolar-Burton, Paul Farber, Mary Jo and Bob Nye, and Lisa Sarasohn. Dear friends Marjorie Sandor, Creighton Lindsay, Deborah Lindsay, Heidi Fischer, Lisa Norris, Anita Sullivan, Kathy Brisker, Chloe Hughes, John and Ellie Larison, Jim Babb, Nick Lyons, Jean Leeson, Greg and Cindy Leeson, Paul Hextell, and George Hopper have indulged me over the years by listening graciously to nightshade lore and making contributions of their own. I am also most grateful to the external reviewers of the manuscript at University of Virginia Press for their valuable suggestions, as well as the savvy comments of readers Jill Rappoport and Bill McKelvy. And without the editorial expertise of Eric Brandt, Angie Hogan, Fernando Campos, Ellen Satrom, and Colleen Romick Clark, this project would never have come to fruition.

I owe a special debt of gratitude to Tracy Daugherty and Henry Hughes, who read the manuscript at various stages of its development and whose comments and encouragement came just when I needed them; their help was invaluable in more ways than they know. Finally, my deepest thanks go to my husband, Ted Leeson, who read, advised, cheered, consoled, and patiently endured without complaint through the years of my life spent under the nightshades' spell.

VICTORIAN NIGHTSHADES

1

A Family Plot

> The nightshades are, in fact, primroses with a curse upon them.
> —John Ruskin, *The Queen of the Air* (1869)

The Vegetable Kingdom

In 1846 the eminent botanist and horticulturist John Lindley published his magnus opus, *The Vegetable Kingdom, or the Structure, Classification, and Uses of Plants*, in his latest effort to modernize botanical science for a British audience. As a proponent of the so-called natural system that had originated on the Continent in the previous century to improve on Carl Linnaeus's "artificial" method of grouping plant species according to the number of their reproductive organs, Lindley aimed to systemize the entire vegetable kingdom by considering every part of the plant, inside and outside, from seed and sap to stem, leaf, and flower. He had already published several versions of his natural arrangement, but the task had become even more monumental by the time he was writing at mid-century, for he was faced with organizing some 82,000 recognized vegetable species, he calculated, the number having increased by 52,000 in just over fifty years.[1]

One species that had arrived in England from South America within the past fifteen years but had already become a beloved garden ornamental was a petunia, which Lindley chose to represent the Solanaceae or nightshade "order," the term he preferred to the more commonly used "family" for the botanical rank above the level of genus (see fig. 1).[2] The *Petunia violacea*, which had been named by Lindley himself, was the last of England's most common Solanaceae species to enter the country; and by giving the petunia the most prominent place in the line drawings by depicting the whole plant as seen above ground, he proclaims its status as a benign and familiar member of a family that had long been anathematized because of the danger posed by

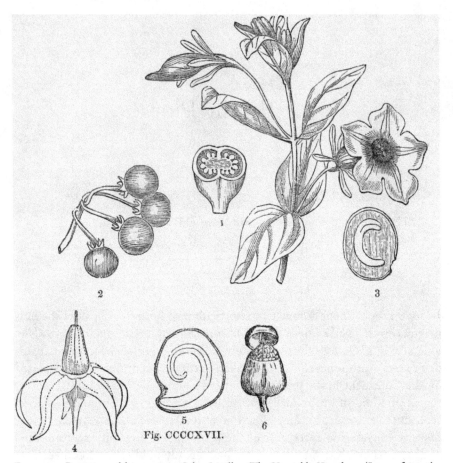

FIGURE 1. Petunia and bittersweet, John Lindley, *The Vegetable Kingdom*. (Image from the Biodiversity Heritage Library; contributed by New York Botanical Garden, LuEsther T. Mertz Library)

some of its poisonous relatives. In fact, the charge of "deadly" had been relentlessly leveled against the flower occupying the lower left corner of the figure, the *Solanum dulcamara*, known as bittersweet or woody nightshade, which was (and is) England's oldest native Solanaceae species, one common in the wild, just mildly poisonous, and the only nightshade actually considered to be truly indigenous to the country. Whatever reasons Lindley had for choosing the flowers of these two plants to illustrate the Solanaceae, the figure suggests the beginning and end of the nightshades' Victorian story, one marked by a radical transformation in the family's botanical reputation, which progressed from suspicion to widespread appreciation by the end of the era.

A FAMILY PLOT

The Solanaceae were among the many plant families whose recognized membership had exploded by the mid-nineteenth century. Lindley counted 60 nightshade genera containing 900 species—not a huge number in comparison, for example, to the 176 genera and 1,814 species of the Scrophulariaceae, or figwort family, to which all botanists agreed the nightshades were closely allied.[3] But there could be no comparison to the collective impact a wider, world view of the nightshades was having on Victorian life and culture. Because there were only five Solanaceae species naturalized in England, all well-known for centuries to be poisonous to some degree, the family had long suffered from an outsized reputation for evil that was still in currency, even while several non-native species had become absolutely vital to England's economy, diet, and the comfort of its citizens. It was as if those homegrown nightshades were pitted against their later-arriving kin, making a reevaluation of the Solanaceae character clearly overdue. This is the drama that forms the plot of this book: how the nightshade "curse" represented in the following chapters by two Solanaceae species native to England, bittersweet and belladonna (deadly nightshade), cast its shadow over the family even as important New World relatives—most dramatically, the potato and tobacco, but also the tomato, pepper, eggplant, and petunia— were succeeding in not only redefining the family character but also revolutionizing Victorian culture.[4] Thanks to advances in medical knowledge and horticulture, evolutionary theory, and crucially, to the impetus of human desire, the nightshade family, historically plagued by fear and hostility in the British imagination, would rise to cultural favor and predominance by the turn of the twentieth century, becoming essential agents in shaping the modern world.[5]

By the time the Great Exhibition of the Works of Industry of All Nations opened in the summer of 1851, the vegetable kingdom had become a national preoccupation, one assisted by the huge influx of new species whose cultivation in climate-controlled environments was made possible by innovations in the construction of glass and its greater affordability. As Jim Endersby observes in his brilliant *Orchid, A Cultural History*, information from the 1851 census revealed that England had become a predominantly urban non-churchgoing nation whose citizens were likely "spending their Sundays in the garden or greenhouse," a pastime that reflected the mid-century obsession with growing exotics;[6] and there could be no better evidence of this development than the magnificent edifice that housed the Exhibition. The mammoth, blazing, sheet-glass and cast-iron prefab structure erected in Hyde Park in less than a year that came to be known as the Crystal Palace gave dazzling proof that British technology and horticulture went hand in hand. The story of the

Palace's creation has become legendary: the brainchild of Lindley's friend and frequent collaborator Joseph Paxton, it was inspired by Paxton's experience, as head gardener for the Duke of Devonshire, in designing greenhouses for the cultivation of tropical plants, most famously the gigantic *Victoria regia* water lily, discovered in the Amazon Basin in 1837 and named (by Lindley) to honor the Queen's accession to the throne that year. Just as the water lily—whose pads could measure over five feet across—went from seed to miraculous bloom under Paxton's care within the Duke's enormous conservatory at Chatsworth, with its large, thin plates of glass ingeniously fitted to allow for maximum light-gathering capability, so the similarly designed Crystal Palace affirmed England's own flowering as a botanical wonderland and horticultural hub, a nation trafficking in and overflowing with rare and exotic floral treasures from every part of the globe.[7]

The fact that the Crystal Palace was in actuality a colossal greenhouse is symbolic of Victorian England's relation to the vegetable kingdom on a local and global scale, literally and metaphorically. The world's great engineers aspired also to be its master gardeners, and England was their greenhouse, site of a victory over the natural world of enormous magnitude; ironically, however, the victory was becoming necessary because of the very technology that made it possible, as smoke from stoves and factories filled the cities, and mines and railroads tore up the countryside, causing large-scale pollution of the air, soil, and water. Like its miniature version the Wardian Case, the small glass container designed to protect plants from salty air on sea voyages or ferns from sooty air in affluent English homes, a greenhouse could shelter and nurture nature. But it also symbolized England's effort to triumph over wildness or nature in the raw—to control and contain it, indeed, to bring nature indoors, as the Crystal Palace had done with several of Hyde Park's stately elms and an abundance of other plants of all kinds, from tropical palms to orchids, which helped to create the opulent ambience of the imposing 100-foot-high central transept and the adjoining exhibit halls. The vegetable kingdom, so full of riches everywhere, was assessed in the same utilitarian terms as the other of the United Kingdom's colonies and dependencies—the terms of conquest, appropriation, and possession—and consequently received the same treatment: at home and abroad, serviceable plants were being protected and transplanted while the intractable ones were chopped down and destroyed. The Victorians were horticulturists par excellence and this was the Age of the Garden, a culture in which even native plants had to earn the right to their existence. This was the challenge the Victorian nightshades faced and ultimately overcame.

The Whore, the Witch, and the Housewife

Edward Forbes's essay "On the Vegetable World as Contributing to the Exhibition," written for the very popular and lavishly illustrated catalogue of the Exhibition published by the *Art-Journal* in 1851, enthusiastically voices the prevailing economic sensibility symbolized by the Crystal Palace—one that viewed plants as items of commerce to be valued for their "beauty and utility" above all other considerations. When Forbes confidently proclaims, "It is an instinct of man's nature to subdue the vegetable kingdom to his service," he expresses a belief about proper human-plant relations at least as old as the Bible; but "subdue" has a peculiarly Victorian ring to it, hinting at the force being mounted against rebellious species as well as the accelerated pace at which vegetable servitude was taking place.[8] Yet Forbes, an eminent naturalist, friend of Charles Darwin, and at the time a professor of botany at King's College, London, wanted also to impress upon his readers that while studying the commercial applications for plants was "part of the science of the Botanist," the primary goal of his discipline was to understand plants on their own terms,[9] a task approached by examining their relations to each other. His essay thus combines an overview of the vegetable kingdom's vast and varied commercial products with an extensive lesson in botany that highlights the role "family" plays in the discipline, including how kinship ties often extend across continents and turn up surprising connections. What he has to say about all plants provides interesting insight into the scope of knowledge about the vegetable kingdom at mid-century, but what he says about the nightshades also hints at their fascinating plot.

Forbes's lesson begins by explaining that thanks to Linnaeus's development of a universal system of binomial nomenclature, every known plant could be precisely identified by a scientific name "composed of two Latin words" that denote its place in the vegetable kingdom: the first indicates "its genus, or relationship with plants very near it in structure and aspect"; and the second, its species, "expressive of some peculiar feature of its own" within its genus.[10] Praising the system as enormously successful if not in eliminating, certainly in reducing, the confusion of plant identities across languages, eras, and cultures, Forbes moves to larger categories of botanical classification, devoting the major section of the essay to describing how the myriad products—the flowers, furniture, fibers, medicine, and food represented in the Exhibition by over fifty families of flowering plants—manifest kinship ties and help shape their respective family personalities. From the structurally simpler grasses and palms to the more complex tribes like the rose and pulse (i.e., legume), member species are enumerated and commended for their usefulness and service—until Forbes arrives at the two closely related families known to include species with unruly dispositions, an unwillingness to serve humanity without posing a high degree of risk: "The foxgloves and the figworts, like doctors, ready to kill or cure, . . . and not far off the nightshades and tobacco-plants, with their associates the capsicums and love-apples [tomatoes]; a strange mixture in one family of man's deadly enemies, with several of his valued friends."[11] At the time Forbes was writing, foxgloves were part of the Scrophulariaceae, the figwort family; the rest of the genera he mentions are all nightshades or Solanaceae, almost predictably the most maligned family in the essay. In fact, Forbes's entire comment about the nightshades' Jekyll-and-Hyde personality, its "strange mixture" of enemies and friends, reflects botanical Britain's old antipathy to the family even while registering the Victorians' dawning awareness of the positive impact nightshades were having on every aspect of personal, social, economic, and political life. Although his plea at the conclusion of the essay that every plant be appreciated "for the plant's own sake" articulates the significant change in attitude taking shape among Victorian naturalists at mid-century,[12] his assessment of the nightshades captures both the affection and the fear the Solanaceae inspired at this defining moment in the family's history.

Forbes rightly claims that his discipline is the most advanced "of the natural history sciences";[13] but to call members of a plant family "man's deadly enemies" would seem to be an unusually anthropocentric and narrowly condemnatory response coming from a botanist who represented the enlightened scientific vision of the day. Nevertheless, it accurately conveys a prejudice against the nightshades going back at least as far as the Renaissance and tied to an age-old habit of attributing moral agency to nonhuman life. The habit arose out of a deeply rooted religious and ethical problem concerning the nature of evil and, more specifically, the perception of evil in nature, one famously addressed just a year earlier by a grief-stricken Tennyson in *In Memoriam* when he fears that "Nature, red in tooth and claw," sets the ruthless law of the universe rather than a loving God.[14] Like Tennyson, Forbes ultimately rationalizes this concern through an appeal to the highest of utilitarian principles, that of providential religion: the belief that all living beings fulfilled "the divine foresight" of the "Creator," and "in accordance with His all-wise designs."[15] But Forbes's conclusion, like Tennyson's celebrated lament about Nature, doesn't soften the impact of his censure: some nightshades were still considered to be "deadly enemies," no doubt in reference to the well-known virulent properties of several Old World species.

Three nightshades considered to be native to Europe—mandrake, belladonna, and henbane—were among the best-known vegetable poisons, ranking with the opium poppy, aconite, and hemlock for arguably the top six most dangerous species of the Old World. A fourth toxic nightshade, *Datura stramonium* or thornapple, was also known in the Old World, but is so cosmopolitan that its place of origin is uncertain. Henbane, belladonna, and, to a lesser extent, thornapple had been growing wild in England for a long time, the first two evidently migrating from southern Europe, Africa, and Asia, and brought in, probably first by the Romans, to be cultivated for medicines. All these nightshades are potent tropane alkaloids with extreme sedating and hallucinating powers that have been used for medical, religious, magical, and/or recreational purposes since antiquity; and not surprisingly, all figure prominently in some of the more lurid tales of European folklore.

For example, mandrake, whose famous forked root, which looks like a man or woman and allegedly screams when pulled out of the ground, either killing or driving to madness anyone within earshot, is probably the subject of as many legends as any known plant. The root's homuncular shape, according to the so-called Doctrine of Signatures, the ancient theory that plants bear signs or "signatures" for God- or nature-given uses in their form or appearance, is no doubt responsible for initiating the character of the superstitions that surround the species. One macabre bit of lore is that it grows under gallows, springing up from the semen of hanged men.[16] Mention of the plant in Genesis as desired by both Leah and Rachel, Jacob's wives and rivals for his affections, suggests that mandrake was used as an aphrodisiac or fertility drug. In any case, the root served as a charm for all sorts of divinatory purposes (the secret to obtaining it is to let a dog do the pulling with a rope tied to its tail; the unfortunate dog of course dies as thanks for its service). Mandrake, however, was extremely rare as a plant in England, but vivid in lore and the imaginations of poets like Shakespeare; and con artists sometimes sold the carved roots of bryony as fake mandrake charms. *Datura* and henbane had ancient roles as hallucinogens in magical/religious rituals. Writing about important religious plants of the classical world, Carl Ruck says that both were (like the poppy and pomegranate) sacred to the dual goddess Demeter/Persephone. In fact, henbane's generic name, *Hyoscyamus,* comes from the Greek for "hog's bean," alluding to the goddess's sacred animal the sow, "a carnivorous beast that responds to the male scent of humans." Henbane seeds were reputedly burned for their intoxicating and oracular effects at Delphos; and in a not unfamiliar move from the ceremonial to the recreational, Ruck reports that by the classical period henbane had become "an abused drug by the younger generation." The Maenads' drug of choice was belladonna, which whipped these revelers into a man-destroying frenzy when mixed with Bacchanalian wine.[17] All these nightshades had long associations with witchcraft and the Devil: according to Maud Grieve's *A Modern Herbal*, some of belladonna's more colorful names are Devil's Cherries, Naughty Man's Cherries, and Devil's Herb.[18] By the Renaissance, it was well documented that ointments made of nightshades, along with other ingredients (some, like the fat of babes, one hopes were more or less spurious), were responsible for witches' hallucinatory flights to the Sabbat.[19]

As the above examples reveal, the ancient world had an extensive knowledge of the most poisonous Old World Solanaceae species, but prized them for their psychoactive properties. The plants were also essential to the materia medica, or treatises on plant remedies, detailed in the many herbals that appeared on the Continent and England throughout European history. But it seems not to have been until John Gerard published *The Herbal, or General History of Plants* in 1597 (revised in 1633 by Thomas Johnson) that his attack on belladonna set in motion what would later become a more sweeping condemnation of the rest of the family.

Gerard's general view of the plants he considered to be nightshades was representative for his time: he knew several species whose similarities in their flowers and other features suggested they were somehow closely related, and he valued them for their properties, as confirmed by his remarks of introduction: "There be divers Nightshades, wherof some are of the garden; and some that love the fields, and yet every of them found wilde; wherof some cause sleepinesse even unto death; others cause sleepinesse, and yet Physicall; and others very profitable unto the health of man, as shall be declared in their severall virtues."[20] In other words, most of these plants had proven themselves to be very useful medicines, especially for their soporific powers (which could account for the name "nightshade"). But the fact that "some cause sleepinesse even unto death" is a warning that those nightshades should not be administered therapeutically except by the most skillful practitioners. This certainly applies to what he calls *Solanum lethale*—commonly known in England at the time as dwale, sleeping, or deadly nightshade—because it is so "furious and deadly." Gerard notes that the "Venetians and Italians call it *Bella dona*," but makes no mention of the usual story that Italian ladies took advantage of its mydriatic properties—its ability to dilate the pupils—to make themselves more sexually attractive. Instead, he emphasizes the plant's narcotic qualities: ingesting it can induce a dead sleep, and even pressing the leaves against the forehead releases its sedative power (hence the name "dwale," which is an Old Norse word meaning "trance" or "sleep-potion," and the epithet "sleeping"). Nevertheless, cultivating the plant for medicinal purposes was not worth the risk, not only because of its extreme toxicity, but also because its berries looked so temptingly tasty—as had been recently demonstrated by the

A FAMILY PLOT 9

fatality of two boys on the Isle of Ely. Thus, in an imperative declaration unprecedented in the *Herbal*, Gerard sternly instructs his readers: "Banish therefore these pernicious plants out of your gardens, and all places neere to your houses, where children or women with child do resort, which do oftentimes long and lust after things most vile and filthie; and much more after a berry of a bright shining blacke colour, and of such great beautie, as it were able to allure any such to eate thereof."[21] Besides the severe tone, what immediately strikes one about this remark is the implication that the plant is to blame for the danger it poses. Because the unwary and unwarned cannot help but succumb to the plant's obvious, though fatal, charms, it must be that belladonna wills their fate, so that death by nightshade is no accident; it is a crime against humanity. The plant was inherently evil, deliberately malicious; it must be destroyed.

Thanks to Gerard, belladonna became England's most notorious nightshade, and his injunction was responsible for initiating the efforts made during the next two and a half centuries toward the plant's wholesale annihilation. Consequently, as his book became a standard British reference work on botany in the seventeenth century—and well beyond, in part owing to the Shakespearean eloquence of his pronouncements—belladonna's reputation spread while the plants themselves began to disappear. Owing to the program of eradication urged by Gerard, what had always been a rather uncommon plant in England became even rarer. Unfortunately, however, because the name "nightshade" was in common use for at least two other quite common species, bittersweet and black nightshade (also called "garden" nightshade for its habit of springing up unbidden in cultivated soil), there was great confusion about which species should be identified as "deadly," thereby extending belladonna's association with evil to its less noxious kin.

Ironically, the nightshades' bad reputation received new attention in the eighteenth century because of the very advancements in science celebrated in Forbes's essay. As Linnaean botany gained widespread currency because of the relative simplicity and clarity of its methods of classification, it also called attention to plant sexuality, which Linnaeus often discussed in vividly anthropomorphic terms: the calyx of a flower was the "marriage bed," and stamens and pistils were the bridegrooms and brides. The nightshades, for example, fell into the large category of *Pentandria monogynia*, plants whose flowers have five (male) stamens and one (female) pistil or stigma—that is, a matriarchal harem of five husbands for one wife. The science historian Ann Shteir explains that even though the Linnaean method of classification appeared risqué in some quarters, botanizing via the system became increasingly popular from the 1760s well into the nineteenth century, especially as part of the female curriculum of study; in fact, it was the first scientific discipline considered suitably ladylike enough for women.[22] Moreover, the growing female interest in "polite botany" dovetailed with a passion for all things related to plants in popular culture, such as floral art and the "language of flowers," which became fashionable during the first half of the nineteenth century.

Regrettably, Linnaeus's specific characterizations of the nightshades—and belladonna in particular—hinted at more violent and evil subjects than just sex, subjects that ladies were better to avoid. Linnaeus assigned the family the name Luridae, which summarizes rather magnificently the pervasively distrustful attitude the nightshades elicited up to about the middle of the nineteenth century, the time of the Exhibition. The root "lurid" is somewhat less pejorative in botany than it has become in popular usage since, originally meaning "sallow" or "wan" in Latin and referring to the color of some species' flowers, like the pale purple-brown blossoms of belladonna or the dull yellow blossoms of henbane. Even so, the new name dramatically reinforced and perpetuated the family's now-established reputation for evil. To emphasize their toxic nature, Linnaeus gave "suspect" as the Luridae's defining character.[23] Even more to the point, in honor of deadly nightshade's deadly ways, Linnaeus created the binomial *Atropa belladonna*, the genus alluding to the last and most fatal of the Greek Fates, Atropos, who cut the thread of human life. Coupled with "belladonna," Italian for "beautiful lady" for its use as a cosmetic, the plant came to represent in the many cultural venues that sentimentalized botany a seductive, but murderous, female—a woman of the worst possible repute. This latter association for "nightshade" was already in currency at least since Jacobean times: Beaumont and Fletcher's play *The Coxcomb*, dating from 1610, refers to one of the female characters as a "nightshade" to mean "a gentlewoman whore."[24]

By the nineteenth century, belladonna's identification as a fatal woman had taken on a life of its own. For example, John Lindley, who early in his career wrote the notes on species for the massive *Encyclopedia of Plants* that John Loudon compiled and published in 1829, reports that one legend explaining the name "belladonna" comes "from its quality of representing phantasms of beautiful women to the disturbed imagination."[25] In the late nineteenth century, the folklorist Hilderic Friend goes further, noting the dangerous femininity inherent in the species' Linnaean binomials: "One of the names of this plant, Fair Lady, refers to an ancient belief that the Nightshade is the form of a fatal enchantress or witch, called Atropa; while the common name Belladonna refers to the custom of continental ladies employing it as a cosmetic, or for the purpose of making their eyes sparkle."[26] One can easily imagine that eye drops made from this notorious plant would have been prized by courtesans, gold diggers, nobility-chasers, and common prostitutes—women for whom sex meant survival. And it is the case that the infamous fruit so maligned by Gerard—that pupil-sized shiny black berry set in a starry calyx—certainly resembles a large, dark, sensual eye whose message, as if derived from some perverse Devil's Doctrine of Signatures, is "come hither." As we shall see, these associations linking belladonna to whores and witches so captured the post-Linnaean imagination that the plant did indeed transmogrify into a living femme fatale in novels, art, and poetry through the Victorian era.

Such were the lurid associations surrounding the nightshade family in the nineteenth century, ones that led the philologist Walter Skeat to define "nightshade" as "a narcotic plant" and to speculate about the name's origin: "compounded of [the Anglo-Saxon words] niht, night, and scadu, shade; perhaps because thought to be evil, and loving the shade of night." However, he goes on to say, "But this may be 'popular' etymology."[27] In any case, it is clear that Forbes's calling some of the nightshades "man's deadly enemies" certainly refers primarily to belladonna's Old World reputation that caused Linnaeus to give the family such a bad name. Not surprisingly, belladonna had only a small presence at the Exhibition as a medicinal plant specimen and an extract exhibited by Spain.[28]

But Forbes's assessment of the nightshades as a "strange mixture in one family of man's deadly enemies, with several of his valued friends" reveals that there is much more to their Victorian story than can be told by the family's allegedly evil Old World species. During the Renaissance when Gerard was maligning belladonna, he was also recording the arrival of several New World Solanaceae species, especially potatoes and tobaccos, which were already launching an invasion of Britain that would have drastic effects on its habits and economy, ultimately leading to major social crises and change during the nineteenth century. Most disastrously, the recent Irish Famine of 1845–48, during which a million people died and another million and a half emigrated from Ireland, called attention to how tragically essential potatoes had become to the economy and diet of an entire nation. According to Redcliffe Salaman, whose exhaustive *The History and Social Influence of the Potato* is the classic work on the subject, potatoes quickly became a primary food source after their arrival in Ireland in the late sixteenth century, thanks to the ease with which they could be grown even in the otherwise crop-unfriendly climate and poor soil of its bogs and mountainsides. Salaman poetically describes the trajectory of the potato's Irish occupation: "Arrived as a foundling, adopted as a favoured child, we see it slowly, but inevitably, dominating the shape of economic life, first of the peasantry, then of the former landowning aristocracy, and later of the industrial workers."[29] Unfortunately, the monoculture of potatoes in Ireland, which led to a population explosion during the late eighteenth and early nineteenth centuries, precipitated the subsequent disaster when the blight caused by the fungus-like *Phytophthora infestans* wiped out what at first appeared to be a healthy crop for two years straight—and with it, a sizeable portion of the Irish population.

In England, potatoes were one of the botanical novelties described by Gerard, who is pictured holding a potato plant in the frontispiece of the *Herbal*. He also started the rumor that they were native to Virginia, and it was not until the nineteenth century that South America was generally known to be their continent of origin. In any case, potatoes became staples of the English diet more slowly than in Ireland and from the opposite social direction, first finding their way to the tables of the affluent before becoming a primary food source for the laboring classes around the middle of the eighteenth century.[30] By the time of the Exhibition, however, potatoes were, next to grains, the most valuable human food plant in England as well as an important source of starch, flour, and distilled spirits. It is surprising that Forbes makes no mention of them along with the capsicums (peppers) and love apples (tomatoes) that undoubtedly represent the "valued friends" among the nightshades he names. In the sentimental language of flowers, "potato" usually means "benevolence," a celebration of the plant's friendliness by way of its association with charity, the most cardinal of virtues.[31] And Victorian history had just demonstrated to the point of tragedy that the potato was a most valued friend, one whose health had become a primary concern of the British Empire. At the Exhibition, potatoes' vital role in the mid-nineteenth-century diet was verified by exhibits from the United Kingdom, Canada, and northern Europe, in which potatoes appeared in the form of preserved specimens, flour, and starch, and vicariously by way of the roasters, steamers, and warmers designed for their preparation and service at table. The German State of Hesse displayed an automated "Braunfels potato mill," which could mash two tons of potatoes in an hour for use in brandy distillation.[32] Here was the quintessentially domestic nightshade and bipolar alternative to belladonna: the New World housewife versus the Old World whore and witch.

What is more, potatoes came to the British table accompanied by two important New World relatives: the peppers and tomatoes mentioned by Forbes had been cultivated for food in South Asia and southern Europe, respectively, for at least a couple of centuries, and they were gaining popularity in England, especially as ingredients for condiments like pepper vinegars and tomato sauce. These comestible nightshades, along with eggplants from southern Europe, were increasingly being grown in Victorian kitchen gardens out of doors and in greenhouses out of season, making possible their eventual status as staples of the British diet and economy, like the potato. Recipes for all of them appear in that indispensable compendium of Victorian housekeeping and cookery, *Mrs. Beeton's Book of Household Management*. As in the kitchen, so in the Crystal Palace: peppers were especially welcome guests, arriving from all over the warmer regions of the globe. Cayenne, bell, bird, bonnet, and Guinea peppers from Barbados, Borneo, Trinidad, Portugal, British Guiana, Java, Tunis, and Van Diemen's Land appeared in the form of raw products, artificial replicas, and powdered spices (Guinea pepper powder was affectionately known in British Guiana as "grains of Paradise").[33] Clearly, the Exhibition was helping to enlarge, qualify, and rehabilitate the nightshade family's shady reputation, which had been primarily formed around a very few Old World plants.

Nevertheless, belladonna and potato represent the extremes of kinship being played out as evil and good family members from two hemispheres pitted against each other: the Old World enemies against New World friends. Appropriately, just as "potato" has become a less pejorative alternative to "nightshade" as the common family name, so its scientific name *Solanum tuberosum* announces the primacy of the genus responsible for the kinder, gentler designation "Solanaceae" given the family by the French botanist Antoine Laurent de Jussieu in the natural system of classification developed in the late eighteenth century and championed by Lindley. The etymology of "Solanaceae," taken from the large genus *Solanum* to which bittersweet and black nightshade also belong (as did belladonna in Gerard's time), is not certain, but there is good reason to assume it comes from the Latin verb *solor*, "to comfort." Just as the more notorious nightshades, carefully dosed, had extraordinarily effective sedative and pain-killing powers, so *Solanum dulcamara*, or woody nightshade, was known in folk medicine for its soothing, pain-relieving effects.[34] The berries of *Solanum nigrum*, black nightshade, were even reputed to be, on occasion, edible. Tomatoes are *Solanum lycopersicum* and eggplant or aubergine is *Solanum melongena*. In the form of mash, chips, and crisps, our old friend *Solanum tuberosum* is arguably still the quintessential British comfort food.

In short, by the time Forbes was writing in 1851, it was already clear that the nightshades' reputation among botanists had advanced far beyond its lurid phase. This greater acceptance and appreciation would accelerate at the end of the decade with the publication of Darwin's *The Origin of Species*, when biological kinship gained increasing attention, blurring the boundaries between "good" and "evil" families and calling into question those very categories. And just as New World Solanaceae species were becoming essential food sources and kitchen garden favorites, stunning ornamentals like petunias, solandras, daturas, brugmansias, and many attractive solanums, brought in for public gardens and by nurseries for sale, were being raised as bedding plants or in conservatories, most arriving from their center of origin, South America, or the warmer parts of the globe like Australia and Africa. The sheer number of so many new and welcome nightshade species would foster the less sentimental and moralistic, more informed and appreciative attitude about plants given lip service by Edward Forbes, ultimately leading to a view of the nightshades from a broader, global, and more dispassionate perspective.

THE WEED FROM HEAVEN AND HELL

Tobacco, the remaining New World nightshade named by Forbes, occupies a pivotal place in this drama as the family's most notorious introduction, quickly becoming a favorite indulgence upon its arrival in England despite evidence that this nightshade more than any other of its kin exhibited both extremes of the family's bipolar personality. As early as 1621, Robert Burton in *The Anatomy of Melancholy* famously proclaimed the good herb/bad herb split in one brief but dramatic paragraph: "Tobacco, divine, rare, superexcellent tobacco, which goes far beyond all the panaceas, potable gold, and philosophers' stones, a sovereign remedy to all diseases. A good vomit, I confess, a virtuous herb, if it be well qualified, opportunely taken, and medicinally used; but as it is commonly abused by most men, which take it as tinkers do ale, 'tis a plague, a mischief, a violent purger of goods, lands, health; hellish, devilish, and damned tobacco, the ruin and overthrow of body and soul."[35] Burton obviously saw and rationalized both sides of tobacco's polarities, but he is still able to convey the fact that whether divine or damned, this plant never fails to arouse a passionate response. And indeed, few plants could have altered Victorian culture and personal life more noticeably and tangibly.

Like potatoes, tobacco came to England from the New World during the Renaissance, and four types appear in Gerard's *Herbal*, including the two most important cultivated species *Nicotiana tabacum* and *Nicotiana rustica*. The generic name given by Linnaeus honors the French ambassador to the Portuguese court in the mid-sixteenth century, Jean Nicot, who was so impressed by stories of the first-named species' reputedly amazing therapeutic powers that he sent tobacco seeds and plants to the French court, where the queen mother, Catherine de Medici, is reported to have given them a hearty welcome. Gerard knew of Nicot and the plant name "*Nicotiana*," which was already in use on the Continent; but he calls the two species henbanes, indicating how similar both tobaccos are to that Old World nightshade genus in some morphological features (all their seed capsules look like teeth) and in its narcotic effects. Gerard says of *Hyoscyamus Peruvianus*, "It bringeth drowsiness, troubleth the sense, and maketh a man as it were drunke by taking of the fume only."[36] Despite what would seem to be these less than desirable qualities, Gerard goes on to list in detail the many medical "Vertues" of tobacco taken from its enthusiastic literary introduction to Europe, the Spanish physician Nicholas Monardes's *Joyfull Newes out of the Newe Founde Worlde*.[37] As the tobacco historian Jordan Goodman explains, praise for the plant's medicinal powers and endorsements from royalty and courtiers (including the celebrated first smoker in England, Sir Walter Raleigh) accelerated tobacco's acceptance in Spain, Portugal, France, Holland, and England. In the latter two countries, smoking for pleasure quickly became as important as strictly medical use or other methods of delivery, like snuffing and chewing. By the mid-seventeenth century, Goodman says, tobacco had become the first non-European "exotic" to be a mass-consumption commodity in England, "well before chocolate, coffee, tea and even sugar."[38] James I's famously condemnatory *Counterblaste to Tobacco*, first published in 1604, registers how quickly England's addiction took hold. In the next two hundred years, England's desire for tobacco, along with cotton and sugar, would lead to its establishment of an American colonial empire and its support of slavery—indications of just how precious and vital to English life and its economy this nightshade had become.

From that point on, with a few rises and falls in fashion, there were always tobacco users of all classes and both genders; but the nineteenth century saw a dramatic increase in consumption with the introduction of the cigar at the beginning of the century and the cigarette in the fifties—their arrival credited to soldiers returning from Continental wars, the Napoleonic and Crimean, respectively.[39] These new ways to smoke made tobacco more portable, popular, and certainly more public, extending it into new segments of the British population. As Matthew Hilton says in his impressive *Smoking in British Popular Culture, 1800–2000*, from the mid-nineteenth century to 1900, smoking tobacco among middle- and upper-class males went from being a still-marginal activity to "an essential social habit" and a form of connoisseurship: knowing one's tobacco, like knowing one's wine, became as important as the habit itself.[40] The female portion of the population took up smoking more slowly, but the New Woman of the late nineteenth century, who would not be denied the liberties enjoyed by the opposite sex, considered smoking to be, as it were, fair game.[41] Tobacco manufacture became increasingly big business with several important firms, like Wills of Bristol and Cope's of Liverpool, that led the way in modern advertising by marketing numerous brands intended to satisfy the taste of every portion of the British population.

Concomitantly, there was as an explosion of literature celebrating the "divine weed" that not only served to normalize the use of tobacco, but to rationalize the dangers of poisonous substances in general—like those found in its solanaceous kin. In fact, the pro-smoking press's appeal to science in defending the admittedly "virulent poison" nicotine worked to cast the whole nightshade family in a new, more positive light. F. W. Fairholt's purportedly "impartial" *Tobacco: Its History and Associations*, published in 1859 and quickly established as the authoritative text on the subject, assuaged his readers' fears about tobacco by pointing out that though deadly nightshade was a relative, so too was the potato, which contained a lethal poison as well.[42] The botanist, science popularizer, and admittedly biased tobaccophile Mordecai Cooke went further, defending the narcotics "thorn-apple and nightshade" for being Solanaceae like tobacco and potatoes, "so that if only from their family connections, independently of any other right, they have a claim on our attention and respect."[43]

Any reader of Victorian novels knows they are rife with tobacco, thus further popularizing its use, even though smokers fell on either side of the hero/villain divide. Female smokers began more frequently appearing in novels in the 1860s, their practice serving as a code for the sexually daring, dangerously seductive "fast" woman, the woman who had liberated herself from stereotypical feminine roles. If this association was a reminder that tobacco trafficked with the witches and whores of the family's dark side, it also revealed how such roles were being redefined, like smoking, as glamorous for a new age.

It isn't clear whether Forbes considers tobacco a friend or foe—despite the hype, public sentiment at mid-century was divided on the subject. Anti-tobacco societies formed in the forties and fifties, and what became known as the "Great Tobacco Controversy" played out in the pages of the *Lancet*, Britain's leading medical journal, in 1857.[44] Depending on whom one consulted, tobacco was either "one of the greatest boons with which man has been blessed," as W. A. Penn would declare at the end of the era, or as John Ruskin claimed, "the worst natural curse of modern civilization."[45] Nevertheless, its violent opponents notwithstanding, tobacco's huge and welcome presence at the Exhibition accurately predicted which side would prevail. The Crystal Palace proudly displayed tobacco products and all sorts of smoking paraphernalia. The British exhibit had an entire subsection devoted to tobacco, with samples of *Nicotiana* species imported from Java, New Orleans, Cuba, South America, Hamburg, Turkey, and Syria.[46] There were cheroots, cigars, and tobacco rolls from the East Indies, Ceylon, the Cape of Good Hope, Canada, Trinidad, and Van Diemen's Land. Related accessories from the United Kingdom and countries throughout the world included every sort of

A FAMILY PLOT

smoking and snuffing accessory imaginable, from tobacco cutters and knives; snuffboxes, tobacco boxes, and pipe boxes; silver pipe tops; meerschaum and clay pipes; pipe cases; tobacco and cigar stands; and amber mouthpieces, to an award-winning newfangled cigarette-making machine.[47] *The Illustrated London News* particularly praised the tobacco from Turkey and the pipes from Germany and Austria, proclaiming, "Tobacco is the vegetable type of the age in which we live."[48] Indeed, there was more than enough evidence to justify the title of a *Chambers's Journal* article about tobacco a few years later, calling it "the most popular plant in the world."[49] In short, by the end of the nineteenth century, tobacco smoke had permeated every corner of Victorian England. Despite its enemies, here was the nightshade that added a twist to the family plot and abetted its change of course: a New World species known and embraced for its poisonous nature—the very quality that had given all of its Old World kin such an evil reputation.

Victorian Nightshades is the story of how the Solanaceae fared as living plants, vegetable commodities, objects of artistic and scientific interest, and cultural ideas in England during the nineteenth century, which marks the turning point in the family's fortunes. Belladonna, condemned as evil and figured as a woman, is at the center of the plot and complicit in the bitter-sweet tale of woody nightshade, the only Solanaceae truly native to England, which suffered, like a Dickensian heroine, because of its family and its name. The potato and tobacco are the New World game changers that permanently altered British history by establishing themselves as essential to daily exis-tence, while petunias, peppers, and tomatoes completed the task of redefin-ing and redeeming the family character, one that wholly reflected the powers of consolation within even its deadliest species—the solace at the root of Solanaceae, its modern scientific name. As Edward Forbes understood, the nightshades' Victorian story is indeed a "strange mixture," but its dark chap-ters shed light on almost every aspect of nineteenth-century life.

2

Bittersweet

The Climbing Nightshade

> Like the autumn wind, when it unbinds
> The tangled locks of the nightshade's hair,
> Which is twined in the sultry summer air
> Round the walls of an outworn sepulchre.
>
> —Percy Bysshe Shelley, *Rosalind and Helen* (1818)

The Tangled Bank

IF YOU WERE to stroll down an English country lane during the fall at any time in the nineteenth century, you might likely notice in the hedgerows, draped among the hips and haws and blackberries, the scarlet berries of woody nightshade, *Solanum dulcamara*, or bittersweet. Woody nightshade berries are beautiful. With their shiny translucence and teardrop shape, they complement the loveliness of the other fruits, but surpass them all in their grace of arrangement: like the starry blue-purple flowers with cone-shaped yellow anthers that preceded them, the berries not so much hang in clusters or "cymes," but rather seem to pose like gymnasts or dancers, asserting their acrobatic skill and artful sense of independence. In his posthumously published botanical journal *Wild Fruits*, Henry David Thoreau lavishly praises this stunning effect with a discerning eye: "No berries that I am acquainted with are so agreeably arrayed, somewhat hexagonally, like a small wasp nest. The cymes are of singular yet regular form, not too crowded, but elegantly spaced, not stiff and flat, but in different stages above and around—finding ample room in the universe." A bit later he waxes even more poetically: "They hang more gracefully over the river's brim than any pendants in a lady's ear."[1] The Victorian novelist R. D. Blackmore, author of the still-popular *Lorna*

Doone, mentions in his 1872 romance *The Maid of Sker* that in Devonshire, bittersweet berries were sometimes called the "Lady's necklace," another indication of nightshade's ability to call to mind the adornments of a beautiful woman, and by extension, to summon an apparition of the woman herself.[2] Although this is the only way *Solanum dulcamara* resembles its relative *Atropa belladonna,* the shadowy, tragic species whose story haunts the whole family, bittersweet was the nightshade most victimized by its confusion with belladonna because of its prevalence, its beauty, and its climbing ways.

Thoreau was writing about botanical observations he made during the 1850s across the Atlantic from England, while on his walks in the environs of Concord, Massachusetts, where bittersweet had probably been introduced at least a couple of hundred years earlier, and which was now naturalized in the shady hedgerows and along the banks of the Concord and other New England rivers and streams. Yet *Solanum dulcamara* is a decidedly Old World plant and as the Victorians discovered, the one member of the nightshade family that could with confidence be considered aboriginal to England. According to the fossil records reported by the late-Victorian paleobotanist Clement Reid, it is one of the oldest known plants in northern Europe, its seeds having been found in the preglacial Cromer Forest-bed deposits in Suffolk and in interglacial deposits in Sussex and southern Sweden.[3] In geologic terms, this evidence indicates that bittersweet was at home and blooming in England at least 450,000 years ago, long before any human inhabitants were around. The great mid-Victorian authority on the geographical distribution of British plants, Hewett Cottrell Watson, also classified bittersweet as native and thoroughly British, with the widest distribution in the United Kingdom of any Solanaceae. In 1849 he estimated that *Solanum dulcamara* grew in a total of seventy-five counties scattered throughout England, Wales, Ireland, and, less frequently, Scotland. More specifically, its choice of local habitats — what Watson termed "situations"—fell into two of the fourteen categories that he devised: Watson classified bittersweet as a septal plant, or one that thrives in hedge banks and hedgerows; and paludal, a marshy-ground plant whose roots are in water or wet ground most or all of the year.[4] As Thoreau's records show, even an ocean away from its homeland, bittersweet naturalized in moist terrain and hedges, sites similar to those where it could be found wild in Britain. Thoreau notes that the berries "in the water at the bend in the river are peculiarly handsome," reliable evidence that this moisture-loving plant had gravitated to its ideal habitat and was thriving beautifully there as it did in England.[5]

Just about the time that Thoreau was recording his botanical observations, bittersweet was no doubt on Charles Darwin's mind as well, for it must certainly have occurred to him in 1859 as one of the plants typically growing in the now legendary "tangled bank" he visualized in his famous conclusion to *The Origin of Species*.[6] Stanley Edgar Hyman referred to Darwin's metaphor as the ideal image for displaying "all the rich complexity of life," choosing *The Tangled Bank* as the title for his masterful critique of Darwin, Marx, Frazer, and Freud "as imaginative writers," since it serves in *The Origin* as the perfect site for representing the struggle for existence and the evolutionary laws of nature.[7] In fact, *Solanum dulcamara* is the only British member of the Solanaceae family whose choice of habitats is the kind of bank that Darwin describes, the only one capable of climbing through the tangle of stems, branches, blossoms, and leaves of the other vegetable inhabitants. Darwin was always fascinated with what his son Francis called the "aliveness" of plants, their capacity for intelligent or sentient responses to their environment; all types of plant movement, which Darwin eventually termed "circumnutation" in *The Power of Movement in Plants* published in 1880, held special interest.[8] But because climbing plants were particularly successful in overcoming an otherwise rooted, stationary existence, they became a subject of his early botanical investigations, published as *The Movements and Habits of Climbing Plants* in 1865. In this work Darwin reports his experiments with many climbers in his home garden and greenhouse in Kent, and he seems disappointed when *Solanum dulcamara* proves to be "one of the feeblest and poorest of twiners," capable of winding around only "a thin and flexible support, such as the stem of a nettle." As Darwin describes his experiment, "I placed sticks round several plants, and vertically stretched strings close to others, and the strings alone were ascended by twining."[9] The tone of the comment may register Darwin's higher hopes for this noticeably sinuous plant, but it certainly is intended to establish *Solanum dulcamara*'s relative place as a climber in the evolutionary scale: Darwin surmises that it has "as yet only partially acquired the habit of twining," which is the first step toward more sophisticated ascent, represented by leaf-climbers (like clematis and another, non-native nightshade, *Solanum jasminoides*, among other plants) and tendril-bearers (like the common pea).[10] Nevertheless, Darwin shows that bittersweet is a highly adaptable plant whose power of movement allows it to conform easily to its surroundings. If isolated from other vegetation or support, it can grow as a shrub; but more often it climbs over other shrubs and vines in banks and hedgerows, searching for both light and shade as it advances. This movement is unusually serpentine, for as Darwin observes, "the stem twines indifferently to the right or left," rather than in the single direction that is more typical of climbing plants.[11] Such flexibility no doubt compensates for bittersweet's weak twining skill while probably accounting as well for that skill's underdevelopment—the plant manages quite satisfactorily to move without such a mechanism.

Darwin's work inspired interest later in the century concerning bittersweet's unusual climbing habits. *Solanum dulcamara*'s ambidexterity earns it a place in the sensationally entitled *Freaks and Marvels of Plant Life, or Curiosities of Vegetation*, which is essentially a retelling of Darwin's botanical research for a popular audience, published in 1882 by the Society for Promoting Christian Knowledge and written by his fan Mordecai Cubitt Cooke (most famous now for *The Seven Sisters of Sleep*, the Victorian classic on drugs published in 1860). *Freaks and Marvels* even includes a graceful black-and-white drawing of bittersweet showing both berries and flowers, recycled from another Society publication, Charles Alexander Johns's very popular *Flowers of the Field*, originally published in 1853 (see fig. 2). Unfortunately, Cooke slightly misquotes Darwin by calling bittersweet "a poor climber" when, in fact, bittersweet has been an adept, successful, and enduring scrambler through England's thickets and tangled banks quite literally for ages.[12]

Two of George Eliot's novels offer further, appreciative evidence that woody nightshade was a common and noteworthy botanical feature in England—in this case of the rural Midlands—throughout the nineteenth century. Eliot was well known for her realistic depictions of people and places, and in her first novel, *Adam Bede* (1859), she nostalgically evokes the towering, massive hedgerows of summer that divided the fields, bordered the lanes, and dominated the visual landscape around the fictional village of Hayslope at the century's turn: "For the hedgerows in those days shut out one's view, even on the better-managed farms; and this afternoon, the dog-roses were tossing out their pink wreathes, the nightshade was in its yellow and purple glory, the pale honeysuckle grew out of reach, peeping high up out of a holly bush, and over all, an ash or a sycamore every now and then threw its shadow across the path."[13] Four novels later in *Felix Holt* (1866), these Midland hedgerows appear again as they would have looked to travelers along the old stagecoach roads in 1831. This time Eliot dwells on their unkempt wildness: "Everywhere the bushy hedgerows wasted the land with their straggling beauty, shrouded the grassy borders of the pastures with catkined hazels, and tossed their long blackberry branches on the cornfields." Their disheveled appearance, however, in no way lessens their appeal: "It was worth the journey only to see those hedgerows, the liberal homes of unmarketable beauty," in which "the purple-blossomed ruby-berried nightshade" figures prominently among the familiar charms of English hedge vegetation, ranking with honeysuckle and roses, blackberries and convolvulus, and the staple of the hedgerows, hawthorn.[14]

FIGURE 2. *Solanum dulcamara*, C. A. Johns, *Flowers of the Field*. (Image from the Biodiversity Heritage Library; contributed by the University of California Libraries)

Yet despite its pedigree, its prevalence, and its beauty, woody nightshade's reputation suffered in Victorian England, largely owing to its family, its name, and a notorious case of mistaken identity—the habitual tendency by novice and expert alike to confuse it with deadly nightshade, *Atropa belladonna*; so that recognition and unqualified admiration for *Solanum dulcamara*, like that of Thoreau and Eliot, are rare in the nineteenth-century literary record. The Victorian botanist William A. Bromfield accurately stated bittersweet's plight when he referred to the plant in 1848 as "the rampant nightshade, gifted with fatal energy in popular imaginings, and one at least of an uncertain and treacherous race, if free itself from the stain of blood-guiltiness."[15] The most inexcusable and possibly the most far-reaching example of this confusion was that of George Bentham, the celebrated botanist whose *Handbook of the British Flora*, designed "for the use of beginners and amateurs," became a standard botanical reference work in England well into the twentieth century. In the *Handbook*'s first edition, published in 1858, Bentham lists "*Deadly Nightshade*" as a synonym for *Solanum dulcamara* (whereas *Atropa belladonna* is listed as "*Deadly Atropa*," "*Dwale*," or simply "*Belladonna*").[16] In a July 21 letter that year to his close friend Joseph Hooker, later the director of Kew Gardens, Darwin writes, "I have ordered Bentham, for . . . it will be very curious to see a Flora written by a man who knows nothing of British plants!"[17] Bentham had done his botanizing in France, which accounts for Darwin's comment, but does not excuse the error, which was corrected in subsequent editions of the *Handbook* (the later ones revised by Hooker). Even so, the original edition perpetuated a mistake that was already ubiquitous throughout the country.

Another egregious example of misidentification occurs in an early short story by the great socialist and polymath William Morris, written for the short-lived *Oxford and Cambridge Magazine*, which he founded with the painter Edward Burne-Jones in 1856 while they were both students at Oxford. Like many of Morris's later works, "The Story of an Unknown Church" has a medieval setting: the narrator is a dead stone mason who speaks from the grave about rebuilding a church for an ancient monastery in a forgotten country. The tale is full of Gothic decor and as lush with plants as one of Morris's later designs; but even though Morris was a close and passionate observer of plants, he is guilty of thoroughly conflating the two nightshade species when he describes one of the wild flowers that have crept into the overgrown churchyard garden and mingled with the roses on the trellises as "deadly nightshade, La bella donna, O! so beautiful; red berry, and purple, yellow-spiked flower, and deadly, cruel-looking, dark green leaf, all growing together in the glorious days of early autumn."[18] Nevertheless, even though he misnames the plant, the comment reveals Morris's appreciation for bittersweet's beauty. In *The Gardens of William Morris*, the authors Jill, Duchess of Hamilton, Penny Hart, and John Simmons identify the flowers appearing in the border of the first tapestry he completed in 1879, called "Acanthus and Vine," as bittersweet—another sign of his appreciation.[19] The flowers in question seem as if they could be attached to the tendriled grapevines that wind through the tapestry, visually connecting them to the clusters of fruit. One would be tempted to think, then, that Morris was intending metaphorical homage to John Keats, whom he considered "one of [his] masters," whose "Ode on Melancholy" famously refers to nightshade as the "ruby grape of Proserpine."[20]

Morris never again features identifiable Solanaceae in his textiles and wallpaper, which combined naturalistic representations of flowers with stylized vegetation imitating designs from the Middle Ages and Renaissance. Nor does he speak explicitly about either belladonna or woody nightshade again in his fiction or poetry. But ironically, "bitter-sweet," often hyphenated or expressed as two words, proves to be one of his favorite terms and formulations, no doubt because it conveys so accurately his private and artistic vision, which was nostalgic, romantic, and melancholy—in short, bittersweet.[21] It is a somber, wise, postlapsarian word whose meaning is tragically realized through hard experience. The witch Medea's knowledge that lovers must "snatch from bitter love the bitter sweet" explains why she is the most sympathetic character in Morris's first major poem, *The Life and Death of Jason*, published to rave reviews in 1867—even despite the fact that she murders her rival for her adulterous husband as well as her own and Jason's children.[22] As we shall see, Medea's association with nightshades was certainly in the air among Morris's Pre-Raphaelite associates at the time he was writing; but it doesn't seem likely that he was thinking about *Solanum dulcamara*, the plant he misidentified so many years earlier. However, Morris could easily have fixed upon the word as a favorite because it originates in the work of the medieval author dearest to him, Geoffrey Chaucer, who uses it in a manner suggestive of nightshade lore. According to the *OED* the earliest recorded usage of "bittersweet" appears in the fourteenth-century *Canterbury Tales*. Like Morris, Chaucer uses the word metaphorically, to mean something like a "fatal enchantment," as one of the *Tales'* translators into modern English, David Wright, proposes. The word appears in the Canon's Yeoman's Tale to describe how the crazed practitioners of alchemy are addicted to their mad pursuit: "Unto them it is a bitter-sweet."[23] Chaucer makes the word sound like a narcotic—a drug or candy, some sort of pretty poison—which brings us back to the plant itself.

The Dictates of Taste

The first recorded instance of "bittersweet" to describe *Solanum dulcamara* occurs in 1568 in the *New Herball*, compiled by William Turner—justly considered to be the "Father of English Botany." As Frank J. Anderson says in *An Illustrated History of The Herbals*, the science of English botany really begins with Turner: this physician, naturalist, and irascible Protestant preacher (he trained his dog "to snatch away bishops' caps at a given signal") was bent upon accurately identifying native plants with their counterparts in European sources rather than simply recycling errors, as had frequently been the case with earlier and even contemporary herbals.[24] Since their primary purpose was, after all, to serve as materia medica for recording the therapeutic uses or "virtues" of plants, accurate identification could literally be a life-or-death matter. In Turner's groundbreaking herbal, what is now, thanks to Linnaeus, officially known as *Solanum dulcamara* appears under the entry "Of the herb called Amara dulcis," and Turner credits himself with the English coinage that exactly translates the Latin, saying it "hath no English name that I know, but for lack of another name it may be called Bitter swete." Turner explains that both the Latin and his English coinage refer to the fact that "the bark of the stem, when tasted of, is first bitter and afterward sweet, and therefore in Dutch *Je lenger je lieber:* The longer the more lovely, that is, the more ye taste it the more sweet it is and the more lovely."[25]

The Dutch name's appreciative emphasis on the pleasure derived from tasting bittersweet twigs suggests that chewing on them must have been a popular activity in the Netherlands in the sixteenth century. We learn from British ethnobotanists David Allen and Gabrielle Hatfield in their comprehensive *Medicinal Plants in the Folk Tradition* that such was the case in one English county during the late Victorian era—perhaps evidence of a long-lived rural practice in a region where, as we shall see, bittersweet has a long and complicated history. They refer to the 1898 *Flora of Cumberland*, in which William Hodgson documents that around his home parish of Workington where *Solanum dulcamara* was abundant, "schoolboys sometimes keep a stock of twigs in their pockets, which they chew as their elders do tobacco."[26] The effect is mildly narcotic, an indication that the twigs contain the glycoalkaloid solanine—the poison, found in highly variable potency throughout the genus (the green sometimes found on potatoes can indicate its presence), that accounts for the first part of bittersweet's oxymoronic name.[27] The sweet taste that follows comes from the glucoside dulcamarin, which is, as the name suggests, peculiar to this *Solanum* species. First isolated in 1820 by a French apothecary named Desfosses experimenting with *Solanum nigrum* (black nightshade), solanine was found to be most highly concentrated in its berries; and in 1828 a Swiss chemist named Peschier demonstrated the same to be true for *Solanum dulcamara*, although to a lesser degree of toxicity than in the former species.[28] John Lindley's description in his 1838 *Flora Medica* is succinct and no doubt apt: "berries oval, scarlet, juicy, bitter and poisonous."[29]

What could be helpfully filled in, though, is that bittersweet berries are also nauseating: they are not only unpalatable to humans but to most animals, although some birds eat them with impunity and thus distribute the seeds encased inside. As the medical botanist Judith Sumner explains, the extremely bitter taste discourages grazing animals and helps to protect the berries against fungi; but vegetable poison as a defense mechanism, so characteristic of bittersweet and to a much greater degree its Old World kin—as well as many other plant families and species—was not fully understood and certainly not appreciated until the late Victorian era. Few of Thoreau's contemporaries at mid-century would agree with his further comment about woody nightshade berries, which offers an aesthetic solution to what was perceived in the Victorian worldview to be a moral problem: "Yet they are considered poisonous! Not to look at, surely. Is it not a reproach that so much that is beautiful is poisonous to us? But why should they not be poisonous to eat? Would it not be in *bad taste* to eat those berries which are ready to feed another sense?"[30]

John Ellor Taylor's provocatively titled *The Sagacity and Morality of Plants*, published in 1884, was among the first to attempt a reversal of the negative opinion about bittersweet in the terms that Thoreau precisely (if inadvertently) hit upon—the berries' bad taste. A popular journalist and lecturer on evolutionary theory, Taylor in fact anticipates twenty-first-century plant philosophers who theorize about plant intelligence and attempt to understand intentionality from a non-anthropocentric point of view.[31] Taylor assigns an honorable and even patriotic motive to plants harboring poisons by calling it "defense, not defiance"—an allusion to the motto of Britain's courageous Volunteer Rifle Corps, which was formed at mid-century as a necessary guard for the homeland during the Crimean War. Taylor argues that bittersweet and other attractive but nausea-producing berries like those of honeysuckle, bryony, and yew serve the greater cause of nature by ensuring that humans and other animals avoid them, thereby providing food for the birds.[32] By the late twentieth century Richard Mabey could say in *Flora Britannica* that bittersweet is one of the less poisonous Solanaceae and ingestion is rarely fatal even for children;[33] but during most of the nineteenth century, the nightshade family's notorious reputation cast its shadow over prevailing botanical opinion, thereby condemning bittersweet in the morally driven canons of middle-class taste. In response to the family's lurid reputation established in the mid-eighteenth century, bittersweet became the representative villain of the family in Gothic and Romantic literature, setting the tone for its reception in the first decades of the nineteenth century.

A stunning example of this censure occurs in Rebecca Hey's anonymously published *The Moral of Flowers* (1833), one of those sentimental flower books, often addressed to a female audience, that were so hugely fashionable among the British bourgeoisie of the gentler sex especially during the first half of the century. Nicolette Scourse, whose engaging *The Victorians and Their Flowers* offers marvelous insight into the scope, variety, and tone of these books, describes their appeal in terms that directly apply to Rebecca Hey: "The sentimental flower writers mirrored maidenly virtues and duties long since passed but they have the magnetism of a lost era. Through their delightful contrived verses and prose, one can still enter the flower-strewn world of the elegant gentlewoman of the nineteenth century."[34] As Scourse makes clear, Rebecca Hey's book gives unusually vivid access to this straitlaced, be-doilied world.[35] *The Moral of Flowers* features charming botanical illustrations by William Clark, a professional artist and engraver for the London Horticultural Society, alongside Hey's prose commentary, which is full of plant lore and a wealth of information from scientific sources, sprinkled with snippets of quoted verse and culminating with one of Hey's own poems. Hey explains her purpose is "to pursue such a train of reflection or draw such a moral from each flower that is introduced as its appearance, habits, or properties might be supposed to suggest,"[36] each floral moral, in turn, informed by Hey's devout Christianity. After suggesting in her preface many religious lessons to be learned from flowers, she adds the most important, no doubt alluding to flowers' ability to rise from the earth from bulb, root, or seed: "But it is when viewed as types of the resurrection that they most vividly affect the imagination and touch the heart."[37] *The Moral of Flowers* covers about fifty native and exotic species of flowers, trees, and shrubs with that "queen of flowers" the rose receiving the most individual attention; and given the book's pious emphasis, it is perhaps surprising that bittersweet appears at all. In fact, it is the only plant in Hey's book subjected to her poetic abuse, one whose moral is chillingly negative. Through the poem's Gothic imagery, bittersweet serves as the token plant to represent the villainous dark side of nature, the botanical serpent in Hey's sacred garden.

Appropriately, Hey opens her entry on "The Woody Night-Shade, or Bitter-sweet" with a comment that reiterates the dilemma faced by the Victorians when considering a beautiful but poisonous plant in moral terms: "Were it allowable for man to desire any thing in nature to be otherwise than it is, one might wish this poisonous plant were clothed in garb less attractive, and more indicative of its deleterious qualities; as the beauty of its blossoms and fruit, known to the peasantry by the name of poison-berries, often proves fatally tempting to children."[38] The accompanying colored plate by Clark certainly conveys bittersweet's beauty with a graceful rendering of both flowers and berries (see fig. 3); but the five-stanza poem that forms the bulk of the entry denounces the plant for its deceptive attractiveness, offering the formula "fair but treacherous" that becomes a byword for bittersweet. Without naming her source, Hey took the name "poison-berries" and her other botanical and scientific information from John Stephenson and James Morss Churchill's *Medical Botany*, published two years earlier—as a long quotation describing the disagreeable smell and taste of woody nightshade makes clear. Hey's comment about the danger the berries pose to children was no doubt inspired by the graphic account that Stephenson and Churchill include from

FIGURE 3. Woody nightshade, Rebecca Hey, *The Moral of Flowers*. (Image from the Biodiversity Heritage Library; contributed by the Chicago Botanic Garden, Lenhardt Library)

one Mr. Wheeler, "Surgeon of Bayswater," who describes four occasions on which he was called in to treat a total of five children poisoned by bittersweet, two of whom died. Considering the regimen of leeches, calomel, castor oil, laudanum, and beef broth that he reports having prescribed in the first case, it's a wonder that there were no more than two fatalities.[39]

As in all the entries, Hey develops her "moral" in verse addressing the flower directly; and in the case of bittersweet, her message to the plant comes as a set of commands designed to mitigate its sinful nature. She begins by ordering bittersweet to find a setting appropriate to its degenerate character, one that is recognizably Gothic in its trappings:

BITTERSWEET

Away, away with thy tempting bloom—
Go seek thee a fitting bower—
In the church-yard drear by the haunted tomb,
Or the falling shrine make thy cheerless home,
Thou fair but treacherous flower;
Or where mandrakes grow by the wizard's cave,
And the adder lurks, let thy garlands wave.[40]

One gets the impression when reading this stanza as it appears on the page that Hey is proudest of "mandrakes" as an atmospheric touch, since what follows is a long footnote recounting lore from "the dark ages"—a time of "ignorance and superstition"—about uprooting mandrakes, recorded here with Shakespearean references and detailed directions for its removal.[41] Although she doesn't mention that mandrake is a Solanaceae relative, she makes the two plants' literary kinship clear: just as the imagery of the church graveyard with ruined tombstones reveals that Hey is recalling references to nightshades that appear in Gothic and Romantic literature (to which I will return), a "fair but treacherous flower" at home with the mandrakes, wizard, and adder can of course be known by the dangerous company it keeps. Thus the opening of the next stanza explicitly states, "For alas! alas! There's a deadly spell / Concealed thy leaves among."[42] The awkwardness of the syntax notwithstanding, "deadly spell" captures the idea of a noxious narcotic and magical potion simultaneously, and "concealed thy leaves among" hints at the plant's sinister resemblance to the lurking adder in the previous stanza.

The next two stanzas use bittersweet as a cautionary tale to warn two types of people susceptible to self-inflicted poisoning—belles and revelers. In stanza 3 it is a beauty who needs to be admonished for her vanity by way of a "graceful wreath" around her mirror, "to show her beneath / What is lovely and bright lurk the seeds of death; / And despite bland flattery's tongue, / She might learn this lesson for after-hour, / That beauty alone is a worthless dower."[43] It seems accurate to say that the wreath has a serpentine quality, especially now that the insinuating verb "lurk" has stolen from the adder to the nightshade's "seeds of death." Without too much of a stretch, the image of a snake and dangerous plant confronting a vain, easily flattered, beautiful woman alludes to Satan's temptation of Eve, for whom an illicit taste, we might say, also led to a sadder but wiser future. In any case, the serpent/plant imagery continues into the next stanza, in which bittersweet is enjoined to be a killjoy at a raucous party, a place "Where the wine flows free and bright," by serving as a memento mori: "Oh! twine, in the reveller's sight, / Round the foaming bowl thy poisonous wreath, / To shew him its draught is linked with death."[44]

The grand finale of this dour characterization of bittersweet comes in stanza 5, in which the plant's "last and fittest shrine" turns out to be "a human brow, / Where aught so baneful and false as thou / May without polluting shine!" We learn in the closing lines of the poem to whom the "brow" belongs: "The skeptic—I tremble to breathe his name— / Thine be the garlands which crown his fame."[45] This conclusion, which connects the plant to "the skeptic"—presumably anyone who questions Christian belief—refers to Hey's epigraph for the bittersweet entry, a well-known and often quoted passage in the Victorian era from Thomas Campbell's enormously popular turn-of-the-century poem *The Pleasures of Hope*, which makes nightshade complicit in science's destruction of faith:

> Oh! Star-eyed science, hast thou wandered there
> To waft us home the message of despair?
> Ah me! The laurelled wreath that murder rears,
> Blood-nursed, and watered by the widow's tears,
> Seems not so foul, so tainted, and so dread
> As waves the night-shade round the sceptic's head.[46]

Campbell himself may well have taken his nightshade-wreath imagery from a more famous Romantic, Samuel Taylor Coleridge, who wrote a poem commemorating the untimely death in 1796 of the otherwise immortal Robert Burns, Campbell's fellow Scot, in an effort to raise money for Burns's impoverished family. Addressed to Charles Lamb, for whom Burns was a hero, the poem "To a Friend, Who Had Declared His Intention of Writing No More Poetry" urges Lamb not to abandon his poetic gifts when "Thy Burns, and Nature's own beloved bard," so deserves to be mourned and memorialized. The nightshade wreath appears toward the poem's end, which is a scathing attack on the Scottish nobility who had refused Burns's bid for their patronage. Coleridge, after instructing "Dear Charles" to honor Burns by wreathing a bough "round thy Poet's tomb," concludes with a final nightshade-laced command:

> Then in the outskirts, where pollutions grow,
> Pick the rank henbane and the dusky flowers
> Of night-shade, or its red and tempting fruit,
> These with stopped nostril and glove-guarded hand
> Knit in nice intertexture, so to twine,
> The illustrious brow of Scotch Nobility![47]

To achieve its effect, Coleridge's invective requires that the reader grasp the disparaging botanical symbolism, which recalls the Solanaceae family's foul and poisonous associations by equating the henbane-and-nightshade wreath with stained honor. Like Campbell's and Hey's condemnation of the skeptic, Coleridge's assault on the Scottish nobility registers the wreath-wearers' sinful perversion of natural gifts, whether in this case of birth or in the former two of intellect. Whereas a laurel wreath bespeaks the glory of righteous achievement, one of nightshade, read correctly, signifies shame—a falling from grace akin to Satan's in the guise of the serpent. That Coleridge's plant, by way of "its red and tempting fruit," can be positively identified with *Solanum dulcamara* is just one of many indications of the prevailing Romantic and Gothic opinion that regarded all nightshade as deadly, whether so labeled or not. Consequently, Rebecca Hey's banishment of bittersweet to "the churchyard drear by the haunted tomb, / Or the falling shrine" is not another instance of confusion like William Morris's, but a deferral to a tradition rather firmly established in the eighteenth century, when woody nightshade's Gothic status matched belladonna's own.

Bittersweet fares slightly better in another popular sentimental flower book of the 1830s, Louisa Anne Twamley's *The Romance of Nature, or The Flower Seasons Illustrated*, which presents the plant in terms that closely echo Hey, but without resorting to lugubrious or imaginary settings. A native of the Midlands like George Eliot and also a close and passionate observer of nature, Twamley writes about the flora familiar to her experience, and her book is also a treasury of her own lovely botanical drawings with sentimental verse and moralistic prose commentary, organized, like Thoreau's journal, phenologically—by the salient botanical offerings of the changing year. Bittersweet appears in late autumn in a poem titled "A November Stroll," in which the poet walks down a "common turnpike road" admiring "the rough tangled mass" of the hedgerow, now sporting autumn foliage and fruit.[48] The few remaining blackberries and the many colors of their turning leaves are her favorites; but the dog-roses, hawthorn, and bittersweet also receive her admiration, even if her praise for the last must include the obligatory warning associated with forbidden fruit. Twamley finds "a long, / Far-creeping, many-angled stalk of that fair plant, / Fair-seeming, yet oft-treach'rous, woody-nightshade" in the thicket. Its leaves are dying away, "But the bright luscious-looking berries hung / In bunches of rich crimson, juicy, ripe, / And tempting e'en to those who know their bale, / Much more to childish lips!"[49] The "far-creeping" nightshade brings Rebecca Hey's serpentine bittersweet immediately to mind. In her prose commentary that follows, Twamley refers to woody nightshade's "most graceful festoons" and again to "its treacherous berries," which are decorously poised above the blackberries, hips, and haws in the accompanying illustration.[50]

Rebecca Hey's and Louisa Anne Twamley's epithet "fair but treacherous" for bittersweet sets the tone for its reputation in sentimental—that is to say moral—botany. As products of the 1830s, however, both books manifest an interest in plants that was becoming increasingly scientific: Hey's allusions to Stephenson and Churchill's *Medical Botany* and Twamley's careful and accurate delineations of plants in their natural settings demonstrate the growing desire on the part of even sentimental writers to accept plants on their own terms by studying them objectively, beyond the moralizing and romanticizing. Yet neither author could escape the influence of the Romantic and Gothic writers who had succeeded in turning woody nightshade into something truly bittersweet: a hideous pejorative on the one hand and a titillating icon of Gothic decor on the other. Besides, Hey's "fitting bower" for nightshade in a "church-yard drear" with its "haunted tomb" and "falling shrine" was not entirely the product of an overcharged sentimental imagination. As we shall see, just such a Gothic habitat for bittersweet famously existed in England's Lake District, around the ruins of Furness Abbey, situated in a secluded, deeply wooded valley once known as Bekansgill and now as the Vale of Nightshade.

The Gothic Sublime

The quintessential example of nightshade's place in the Gothic literary tradition occurs in references by Ann Radcliffe, the doyenne of the eighteenth-century Gothic novel, whose *Mysteries of Udolpho* will remain unforgettable, if for no other reason than that Jane Austen satirized it in *Northanger Abbey* for its melodramatic plot and highly impressionable heroine. When Emily (against her will) arrives within the inner courtyard at Udolpho, a remote, ancient castle in Italy's Apennines belonging to the villain Montoni, the plants growing there have been chosen as much for symbolic effect as for naturalistic plausibility: "As she surveyed through the twilight its desolation—its lofty walls overtopped with bryony, moss, and nightshade, and the embattled towers that rose above—long suffering and murder came into her thoughts. One of those instantaneous and unaccountable convictions, which sometimes conquer even strong minds, impressed her with its horror."[51] Ignoring the unintentionally amusing effect of the allusion to Emily's mental prowess, the reader can still appreciate Radcliffe's forte for what might be called telling description, in which the plants reinforce the scene's message of danger, wildness, and neglect—of civilization gone feral. The bryony, a poisonous tendril-bearer, and the easily attachable, moist-stone-loving moss give evidence that all three plants must be capable of scaling walls or hanging from them, thus identifying the nightshade as bittersweet. But its place in the list of flora has arguably the most emblematic resonance as a cue that we have entered the realm of what Edmund Burke had designated in 1757 as the "sublime," a scene or setting of the most compelling or overpowering type because "terror" is its "ruling principle."[52] As Burke pointed out, the strongest emotions are those aroused by the need for self-preservation;[53] and read from the perspective of Emily, who is overcome with fear, the very word "nightshade" reemphasizes the sense of foreboding created by the dying light and tall dark towers that are closing in upon her. Burke, the brilliant conservative statesman best known for his *Reflections on the Revolution in France,* was also one of the great eighteenth-century theorizers about aesthetics, and his *Philosophical Enquiry into Our Ideas on the Sublime and the Beautiful* coincided with and helped to foster a rage for the ruins, haunted castles, violence, darkness, craggy precipices, and storms that were the staples of Gothic fiction. Of course, that Udolpho's vegetation grew from the hotbed of Radcliffe's imagination is unquestionable since she hadn't set foot on the Continent before the novel's publication in May 1794. However, in a book recording her trip to Europe shortly afterward, *A Journey Made in the Summer of 1794,* we learn that Radcliffe returned home by way of England's Lake District, where she found a real setting eminently worthy of her fiction, the ruins of St. Mary's Abbey of Furness, a Cistercian monastery established in 1127 and dissolved by Henry VIII in 1537—appropriately located in the Cumbrian Vale of Nightshade, whose very name advertised its somber Gothic identity.

Radcliffe introduces the reader to the Furness Abbey very much as if it were Udolpho—but without the terror felt by the heroine, Emily, who necessarily from a generic point of view could only sense danger where Radcliffe sees delight:

> The deep retirement of its situation, the venerable grandeur of its gothic arches and the luxuriant ancient trees, that shadow this forsaken spot, are circumstances of picturesque and, if expression be allowed, of sentimental beauty, which fill the mind with solemn yet delightful emotion. This glen is called the Vale of Nightshade, or, more literally from its ancient title Bekangsgill, the "glen of deadly nightshade," that plant being abundantly found in the neighborhood. Its romantic gloom and sequestered privacy particularly adapted it to the austerities of monastic life.[54]

The key to Radcliffe's wholly appreciative response to "this forsaken spot" of "romantic gloom" is contained in the word "picturesque," which was as heavily weighted in eighteenth-century aesthetic theory as "sublime," but which indicated a crucial shift in point of view. In distinguishing the sublime from the beautiful, Burke had argued that the two qualities were actually opposed to each other: whereas the sublime moved in the realm of painful emotion, beauty was all about pleasure, so the two ideas were in essence mutually exclusive and thus could not inhere in the same scene or object.[55] In the aesthetic debate that Burke's definitions set in motion, a Cumbrian native, William Gilpin, offered a solution via the mediating term "picturesque," which could accommodate the idea of the sublimely beautiful or beautifully sublime by completely removing the perceiver from the scene, as if he or she were looking at a picture. Like the imaginary fourth wall in the theater separating the stage from the audience or the frame of a painting, the "picturesque" attitude offered the observer immunity from any pain and terror aroused by the scene itself.[56] By viewing the ruins in a painterly way, Radcliffe could therefore appreciate the scene's "sentimental beauty" and experience a "solemn yet delightful emotion"—despite the inherent dreariness of uninhabitable, crumbling buildings and monks' graves set in a dark glen filled with poisonous vegetation. Gilpin himself had written about Furness Abbey in his third book recording his aesthetic travels; and although he didn't actually visit the ruins, he commented, "From the drawings I have seen of it, it seems to have been constructed in a good style of Gothic architecture, and has suffered, from the hand of time, only such depredations as picturesque beauty requires."[57] To prove his point, Gilpin included "a very pleasing drawing from Mr. Smith" showing the Furness Abbey school—one that displays the picturesque qualities of the ancient building being taken over by wild nature, which results in an overall softening or beautifying effect.[58]

In short, Ann Radcliffe's thrilled reaction to the ruined abbey and its environs not only demonstrates how a theory of the picturesque served to domesticate the sublime as part of the Gothic aesthetic; it also articulates how the Vale of Nightshade became a local habitation and name for "romantic gloom" in fiction, poetry, and fact during the late eighteenth and early nineteenth centuries. Unfortunately, this literature shaped bittersweet's "fair but treacherous" reputation while reinforcing its mistaken identity as deadly nightshade. Radcliffe was not the first Gothic writer to respond to the site's gloomily titillating possibilities. William Beckford, author of the horrific "Arabian Tale" *Vathek,* had toured Furness in 1779; and we learn from the author of his *Memoirs,* "The Abbey in the Vale of the Deadly Night Shade delighted the sensitive mind of young Beckford so much, it was difficult to detach him from it, and return to [nearby] Dalton."[59] And in the years immediately following Beckford's trip to the ruins, probably the most celebrated visits in literature began taking place: William Wordsworth, arguably the most influential of the Romantic poets, records in *The Prelude* his joyous trips on horseback with his fellow schoolmates to

the antique walls
Of that large abbey, where within the Vale
Of Nightshade, to St. Mary's honour built,
Stands yet a mouldering pile with fractured arch,
Belfry, and images, and living trees,
A holy scene![60]

These lines and the following register Wordsworth's sense of picturesque beauty:

Trees and towers
In that sequestered valley may be seen,
Both silent and both motionless alike;
Such a deep shelter that is there, and such
The safeguard for repose and quietness.[61]

The following section, however, captures the raw animal spirits of adolescent boys indifferent to aesthetics and picturesque detachment as they irreverently romp over the grounds:

With whip and spur we through the chauntry flew
In uncouth race,[62]

riding their horses through what was formerly the most sacred place of all, the altar, and past the remaining effigies in the church.

On one of these occasions, however, Wordsworth recalls being moved profoundly by the song of a "single wren" from the church's nave:

So sweetly 'mid the gloom the invisible bird
Sang to herself, that there I could have made
My dwelling-place, and lived forever there
To hear such music.[63]

This fleeting moment of exaltation is a type of the sublime, but one more akin to its original, classical meaning of transcendence, grandeur, or elevated thought than to Burke's definition, with its connotations of danger. The Victo-

36   VICTORIAN NIGHTSHADES

rians would come to prefer Wordsworth's sublime to Burke's, but this change in aesthetic taste had to wait until the first publication of Wordsworth's poetry at the turn of the nineteenth century. Moreover, *The Prelude*—and thus Wordsworth's account of his visits to the abbey—would not be published until after his death in 1850.

In the meantime, the cultural obsession with picturesque Gothic ruins prompted other publications about the abbey and the Vale of Nightshade that promoted the plant family's notoriety as well as the confusion between deadly nightshade and bittersweet. The writings of antiquarians became especially important as purveyors of Gothic lore, and often became the guidebooks for visitors like Radcliffe. Her introduction not surprisingly echoes Thomas West's *Antiquities of Furness*, first published in 1774, two years after Gilpin's tour. This book achieved some fame among the Gothic novelists (Horace Walpole, author of *The Castle of Otranto*, mentions reading it in his correspondence),[64] but it either began or furthered a misreading responsible for affixing the "deadly" to the Vale of Nightshade's name. As West describes the place, "The situation is gloomy and romantic, and formerly produced abundance of Lethal Bekan, the Solanum Lethale, or deadly nightshade, from which circumstance the vale first obtained the name Bekangs-Gill."[65] West gives as his source for this juicy morsel the relevant lines from a fifteenth-century Latin poem included in the abbey's repository of official records;[66] but no one seems to have bothered looking closely at the Latin until late in the nineteenth century, when a flurry of exchanges took place in *Notes and Queries* about the meaning of "Bekan," initiated by a plant geographer compiling a flora of Lancashire. Thomas K. Fell responded by identifying bittersweet as the plant in question via his translation of the Latin poem: "This valley took its name a long time ago from the herb Bekan, the bittersweet, where it flourished. Thence the name of the house 'Bekan's gill' was known aforetime. Now it receives the auspicious name for so important a dwelling-place."[67] According to the rest of the *Notes and Queries* exchange, "Bekan," which was the valley's name before the monks arrived, turns out to be the Scandinavian proper name that later became "Bacon" (the information prompting a few jokes about pigs). But the phrase in the original Latin, "*dulcis nunc tunc acerba*" (literally, "sweet now then bitter"), that describes the herb certainly seems to refer to bittersweet, even if it reverses the order of the tasting sensation (*tunc* usually implies "next")—which the Latin species name *Dulcamara* does as well.[68] How and when the plant's being called "Bekan" entered the record must remain one of those Gothic mysteries, but its being "lethal" adds a good deal of sensational detail to the abbey's history.

The last bit of evidence about the identity of the plant in question, however, is the Seal of Furness Abbey, which allegedly pictures "Bundlets of Nightshade" as the decorating greenery; but the leaves are too stylized for positive identification.[69] In any case, there is a better-than-average chance that deadly nightshade was cultivated by the monks as a medicinal plant, and in an appendix to the 1805 edition of West's *Antiquities*, William Close includes *Atropa belladonna* in a list of the local flora. Nevertheless, he concludes, "It used formerly to grow very plentifully and luxuriantly amongst the ruins, but is now almost exterminated."[70]

Ann Radcliffe on tour, however, doesn't trouble herself with niggling details of plant identification when she describes her first sight of the ruins from the winding road leading to the abbey grounds: "A sudden bend in this road brought us within view of the northern gate of the Abbey, a beautiful gothic arch, one side of which is luxuriantly festooned with nightshade." And not long after, "We made our way among the pathless fern and grass to the north end of the church, now, like every other part of the Abbey, entirely roofless, but shewing the lofty arch of the great window, where, instead of the painted glass that once enriched it, are now tufted plants and wreaths of nightshade."[71] If Radcliffe is describing what she actually saw—and there is no reason to think otherwise, except that West mentions only "picturesque appendages of ivy" decorating the gate—the festooning, wreathing nightshade has to be *Solanum dulcamara*. Even today it's possible to take the "Vale of Nightshade Walk," where according to the brochure the visitor can find that "here and there, in season, the thickly vegetated path is indeed hung with growths of woody nightshade (bittersweet), its purplish flowers of summer succeeded by red berries in autumn."[72]

Three years after Radcliffe's visit, the soon-to-be greatest of Romantic landscape painters J. M. W. Turner would sketch the ruins, and in 1837 his most famous critic and admirer, John Ruskin—already an extremely talented artist at the age of eighteen—would follow suit.[73] Ruskin's drawing of one of the abbey's ornate windows, framed by a tree whose trunk leans away from the frame as if to invite the viewer inside the ruin, suggests his abiding love for both nature and Gothic architecture, subjects that would become preoccupations of his aesthetic criticism for the rest of his life. But unlike Radcliffe and her fellow writers, Ruskin had little sympathy with the nightshades, despite their impeccable Gothic credentials; and his assessment of bittersweet throughout his career articulates his extremely moralistic Victorian sensibility, one that could not divorce the plant's beauty from his perception of its evil nature.

The most influential critic of the Victorian era, Ruskin would make his reputation with the publication of *Modern Painters*, the five-volume work writ-

38   VICTORIAN NIGHTSHADES

ten from 1843 to 1860 that began as a defense of Turner's late-career paintings and ended as a pulpit for pronouncements about all art, nature, and society. During the same period two works celebrating Gothic architecture—*The Seven Lamps of Architecture* (1849) and *The Stones of Venice* (1851–53)—would help to foster a cultural revival of Gothic design, establishing Ruskin as its foremost Victorian authority. Thus, like Burke and Gilpin, Ruskin took up the eighteenth-century debate about the sublime's relation to the beautiful, ultimately arriving at his own conception of the "noble picturesque" as the superior version of that quality embodied in Turner's paintings, which expressed the artist's deep sympathy with his subject rather than the surface effects that Ruskin claimed characterized the typical "picturesque" painting. As George Landow explains in his incisive discussion of the evolution of Ruskin's aesthetic theory, the "noble picturesque" invested nature with a deeply ethical sensibility: because Ruskin "could not avoid the human implications" of picturesque subjects, he could not escape awareness of their social—and thus their moral—connotations.[74] The nightshades are therefore left to suffer the consequences.

For example, in volume 3 of *Modern Painters*, after Ruskin half-jokingly moves from ranking paintings to fruit, he asserts, "It is, indeed, true that there *is* a relative merit, that a peach is nobler than a hawthorn berry, and still more a hawthorn berry than a bead of the nightshade,"[75] a critique that seems as ludicrous as it is cryptic. But a later comment in volume 4 more clearly reveals the moral character of Ruskin's aesthetic judgment as applied to the nightshades. To support his praise for Turner's depiction of light in paintings, in which even the shadows are full of "lovely color," Ruskin pronounces his belief in "the fact of the sacredness of color, and its necessary connection with all pure and noble feeling."[76] Ruskin instructs his reader to "observe how constantly innocent things are bright in color" before qualifying, "I do not mean that the rule is invariable, otherwise it would be more convincing than the lessons of the natural universe are intended to be; there are beautiful colors on the leopard and tiger, and the berries of the nightshade."[77] The obvious implication is that because leopards, tigers, and nightshade berries are predatory and dangerous, they are all examples of the treacherous beauty that Rebecca Hey and Louisa Anne Twamley had warned about. It's also clear that like Hey and Twamley, Ruskin is talking about bittersweet, since its berries are the only ones of the nightshade family that could be considered colorful, as they ripen from green to shades of orange to red. Ruskin would eventually describe "the purple and yellow bloom of the common hedge nightshade" as "deadly and condemned," and "the whole poisonous and terrible group" as a "tribe set aside for evil" in *The Queen of the Air* (1869),[78] a verdict that is a more extreme version of the morally driven taste of Rebecca Hey and Louisa Anne Twamley, rather than the more appreciative view of Ann Radcliffe, whose aesthetic sensibility allowed her to relish treacherous beauty as a thrilling aspect of the Gothic sublime.

In the first decades of the nineteenth century, the second-generation Romantics certainly profited from bittersweet's notorious reputation, using it brilliantly for sensational, Gothic effect. John Keats flirted with vegetable poisons in his "Ode on Melancholy," warning the melancholic not to take "wolf's-bane" or "yew-berries," "Nor suffer thy pale forehead to be kiss'd / By nightshade, ruby grape of Proserpine."[79] The metaphor may suggest using nightshade as an external compress to promote a deep sleep, as Gerard reported was done with belladonna; but the color of the berries identifies this fruit of Proserpine, the queen of the underworld, as bittersweet. In *Romantic Medicine and John Keats*, Hermione de Almeida makes much of the poet's having studied medicine, and thus of his awareness of plants' therapeutic and deleterious effects; however, she, perhaps like Keats himself, doesn't make clear distinctions among Solanaceae species, and mistakenly refers to belladonna's "scarlet berries."[80] In any case, we once again see nightshade's connection to the brow of someone with a troubled mind, like Thomas Campbell's skeptic and, later, Rebecca Hey's.

Finally, Percy Bysshe Shelley probably referred to nightshade for Gothic effect more than any other of the great Romantic poets; and the lines that serve as the epigraph for this chapter, taken from *Rosalind and Helen* and referring to "nightshade's hair" twining "Round the walls of an outworn sepulchre," epitomize the sometimes affectionate, if rather gloomy, treatment the plant—usually identifiable as bittersweet—received in Shelley's works.[81] His most famous reference, however, has unequivocally treacherous connotations of a woman poisoner. In *Epipsychidion*, a poem written in 1821 and generally accepted to be autobiographical, the speaker, who is searching for the soul mate cryptically alluded to in the title, at one point meets a terrifying femme fatale instead:

There, —One, whose voice was venomed melody
Sate by a well, under blue nightshade bowers;
The breath of her false mouth was like faint flowers,
Her touch was as electric poison—flame
Out of her looks into my vitals came,
And from her living cheeks and bosom flew
A killing air, which pierced like honey-dew
Into the core of my green heart, and lay

Upon its leaves; until as hair grown gray
O'er a young brow, they hid its unblown prime
With ruins of unseasonable time.[82]

Shelley's lines are beautiful, terrifying, and enigmatic; but whatever else is going on here, the nightshade is clearly an accomplice: a necessary, villainous accessory, in both ornamental and legal terms. We can say with certainty that the mysterious woman sitting beneath "the blue nightshade bowers" is connected to the flowers by more than proximity—they are even on her breath and in her melodiously poisonous voice. The color blue positively identifies the nightshade as *Solanum dulcamara*, and the fact that it forms "bowers" is further evidence of the woody, climbing, wreathing plant that winds through the tangled banks of England and that had gained notoriety among the Gothic and Romantic writers in the previous century. It seems fair to identify the plant itself with the dangerous enchantress who has somehow infected the speaker, who in turn comes to represent the untimely "ruins" resulting from her toxic contact. Here is the Gothic sublime writ large, with no mediating barrier to protect the poet himself from becoming the "ruins." Some literary critics speculate that this passage refers to Shelley's encounter with a prostitute from whom he contracted syphilis; and in *Shelley's Venomed Melody*, Nora Crook and Derek Guiton note that *Solanum dulcamara* was in fact used in treating the disease;[83] but Shelley's lines seem to be more about the act of infection rather than its cure. At any rate, the association of "nightshade" with whores and witches as noted in chapter 1 certainly seems pertinent here.

Writing about Shelley's poetry in 1856, one of his most astute critics, the economist and essayist Walter Bagehot, remarked that the poet had "a perverse tendency to draw out in keenness the torture of agony," a comment borne out by the above passage. A bit later Bagehot continues: "The nightshade is commoner in his poems than the daisy. The nerve is laid bare; as often as it touches the open air of the real world, it quivers with subtle pain. The high intellectual impulses which animated him are too incorporeal for human nature; they begin in buoyant joy, they end in eager suffering."[84] It's clear from the context that "daisy" and "nightshade" represent the poles of "joy" and "suffering," and equally clear which is which. By the time Bagehot is writing, just as the daisy was immediately understood to mean innocence and "buoyant joy," "nightshade" had been well established in the Romantic literary tradition as a botanical cliché to mean intellectual angst, melancholy, treacherous beauty—or in Bagehot's words, "eager suffering," which is also a perfect definition of the emotion that informs the Gothic sublime. Were it not for the context, however, nonspecific references to "nightshade" obscure *Solanum dulcamara*'s identity. As we shall see in chapter 3, such is not the case for the Language of Flowers, in which bittersweet is unambiguously identified and assigned the meaning "truth," a verification of the eager suffering that characterizes a bittersweet romance.

3

Dulcamara

Affairs and Elixirs of Love

> If you have any doubt about the truth of Dr. D.
> You've only got to cast your eye upon the face of me.
> A plainer man than I was once you very seldom see,
> But now I am as beautiful as you could wish to be.
> My sham complexion brings me fame,
> My sham complexion brings me fame;
> I lay it on the latest thing at night, boys —
> In the morning I am fascinating quite, boys.
>
> — W. S. Gilbert, *Dulcamara, or The Little Duck and the Great Quack* [1866]

True Romance

IT WAS THE wonderfully suggestive common name "bittersweet" rather than the species' lurid family connections that accounts for *Solanum dulcamara*'s more positive treatment in the most sentimental of all Victorian floral obsessions, the language-of-flowers craze that took hold first in France and later in England and America, especially through the first half of the nineteenth century. At its simplest, this "language" took the form of a dictionary: an alphabetical list of flower and plant names, each followed by a definition that most often related, in the words of Beverley Seaton from her fine history of the subject, to "the conduct of a love affair."[1] The practice of lovers communicating secret messages through bouquets or flowery poems supposedly could be traced to Turkey, the harem, and clandestine courtship, thus imbuing "floriography," as it was also called, with all the Oriental exoticism that such an origin implies; and learning this botanical code had obvious appeal to a young female audience.[2] But as Jack Goody explains in *The Culture of Flowers*, his fascinating account of symbolic and ceremonial uses of flowers through history and around the globe, its invention had more to do with money than love, for it coincided with the burgeoning European market for new horticultural varieties and the development of the florist trade, which, like the publishing business, profited immensely from saying it with flowers.[3]

Thus, as the flowers themselves became major commodities and status symbols in Victorian England, books containing floral lexicons enjoyed a similar popularity, reflecting an attempt by the English bourgeoisie to imitate French elite society, whose interest in such works had reached its height in the early 1820s.[4] The often lavishly produced gift annuals and albums that were filled with illustrations of flowers, verses about flowers, and floral symbolism—occasionally hiding their fair owners' blossoms pressed between the pages—so frequently included a language of flowers that the lists became an almost essential feature of nineteenth-century decor, and certainly of its botanical lore. Kate Greenaway's *Language of Flowers*, first published in 1884 with her own sentimental illustrations, represents the type of the genre for twenty-first-century readers because of its continued availability, with a facsimile in print from Dover Publications since 1992. But Greenaway's *Language* is a stripped-down affair that offers no explanation for the flower definitions in the lexicon or the reverse dictionary that follows; and the extended floral associations, commentary, and lore that were formerly a major feature of such works are represented only by twenty-seven poems at the book's end, presumably added to fill out the tiny volume. Its real interest for Greenaway and her original readers must have been the dainty flower drawings and vignettes of well-dressed late-Victorian ladies and children that frame the pages. Not surprisingly, Dover markets it as a children's book.

As Greenaway's version suggests, in the many permutations of the language that were produced through the century the emphasis on romance varied widely, and it was more obvious in some individual definitions than in others. Rose could mean "love," as one would expect, but Greenaway and others list as many as thirty kinds of roses, each with its own definition. Obviously, the assigned meanings were often completely arbitrary, although many drew on traditional associations from folklore or literature. For example, narcissus maintains its connection to classical mythology and predictably means "egoism" or "self-love" in the lexicons, and rosemary often carries its Shakespearean meaning of "remembrance" in honor of Ophelia's definition in *Hamlet*. There are variations in the definitions among the lists, but it's clear there was quite a bit of borrowing among sources. Originality, of course, was not the issue; as the French would say, the raison d'etre for the genre was the celebration of the flowers and plants as aesthetic objects, so that its emphasis was on horticultural and/or ornamental species—that is, those with a high commercial value.

Considering the fashion's genesis and emphasis, it is not surprising that the Solanaceae figure less prominently in the lists than other plant families; even so, bittersweet has a steady, if relatively modest presence from the beginning, appearing in B. Delachénaye's early *Abécédaire de flore ou langage des fleurs*, published in Paris in 1811. There and with only a few exceptions, bittersweet (or *douce-amère* in French) means "truth,"[5] including in the slightly later but by far most influential, the pseudonymous Charlotte de Latour's *Le langage des fleurs*, initially published in December 1819, and instantaneously reproduced in many editions, translations, and imitations into the twentieth century (with some nostalgic resurgence of interest today).[6]

Who originally assigned *"vérité"* to *douce-amère* is impossible to determine. It should be noted, however, that the French common name, like the Latin species designation *dulcamara* for which it is an exact translation, reverses the order of bittersweet's attributes, so that the bitterness associated with the term is its final, lingering sensation. Interestingly, this idea seems to be borne out in a colored plate appearing in the first and many later editions and translations of Latour's *Langage* illustrated by the famous French botanical artist Pancrace Bessa.[7] The plate features sixteen separate flower drawings arranged above the words of a verse to suggest their individual meanings (or, with the two upside-down drawings, their meanings' negatives), serving, as Frederic Shoberl explains in his English adaptation of Latour's work, as an "illustration of flower-writing" (see fig. 4).[8] The verse, by "the Chevalier Parny" (a popular love poet of the Napoleonic period), is a simplified rendition of a "song," which Shoberl quotes in the original French and then translates as follows: "To love is a pleasure, a happiness, which intoxicates: to love no longer, is to live no longer; it is to have bought this sad truth, that innocence is falsehood, that love is an art, and that happiness is a dream." The accompanying illustration changes the second major clause so that it reads "to cease to love is ceasing to exist";[9] but the message remains, as in the French original, a melancholy one, because the "truth" seems ultimately to be a recognition of love's inherently deceptive nature. The sentiment itself is a sweet-bitter expression of the move from innocence to experience, the theme of so many nineteenth-century French novels of "realism" by writers like Stendhal and Balzac (*Lost Illusions* is the title of the latter's masterpiece in this vein). Of course, it's reasonable to interpret the poem as an exhortation to love always and at all costs, since the end of love has such depressing, even fatal, consequences; but

FIGURE 4. Illustration of flower writing, Frederic Shoberl, *The Language of Flowers*. (Image from Biodiversity Heritage Library; contributed by Cornell University Library)

the "truth" of the matter doesn't change. The song ends as a sober reminder of Cupid's legendary blindness and the cruel reality beyond love's intoxication. Like the French realists, Parny celebrates love, but acknowledges the folly of doing so.

As Beverly Seaton points out in her history, it would be difficult to decipher any coherent message solely from the colored plate because not all the words have a floral counterpart, nor do all the flowers appear in the accompanying dictionary. But "truth" is unambiguously aligned with the tenth drawing representing bittersweet, while the ninth drawing of a sprig of yew appears above the words "this sad" to convey truth's essential quality.[10] Taking the verse as a whole, then, the move from joy to sorrow would seem to be the

46 VICTORIAN NIGHTSHADES

inevitable trajectory of love; for the "truth" about it resides in the last, decisive, bitter realization, one which the French plant name *douce-amère*—literally, "sweet-bitter"—communicates in a single word.

In the dictionary that follows the plate, Latour's justification for the *vérité-douce-amère* association, as translated by Shoberl, follows a similar trajectory: "The ancients thought that Truth was the mother of Virtue, the daughter of Time, and Queen of the world. It is a common saying with us that Truth conceals herself at the bottom of a well, and she always mingles some bitterness with her blessings: we have given for her emblem this useless plant that, like her, delights in shade and is always green. The bittersweet Nightshade is, I believe, the only plant in this country that loses and re-produces its leaves twice a year."[11] The entry begins auspiciously with allusion to an amalgam of proverbs that give homage to Truth in the guise of a goddess—known as Aletheia to the Greeks and Veritas to the Romans—by way of honoring her mythic family credentials and royal status. The reference to the well, her symbolic hiding place, is also of classical origin, usually attributed to the skeptical philosopher Democritus, whose doubts about the reliability of the senses to perceive truth explain her legendary elusiveness and the metaphorical depth and inaccessibility of her dwelling. But finding Truth has yet another downside, one that calls to mind Parny's song and calls into question the desirability of the pursuit in the first place. "Her blessings" are for the philosopher, not the lover; for romance exists in an unusually delusional, sensory realm that wants nothing to do with truth. Given this interpretation, it is most appropriate that Truth's emblem be a "useless plant," although this curious designation probably refers to the fact that woody nightshade was a bank and hedgerow weed "of unmarketable beauty," as George Eliot said: that is, a plant not worth cultivating for ornament or other horticultural purpose.[12] The final remark about bittersweet's semiannual leaf production seems like a last-ditch effort to ascribe to the plant a unique character—even though the reference to "this country" originally applied to France rather than England, as Shoberl's entry implies.

Latour's influence was immediate and widespread. It's even possible that Shelley was thinking about bittersweet's description from Latour's *Langage* when he wrote of the dangerous but seductively attractive woman in *Epipsychidion* who "Sate by a well under blue night-shade bowers," thereby serving as the embodiment of the ill-fated truth that bittersweet represents.[13] In any case, the exact definition surfaced not only in Frederic Shoberl's many English translations, but also in variations published by other floriographers throughout the century. For example, an abridged version of the bittersweet-truth pairing appears in another early English rendition of the genre by Henry

Phillips, a well-known horticultural writer and landscape gardener whose partiality to garden varieties of plants probably explains the lackadaisical treatment bittersweet receives in *Floral Emblems*, published in 1825. Phillips drastically shortens the entry on "Truth" and except for naming bittersweet, says nothing about the plant itself. Beginning with an epigraph by the seventeenth-century moralist Roscommon ("The first great work / Is, that yourself may to yourself be true"), the brief explanation follows: "However delightfully sweet truth must appear, it is frequently found a bitter draught to those to whom it is presented; and from this cause we presume the emblem has been established."[14]

Bittersweet gets fuller treatment in the language-of-flower adaptations from Latour composed by the prolific Robert Tyas, a clergyman whose mid-century botanical works reveal an eclectic interest in plants, from *Favourite Field Flowers* (1850) to *Flowers and Heraldry* (1851) to *Flowers from the Holy Land* (1851). In his first entry in the Latour category, *The Sentiment of Flowers, or Language of Flora*, which appeared in 1836 and went through many editions and variations in the next four decades, Tyas acknowledges in the preface his debt to his source, but insists that his work is not a "mere translation," a claim that the later editions partly bear out. He also removes Parny's perhaps too-cynically French "song" along with its rendering in "flower-writing," which must be what Tyas means by noting that "a discretion has been exercised in the rejection or alteration of those passages which were not suited to English taste."[15] However, Tyas's major contribution to the genre is the addition of more actual botanical information about the plants than the Latour imitators usually provide. For instance, the ninth edition (1842) not only includes a "Glossary of Technical Terms";[16] the entries begin with the plant's Latin and common names, its class and order in both the Linnaean and natural systems, and a brief physical description, keyed to the colored illustrations by the botanical artist James Andrews. Thus "bitter-sweet nightshade" once again appears under "Truth," with Latour's (and Shoberl's) explanation given almost verbatim; but the emphasis of the entry is weighted toward the plant itself rather than its assigned "sentiment." After locating bittersweet's place in the natural and artificial (Linnaean) systems, Tyas gives an accurate, if brief, verbal sketch of the plant as "a poisonous climber, common in hedges. Flowers, which appear in June and July, purple. Anthers large, yellow, united into a cone-shaped figure. Berries ovate, red."[17] Finally, Tyas attaches the following information about bittersweet to the original entry's conclusion: "Its roots smell somewhat like the potato, and being chewed, produce a sensation of bitterness on the palate, which is succeeded by sweetness. From this singular fact it derives its specific name 'bitter-sweet.'"[18] All these additions certainly make bittersweet more interesting as a living plant; and by explaining at last the derivation of the English common name, Tyas replaces the bitterness that lingers in the original Latour version with a sweet taste—a fact that seems to temper the criticism his entry retains that it is "an useless plant."[19]

In later floriographic offerings, Tyas chose to omit the scientific apparatus that appeared in 1842, instead giving priority to poetry and to James Andrews's lovely illustrations of ornamental "Floral Emblems," each portraying a few plant species to convey a single message or aphorism. In an 1869 version of *The Sentiment of Flowers*, nightshade (truth), heath (solitude), and bindweed (humility) illustrate that "Truth is humble and retiring," a sentiment that casts all three plants in as flattering a light as the illustration itself.[20]

THE LURID LEGACY

There is no indication why Robert Tyas decided to downplay the emphasis on science and plant identification in his later floriographies; but it's quite possible that he felt such information, even for a predominantly female audience, had become unnecessary. By the second half of the nineteenth century, there were scores of works available offering instruction in botanical science and written at every level of technicality. Moreover, the sheer number and variety of the earlier publications—many written expressly for, as well as by, women—attest to the fact that women's desire to know about plants went far beyond the frivolous. Hey's and Twamley's sentimental flower books had incorporated a wealth of scientific information with the sentiment, more evidence, like Tyas's publications, that the lines between these two approaches to the study of flowers were by no means distinct. It is no surprise, then, that during the same period of the language-of-flowers vogue, the study of botany for ladies was also in high fashion, and even undergoing a revolution, one that would eventually lead to a more enlightened view of the nightshades once the family could escape from its lurid Linnaean legacy.

In *Cultivating Women, Cultivating Science: Flora's Daughters and Botany in England, 1760 to 1860*, Ann Shteir has done a remarkable job of documenting the crucial role women played as botanical students, collectors, artists, writers, and teachers in the development of the discipline, the period when the Linnaean "artificial" system of plant classification, deemed easy enough for a female to understand, helped to turn botany into what could be considered the gateway science for women.[21] But with the takeover of the newer "natural" system that John Lindley pioneered in England beginning in the 1820s, there was a movement to modernize and professionalize botany, ultimately shifting it away from what had become its fashionably feminine—and sentimental—associations.[22]

Although the idea that there is a special affinity between women and flowers is ancient and cross-cultural, surfacing long before the eighteenth century, the great Enlightenment philosopher and guru of the Romantics Jean-Jacques Rousseau helped to steer the female-flower connection in a scientific direction via his *Letters on the Elements of Botany, Addressed to a Lady*, penned from 1771 to 1773 to a "dear cousin" who had enlisted his aid to instruct her young daughter. In the first letter Rousseau applauds the mother's project of "amusing the vivacity" of her child with botany and agrees to assist because of his conviction that botanizing is a character-honing activity beneficial at any age. He declares, "At all times of life, the study of nature abates the taste for frivolous amusements, prevents the tumult of the passions, and provides the mind with a nourishment which is salutary, by filling it with an object most worthy of its contemplations."[23] For Rousseau nature is the great teacher, and a student of plants will of necessity be led to discover valuable lessons about existence. Nevertheless, botany is an occupation for the amateur, not really a discipline to be pursued with serious intentions, as Rousseau insists in Letter 7: "You must not, my dear friend, give more importance to Botany than it really has; it is a study of pure curiosity, and has no other real use than that, which a thinking sensible being may deduce from the observation of nature and the wonders of the universe."[24] Evidently a reaction to his cousin's too-pressing botanical questions, this odd comment seems to suggest that botany's crucial lessons are simple and accessible without effort for the "thinking sensible being": hence, their status as an appropriate and innocent recreation for the gentle sex and even its very young (if precocious) members.

Originally comprising a brief history followed by eight letters explaining basic plant structure and focusing on the characteristics of six common families, Rousseau's botany gained in both size and popularity in England thanks to Cambridge professor Thomas Martyn, who in 1785 published a translation that augmented Rousseau's text with twenty-four additional letters presenting the Linnaean classes. Like Rousseau, Martyn intended the book as an introductory study written primarily for females, retaining the now-fictional "cousin" as the addressee; but in his preface he invited a larger audience, expressing his hope that the letters "might be of use to such of my fair countrywomen and unlearned countrymen as wished to amuse themselves with natural history."[25] We might call this the eighteenth-century equivalent of "Botany for Dummies": but even though cast as a leisure pursuit that would not be too intellectually taxing, botany nevertheless required vigorous participation.

Martyn exhorted his readers: "I beg leave to protest against these letters being read in the easy chair at home; they can be of no use but to such as have a plant in their hand. . . . Botany is not to be learned in the closet; you must go forth into the garden or the fields, and there become familiar with Nature herself."[26] Given the evident popularity of the Rousseau/Martyn *Letters* (the book went through eight editions in thirty years), this hands-on approach and emphasis on close examination certainly raised the level of botanical literacy in England, particularly among women. However, even with the letters' scientific emphasis, both Rousseau and Martyn promoted the sentimental female-flower tie by assuming a patronizing and protective voice—one unfortunately suited to a disparaging, if dramatic, presentation of the nightshades that would reinforce the family curse well into the next century.

Using the Linnaean designation "Luridae" in Letter 16 on the fifth class, Pentandria, and its first order, Monogynia, Martyn begins his account with unusual trepidation, as if he intends to dissuade both the mother and "our young cousin" from even knowing about the family:

> I am almost afraid to present you with a set of plants, which from their lurid, gloomy, appearance, are kept together under the title *Luridae*. They have also most of them a disagreeable smell, which, with their forbidding look, will deter our young cousin from examining them, she not being yet sufficiently tinctured with enthusiasm to go on in spite of such circumstances. Indeed I would not wish her to be too busy with some of these *insane roots that take the reason prisoner*, and which I can never collect and examine myself, without their affecting my head. You will consider that nature has kindly given us notice in general of approaching danger, by means of our senses; and accordingly some of these Lurid plants are highly poisonous; most of them to some degree, though soil and climate may mitigate the poison, and even render them wholesome. I will select some of the least disagreeable in smell and appearance; or, if they be otherwise, will announce it to you.[27]

When Martyn wrote this off-putting introduction, he evidently had henbane (*Hyoscyamus niger*) uppermost in mind because it manifests all of the unsavory characteristics he mentions. It is smelly, sticky, and arguably rather sinister-looking, qualities recorded in a former common name in England, "stinking nightshade," and in Germany, "Devil's Eye," an apt depiction of the flower, which looks a bit like the diseased and bloodshot sclera ("white") of an eye with a deep purple iris.[28] The flower's main color is a pale shade of yellow, which "lurid"—derived from the Latin *luridus*, meaning "pale yellow, wan, ghastly"—literally designates. Linnaeus characterized the Luridae as "suspect,"[29] so he must have thought that "ghastly" was an appropriate description as well for a family that included henbane, belladonna, datura, and mandrake. But only the first two of these species have flowers that could be called "lurid" in the sense of color: henbane's yellow streaked with purple and belladonna's pale brownish-purple flowers exhibit the dingy yellow or yellow-brown that the color now signifies in scientific terminology, thanks to Linnaeus. Henbane is also by some accounts considered to be the "insane root," the hallucinatory plant referred to by Shakespeare in the italicized quotation, from *Macbeth*—although belladonna was more frequently given this honor during the nineteenth century. Henbane and mandrake were the two most important medical nightshades since ancient times, known for their sedative and painkilling powers; but they were also hallucinogens and thus subject to abuse, a circumstance to which Martyn somewhat cryptically alludes when he describes the former a few pages later: "Henbane is a very common plant, and has often done mischief to such as will not suffer their appetites to be corrected by their senses." He continues, "You will agree with me that the smell is sufficient to deter any person from eating it." Then, in a surprising reversal of his original warning about the family's "forbidding look," he praises the flower's appearance, but does so in terms implying that his is the minority opinion: "I cannot however dispense with your examining the flower, which is really beautiful on a near view. The corol is funnel-shaped, and obtuse; of a pale yellowish colour, beautifully veined with purple."[30]

Martyn's aesthetic appreciation of henbane notwithstanding, the presentation of the rest of the family, covering about twenty wild and cultivated nightshade species and a couple of mulleins (now considered figworts), is most striking for the lurid details that punctuate this entire section. In fact, the lion's share of the discussion here is devoted to mandrake, an emphasis especially glaring since a non-native, non-ornamental plant would not have been available for actual inspection, which was one of Martyn's requirements as prescribed in the preface. What becomes clear is his pleasure in recounting mandrake's titillating lore—and in quoting Shakespeare, for whom it was the nightshade of choice. Martyn begins dramatically, recalling passages from *Henry VI, Part 2* and *Romeo and Juliet*, respectively: "You have heard of the *Mandrake's Groan*, and 'of shrieks, like Mandrakes torn out of the earth;' superstition having endued this plant with a sort of animal life, fatal to whoever presumed to destroy it, by digging up the root." Martyn continues with two more Shakespearean references, noting that the "*Mandragora*" desired by the heroine in *Antony and Cleopatra* and mentioned by Iago in *Othello* was once prized as much as opium for its sleep-inducing properties. He then returns to the magical lore about its anthropomorphic qualities with a tongue-in-cheek explanation that makes the myth seem almost believable: "Since Mandrake groans and shrieks when injured, it must needs have a human form; and accordingly, such have been carried about for sale, notwithstanding the danger that attends the procuring it; but this is cunningly avoided by tying a dog to the root, and thus making the blind fury of the poor Mandrake fall upon the innocent dog instead of the aggressor." Martyn describes how purveyors of fake mandrake roots made them from angelica or bryony, "either cut into form, or compelled to grow through earthen moulds put into the ground for this purpose; they were used in magical incantations; and though these are now pretty much out of fashion, yet I have had them very gravely offered me for sale." This last is a valuable piece of late eighteenth-century ethnobotanical information from an eminently reliable source; but even though Martyn was apparently not tempted to make a purchase, he could not resist trafficking in mandrake's lurid and fabulous history. No other plant—or family, for that matter—gets such sensational treatment in the *Letters*. "The root and leaves are stinking, and the whole plant is poisonous, though, in small doses, it is used medicinally" are Martyn's last words on mandrake.[31]

These theatrical accounts of two of the family's most "insane roots," along with some provocative comments about belladonna (of which, more, in the next chapter), make the rest of Martyn's discussion of the Luridae literally anticlimactic. The two genera containing most of the edible, "wholesome" species, the *Solanum* and the *Capsicum*, are relegated to the end of the section, where the English native and exceedingly common *Solanum dulcamara*—a species so available for examination that one would think it deserved more attention—gets only one brief paragraph: "Another shrubby sort, without spines, is the *Woody Nightshade*, or *Bitter-sweet*, which grows commonly wild in moist hedges. This has a climbing, flexuous stalk; the lower leaves lance-shaped, the upper ones sometimes trifid; the flowers in bunches, or branched cymes, coming out from the axils of the leaves; the corol revolute, purple, marked with two shining green spots at the bottom of each segment; and the berries red."[32] This close and accurate description would certainly facilitate identifying woody nightshade *in situ*; but compared to the highly dramatic presentations witnessed earlier, the clinical tone makes the plant rather unremarkable, even of less interest than the garden nightshade (*Solanum nigrum*) described next as a "common weed" whose frequent appearance "on dunghills, in gardens, and other richly cultivated places" invests it with more personality. Nevertheless, thanks to their identification as nightshades, these two *Solanum* species would be falsely considered to be as dangerous as their most notorious Luridae kin.

The Rousseau/Martyn *Letters* inspired two important epistolary botanies written primarily for females, the convention of lessons couched as a one-sided correspondence to a real or fictional person proving to be a highly effective pedagogical method useful for customizing information and conducive, as we have seen, to a chummy familiarity of tone. The first, Priscilla Wakefield's *An Introduction to Botany: In a Series of Familiar Letters* (1796), which was extremely popular for the first four decades of the nineteenth century, advanced the same "get out of the parlor and head for the gardens and fields" approach advocated by Martyn. But as an enlightened Quaker with a Girl Scout, no-nonsense sensibility, she easily dispensed with titillating lore and the sensational bits from Shakespeare when describing the Luridae, leaving only the disagreeable and dangerous and furthering the confusion over nightshade identification—a problem, as we already know, that would continue to plague the family at large and bittersweet in particular.

To appeal to the "young persons" designated in the preface as her intended audience, Wakefield devised a fictional correspondence between two sisters (ages unspecified): Felicia, saddened by a six-month absence from her sibling Constance who is visiting an aunt, writes to keep her own spirits up by rehashing the lessons in Linnaean systematics she is receiving from her governess. In the first letter to Constance, Felicia proposes an epistolary buddy-study plan: "You may compare my lessons with the flowers themselves, and, by thus mutually pursuing the same object, we may reciprocally improve each other."[33] Despite the fiction of the sisterly exchange, however, the voice that speaks in the following description of the Luridae as one of the families representative of the Pentandria seems undisguisedly to be Wakefield's, although Martyn's influence is indelible in the substance and structure of her comment, which is a succinct and accurate rendering of his information, sans poetry: "The same division of the first order of this class contains a tribe of plants, called Luridae, a name expressive of their noxious appearance and strong scent; marks kindly impressed by nature, to warn the incautious against their baneful effects, most of them poisonous in the wild state. But change of soil and cultivation have rendered even some of these eatable; others yield to the skill of the physician, and under proper management, are useful in medicine."[34]

After briefly acknowledging some of the Luridae's beneficial qualities above, the remainder of Wakefield's account of this "noxious family" is intended as a warning, she says, "to guard you against approaching the rest too familiarly."[35] Omitting discussion of all the cultivated ornamentals and edibles, "Felicia" describes only the five species most likely to be found growing wild in England—thornapple (datura), henbane, and the three species commonly known as "nightshade," *Solanum dulcamara*, *Solanum nigrum*, and *Atropa belladonna.*[36] Wakefield has especially disdainful things to say about henbane and belladonna (or as she calls it, "Dwale, or deadly Nightshade," evidently preferring not to mention its sexier name). With regard to the former, henbane's smell, "though very disagreeable, has not always been sufficient to deter ignorant persons from suffering the fatal consequences of its poisonous qualities," which are "madness, convulsions, and death." And there is none of Martyn's praise for the flower's beauty here; rather, the plant is irredeemably repugnant: "The whole is hairy, and covered with fetid, clammy juice designed, perhaps, to drive away insects, which would otherwise be injurious to it."[37]

Belladonna predictably gets the roughest treatment; but what is most troublesome about the presentation is the possibility of the reader's inferring that the three nightshade species are members of the same genus, with similarly shaped flowers and fruit, since the section introducing the three begins with the information that "the Nightshade [i.e., *Solanum*] is a principal genus in this forbidding order," whose "wheel-shaped corolla" and "round glossy berry" are the features that "readily distinguish the plants that belong to it."[38] Dismissing bittersweet and garden nightshade in one sentence, Felicia immediately moves to deadly nightshade and its dark terrors, making the two solanums seem complicit in the last species' crimes: "The berry of the woody Nightshade is red, and its blue blossoms sometimes change to flesh-colour or white, whilst the garden Nightshade is known by its black berries and white blossoms. The Dwale, or deadly Nightshade, is most fatal in its effects. The leaves are egg-shaped and undivided, the blossoms a dingy purple. Woods, hedges, and gloomy lanes mostly conceal this dangerous plant; though it too frequently lurks near the husbandman's cottage, whose children are endangered by the tempting appearance of its bright shining black berries."[39] The arrangement of information could lead one to think erroneously that bittersweet's berries are round and belladonna's flowers are wheel-shaped; and worse, if there are no berries present, all sorts of confusion could occur in trying to identify the species by the color of its flowers. Wakefield (or her editor) does name the separate genera in one-word footnotes, but the text itself makes no such distinction. What is abundantly clear, however, is that Wakefield and Felicia are one in urging that Constance and the reader leave these plants alone: the letter ends with this "account of the Luridae, from whose poisonous influence, I hope you will always be preserved."[40]

Although Rousseau/Martyn and Wakefield were widely read into the early decades of the 1800s, significant advances in botanical science were by that time already rendering their books outdated. The more sophisticated "natural" system devised by Antoine de Jussieu and modified by Augustin Pyramus de Candolle, which categorized plants based on what we now know to be evolutionary characteristics, led to a greater understanding of relations not only within families but between them, ultimately resulting in the discipline's profound respect for the nightshades late in the century. The natural system advantageously eliminated the derogatory designation *Luridae*, finally replacing it with its Jussieuvian name *Solanaceae* in honor of its largest, and in England its friendliest, genus. But in by far the most important epistolary botany of the nineteenth century, the family retains a purgatorial position, which its designation simply as "the Nightshade Tribe" reinforces.[41]

Ladies' Botany was written by the great Victorian botanist John Lindley himself who, despite his efforts to make the science a serious (and thus male) profession, worked to bring women up to speed with the natural system by publishing in 1834 and 1837 the two volumes of this "Familiar Introduction" addressed, following the example of the Rousseau/Martyn *Letters*, to a mother wishing to teach her children. Lindley had published a technical *Introduction to the Natural System of Botany* in 1830, but he considered the earlier book to be "far too difficult" for women and children, or as he puts it in *Ladies' Botany*, "for those who would become acquainted with Botany as an amusement and a relaxation."[42] In his introductory remarks from Letter 1, Lindley acknowledges his debt to Rousseau, whom he would have recommended, "if his inimitable Letters were not both incomplete and obsolete."[43] Still, the Rousseau/Martyn influence can be deduced not only from the chatty, so-called familiar tone, but also from Lindley's presentation of the nightshades, which dispenses with the associated Linnaean term "lurid," but not with the lurid details. In Letter 15, following his discussion of the borages, Lindley begins with what by now has become the obligatory voice of foreboding:

> From this harmless natural order, let us turn to one, the properties of which are too often dangerous. Henbane, Nightshade, and Tobacco, the narcotic Thorn-apple, with the half-fabulous Mandrake, whose roots were said to shriek as they were torn out of the earth to give effect to magical incantations, form, with a number of other plants, a large natural order, the prevailing quality of which is poisonous. Many of them are common wild plants, and none more so than the species called *Black Nightshade* (Solanum nigrum), which is sure to spring up wherever a spot of ground is neglected, and suffered to become waste.[44]

Lindley uses black (garden) nightshade as the representative species for illustrating the morphological characteristics of the family in detail; and after the rather sensational beginning quoted above, he generally sticks to brief, straightforward presentations of a few more solanums, plus belladonna, mandrake, henbane, datura, and tobacco. But he never lets his readers forget about the threat that these plants pose. The worst is *Atropa belladonna*, whose fruit "is the most venomous of all our wild berries; it is of a deep shining black, and follows a livid brown corolla" ("livid" is Lindley's substitution for "lurid" as well for henbane flowers, which "have a large dirty-yellow corolla, veined with brownish purple, which gives them a peculiar livid appearance").[45] Bittersweet, "whose red and tempting berries present a dangerous decoy to children," also falls into the threatening category.[46]

In fact, the most striking feature of Lindley's description of the family comes in his summary remarks about the solanums. After mentioning the culinary uses of tomatoes, eggplants, and potatoes, he ends his account of the genus with a long and prejudicial warning about those species Martyn considered as "wholesome":

> Here, you will imagine, is a singular assortment of eatable and poisonous plants in the same genus; but in truth, the fruit of these is in all cases deleterious till it is cooked; Tomatoes are stewed, Egg-plants are washed and fried before they are eaten, and it is not to be doubted that they all would prove injurious, if used in a raw state. The fruit of the Potatoe [sic] is notoriously unwholesome; and if the roots are not so, that circumstance is to be ascribed in part to their being cooked, and in part to their being composed almost entirely of a substance like flour, which in no plant is poisonous, if it can be separated either by heat or by washing, from the watery or pulpy matter it may lie among.[47]

With this kind of caution about the family's friendliest plants coming from one of the century's most eminent and influential botanists, it is no wonder that "nightshade" was so frequently a Victorian pejorative.

Moreover, the effect of these comments surfaces as late as 1869 when in *The Queen of the Air* John Ruskin describes the potato as "the scarcely innocent underground stem of a tribe set aside for evil."[48] Ruskin doesn't acknowledge *Ladies' Botany* as the inspiration for this comment, but his dramatic claim a few sentences later that "the nightshades are in fact, primroses with a curse upon them"[49] is based on the following observation by Lindley that concludes the botanist's discussion of the family:

> You will scarcely suspect that those prettiest of spring flowers, *Primroses, Oxlips,* and *Cowslips,* can be in any way related to the venomous [nightshade] plants I have just mentioned; nor do they in fact belong to the same tribe; but they are so similar in many respects that I shall have no oppor-

DULCAMARA   57

tunity more fitting than the present, to say a word to you about them. . . .
They are distinguished by one circumstance in particular, by which they
may be at all times known among our wild flowers with certainty,—their
stamens are not placed between the lobes of the corolla, as in the Nightshade
tribe, *but are opposite to them,* a very curious and permanent difference.[50]

This technical differentiation between two otherwise similar-looking families
is an illustration of the concept of "affinity," which Lindley defined in Letter 5
of *Ladies' Botany* as "resemblance in *most* characters of importance." At that
point Lindley says that the concept is "one of the great practical difficulties in
the way of the student of the Natural System of Botany," and even prefaces
his extended explanation of the term with the disclaimer, "If you do not understand me, you may skip all that follows upon the subject."[51] Nevertheless,
Lindley continues to employ the principle by calling attention to exacting
distinctions like the one above concerning the respective position of the two
family's stamens—a distinction that Ruskin repeats when he differentiates
bittersweet's flower (or as he puts it, "the purple and yellow bloom of the common hedge nightshade") from that of the cyclamen: "A sign [is] set in their
petals, by which the deadly and condemned flowers may always be known
from the innocent ones,—that the stamens of the nightshades are between
the lobes, and of the primulas, opposite the lobes, of the corolla."[52] Although
the difference in the arrangement of the floral parts asserted by both Lindley
and Ruskin would likely be lost on most of their readers, Lindley's accompanying plate helpfully displays the black nightshade's "corolla laid open" for his
readers, revealing the placement of the stamens (see no. 1 in fig. 5).[53] For both
authors, however, the contrast between archetypal innocence and incarnate
evil couldn't be more explicit. Ruskin's sensational characterization of the
nightshades as "primroses with a curse upon them" is a graphic restatement
and reminder of the family's lurid eighteenth-century reputation.

Not all of Lindley's successors treated the nightshades with Ruskin's animosity. Jane Loudon is particularly noteworthy for her measured assessment
of the family. The wife of the celebrated horticulturist and garden-design
theorist John Claudius Loudon, whose groundbreaking *Encyclopedia of Plants*
(1828) is the definitive Linnaean tome for flora in Britain (with the text written
by Lindley), "Mrs. Loudon," as she was known to her readers, helped to define
the genteel Victorian female-flower connection with her several gardening
books, which, like her magazine the *Ladies' Companion at Home and Abroad,*
promoted the gender association in their titles. She also carried on the lay
scientific tradition with her very accessible *Botany for Ladies, or A Popular Introduction to the Natural System of Plants,* published in 1842. Although Loudon

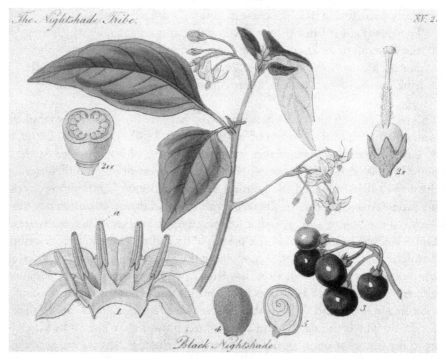

FIGURE 5. Black nightshade, John Lindley, *Ladies' Botany*. (Image from the Biodiversity Heritage Library; contributed by University Library, University of Illinois Urbana Champaign)

dropped the epistolary format, she assumed a personable, confiding voice as a means of allaying the technical difficulties she admits to having struggled with herself. Loudon speaks despairingly of her effort to learn botany from books, saying in the preface, "With the exception of Dr. Lindley's Ladies Botany, they were all sealed to me; and even that did not tell half I wanted to know, though it contained a great deal I could not understand."[54] To correct this problem, Loudon emphasizes the broader similarities defining plant groups rather than dwelling on the subtler—if more exacting—distinctions that we saw with Lindley. This is the case with the Solanaceae, the term Loudon uses to designate both a large order comprising four tribes as well as the name for the largest one within it, whose characteristics she describes charmingly: "The plants within this tribe are easily recognized by their flowers, which bear considerable resemblance to each other, and by their berry-like fruit, which always has a persistent calyx. The corolla is also always folded in the bud; and the folds, like those of a country woman's clean apron, are often so deeply impressed as to be visible in the newly opened flowers."[55] Loudon chooses a bittersweet flower and its berries to illustrate the genus *Solanum*, and comments with unusual appreciation and fairness, "The berries of the Bitter-sweet are red, and they have a very pretty effect in hedges and coppices, where they are produced in great abundance during the latter part of the summer and autumn; and those of the Garden Nightshade are black. Both of these plants are poisonous; but this is by no means the case with all species of the genus."[56] This balanced assessment characterizes Loudon's entire discussion of the Solanaceae, and considering that she was one of the most widely read — and prolific — botanical writers of the period, her objective treatment surely helped to rehabilitate the family reputation. The second edition of this work, published in 1851, was tellingly retitled *Modern Botany*, certainly to broaden its audience, but probably also to alert her readers that this book valued science over ladylike sentiment, with all its attendant moralizing.

The lurid legacy was not destined to die unsung, however. Anne Pratt, who was certainly Jane Loudon's equal as a prolific popularizer of botany,

FIGURE 6. Nightshade tribe, Anne Pratt, *Flowering Plants, Grasses, Sedges, and Ferns of Great Britain*, vol. 2 [1905]: plate 151. (Image from the Biodiversity Heritage Library; contributed by New York Botanical Garden, LuEsther T. Mertz Library)

but who wrote and illustrated mostly descriptive floras of indigenous plants rather than gardening or botanical textbooks, obviously took pleasure in recounting—and thus preserving—the nightshades' sensational story. Her most important work, *Flowering Plants, Grasses, Sedges, and Ferns of Great Britain*, a massive, beautifully illustrated, originally four-volume publication that went through many editions from the mid-1850s into the twentieth century, not only offered technical botanical information, but also served as a fascinating and valuable repository for all kinds of plant knowledge, lore, uses, and associations—which included the fact about the Solanaceae that "Linnaeus gave to it the name of Luridae, from the dull, lurid appearance of the flowers of many of the plants, which he regarded as indicative of their noxious properties."[57] But Pratt's account of the family here and elsewhere manifests none of the aversion to discussing its members that was apparent in earlier sources, and her illustration of the family seems to convey her appreciation (see fig. 6);[58] yet she persists in calling bittersweet's flowers a "lurid purple," a description that seems inaccurate except as Linnaean moralizing.[59] In any event, her opening remark for the entry on woody nightshade in *Wild Flowers* (1857)—"The lurid purple blossoms of this plant would lead the botanist to infer, at first glance, that poison lurked there"—may be the most dramatic assessment of the flower's color found in a Victorian botanical work.[60]

Dr. Dulcamara

During the same period that the language of flowers and "familiar botanies" were introducing a female audience to *Solanum dulcamara*, the species' scientific name, "dulcamara," was also gaining a new identity in Victorian popular culture, and as in floriography, one beginning with affairs of the heart. In this case, however, the surprising—and satiric—association developed from medical science's interest in those alkaloidal properties that gave bittersweet its romantic name, as well as its long medical history. Such an old and available British plant with easily discernible active properties must have been put to therapeutic use since ancient times; but references to it in the earliest medical manuscripts composed in England are extremely difficult to locate, no doubt owing to what seems to be the eternal confusion over its identity. The British scholar Alaric Hall has convincingly argued in *Elves in Anglo-Saxon England* that the Old English word translatable as "elf-vine," referring to a plant mentioned in two tenth-century "leechbooks" (physicians' manuals) as remedies "for fevers, madness, or ailments" attributed to elves, is in fact woody nightshade; but it had been misidentified as hellebore for centuries because of its being confused with a different plant in Dioscorides's first-century *De materia medica*. Hall postulates further that the plant could have been used to induce a mind-altering state. Given the rest of the family's known narcotic properties, these are fascinating possibilities.[61] In any case, by the early modern period the English herbals generally agree regarding bittersweet's primary virtues, which are described as being of a diuretic, laxative, and anti-inflammatory nature. Gerard's *Herbal* records that its leaves, twigs, and berries were made into "decoctions" to treat a variety of ailments, from "stoppings of the liver and gall," to bruises from falls (he calls this condition being "dry-beaten"—a term that especially amused Thoreau) and difficulty breathing.[62] Gerard also includes this choice recipe from the *Kreuter Buch* (1546) of a self-taught but talented and important German botanist (whose last name was actually Bock): "*Hieronymous Tragus* teacheth to make a decoction of Wine with the wood finely sliced and cut into small pieces; which he reporteth to purge gently by urine and siege those that have the dropsie [edema, or swelling] or jaundice [yellowish skin]." And finally, Gerard says, "it thoroughly cleanseth women that are newly brought a bed."[63]

We know that similar therapeutic uses for bittersweet continued to be in vogue in the nineteenth century among lay medical practitioners, since Gerard's catalogue of virtues appears in Nicholas Culpeper's *Complete Herbal and English Physician*, first published in 1653, but still touted as a reliable compendium of medical knowledge in reprints up until at least 1840.[64] In truth, many of Culpeper's recommendations were considered trustworthy in Victorian times by trained medical practitioners as well, but Culpeper had a reputation for being a quack even in his own day because of his astrological method of assigning occult properties to plants according to alleged planetary influence. In his wonderful *The Art of Simpling* (1656), Culpeper's rival William Coles launches a diatribe against the practice of "the gathering of Simples" by the stars that begins: "If Mr. Culpeper had but in a moderate measure understood this doctrine, or known but a tithe of what he pretended unto, the world had not been abused with such lame and imperfect directions, as he (in his English Physician enlarged) has left unto it."[65] Culpeper's approach did not, however, lessen the popularity of his book in rural Victorian England, where, as Owen Davis explains, magic and medicine were still a time-honored combination practiced by so-called "cunning-folk," whose literacy accounted for their status both as magicians and healers.[66]

In fact, Culpeper praises bittersweet for its magical properties, claiming that since it is governed by the planet Mercury and thus an herb "of very subtle parts," Amara Dulcis is "excellent to remove witchcraft both in men and beasts, as also all sudden diseases whatsoever." He recommends tying the herb "round about the neck" as a cure "for the vertigo or dizziness in the head," and notes that according to Tragus, "the people in Germany commonly hang it about their cattle's necks, when they fear any such evil hath betided them"[67] (William Coles includes the same remedy for cattle, calling this condition "the Staggers").[68] Interestingly, the botanical eminence David Mabberley notes in *The Plant-Book* (1997) that woody nightshade berries were strung "on strips of date-leaf" around the decorative collar of the Egyptian pharaoh Tutankhamun's third coffin, an indication that wearing a bittersweet necklace to protect against evil may well have been a widespread and ancient practice.[69] Perhaps it was for a similar effect, as Roy Vickery reports in *A Dictionary of Plant-Lore* (1995), that a Norfolk manuscript recommends necklaces of the dried berries to prevent convulsions in teething babies.[70] But even though Culpeper credits bittersweet with supernatural powers, he also reports on a less sensational but probably more effective external application in folk medicine for which the plant was well known: "Country people commonly use to take the berries of it, and having bruised them, they apply them to felons, and thereby soon rid their fingers of such troublesome guests." Coles notes this remedy as well and adds that the leaves can also be used, "stamped together with refry Bacon."[71] A "felon" in this case is an inflammation around the fingertip, and the condition and treatment were evidently common enough that, according to David Allen and Gabrielle Hatfield's research on British folk medicine, one of bittersweet's local names in several English counties was still "felon-wort" even into the twentieth century.[72] Vickery reports a related folk remedy in use in the Cotswolds of rubbing berries into the skin to cure chilblains; he says the berries were bottled and preserved for this use in winter. Allen and Hatfield note that in several English counties, a salve made from the juice mixed with lard served a similar purpose.[73]

As the record demonstrates, bittersweet's traditionally perceived benefit to humankind was certainly of a medicinal nature; and in the explosion of medical research that occurred from the early modern period into the nineteenth century, there was much experimentation to refine and extend its traditional applications. *Solanum dulcamara*'s medical heyday seems to have been the latter half of the eighteenth century after the great Linnaeus himself—rather ironically, considering his view of the whole nightshade family as "suspect"— brought it into general medical practice in Sweden for treating rheumatism, syphilis, and scurvy, among other complaints. A Linnaean dissertation on dulcamara written by George Hallenburg in 1771 praises its effectiveness as a diuretic and its success in relieving "violent ischiatic [sciatic] and rheumatic [joint] pains."[74] In England during the mid-eighteenth century, there was even a feud over the nightshades waged in print between Doctors Thomas

Gataker and William Bromfeild, the former promoting the medical uses of the nightshades in general and the internal use of *Solanum nigrum* in particular, while the latter warned of the dangerous consequences of these applications.[75] In his *Medical Botany*, first published in 1790 and still considered authoritative when an expanded third edition appeared in 1832, William Woodville summarizes the prevailing medical opinion at the turn of the century regarding *Solanum dulcamara*, saying it is "very generally admitted to be a medicine of considerable efficacy," recommended for an "extremely various" assortment of diseases, including rheumatism and "obstinate cutaneous affections." Woodville gives directions for administering the plant via decoctions of the "stipites, or younger branches" diluted with milk to prevent nausea, with a warning not to overdose. Even so, there is evidence in Woodville's comments to register doubt about bittersweet's ultimate therapeutic usefulness. First of all, "dulcamara does not manifest those narcotic qualities, which are common to many of the nightshades," a fact that makes the plant less dangerous, but also less potently therapeutic. And he reports at length that William Cullen, the esteemed physician and professor of the Edinburgh Medical School, had had less than satisfactory results in its use, in part because the potency of the twigs proved to be so highly variable.[76]

In 1809 dulcamara was admitted to the London Pharmacopoeia, the list of authorized drugs approved by the city's College of Physicians; but after the French botanist Michel Felix Dunal, who performed extensive experiments administering the plant to animals as well as to himself, discredited its therapeutic efficacy in his exhaustive treatise on the genus *Solanum* published just four years later, its use significantly declined. By 1829 the physician George Spratt, who reported Dunal's and other French botanists' experiments with dulcamara in his *Flora Medica*, could summarize its place in mainstream British medicine as follows:

> Dulcamara acts on the animal economy as a stimulus, exciting the action of the heart and arteries, and it is said to increase all the secretions and excretions, hence it has been recommended in a variety of diseases, by different authors: namely in rheumatism, scrofula, jaundice, dropsy, obstructed menstruation, and many cutaneous diseases, particularly lepra, for which it has been recommended as one of the most effectual remedies, taken internally and applied externally in the form of a lotion. Dulcamara is now, however, little used in this country.[77]

Spratt's concluding sentence, which seems to undercut everything that preceded, can probably be accounted for by the sheer number of other remedies available at the end of the nineteenth-century's second decade. By that time pharmaceutical science had advanced far beyond dependence solely upon the plant "simples" that had been the basis of the first medicines. Most importantly, the isolation of morphine from opium shortly after the turn of the century launched a revolution in plant chemistry leading to the discovery of numerous alkaloids and thus to the creation of both stronger and more controllable drugs. Consequently, Spratt could also report on solanine's having been recently isolated from both *Solanum nigrum* and *Solanum dulcamara;* and although he surmises that the pure alkaloid ought to be as useful in treatment as the plants themselves, experiments with animals were beginning to locate the therapeutic liabilities of the solanums in any form. Given what we know today about the ascendancy of morphine as the world's foremost painkiller, Spratt's concluding remark is prophetic: "It appears that Solanine, like opium, produces vomiting and sleep; that its emetic properties appear to be more violent than opium, but its narcotic properties much less so."[78] An appendix on solanine written by Spratt for the 1832 edition of Woodville's *Medical Botany* included the same information, but with the addition that "on man, a very small quantity of Solanine occasions great irritation in the throat, and excites a nauseous bitter flavour in the mouth"—results that could easily be produced, if desired, using any number of alkaloids, with more interesting and certainly more potent effects otherwise.[79]

At the same time that Spratt was prophesying dulcamara's demise in mainstream medical practice, however, the plant was establishing itself in what would prove to be a long-lived alternative approach to treatment, the homeopathy of the German physician Samuel Hahnemann, which had achieved both considerable popularity and notoriety throughout western Europe by the early 1830s. Hahnemann was a brilliant man, unquestionably a gifted healer as well as a dogged researcher and experimenter, distinguishing himself even among the Germans and French who were then the leaders in most aspects of medical science. But Hahnemann parted ways with the majority of his peers in his medical philosophy, most fully articulated in the *Organon of Rational Healing,* first published in 1810 and expanded several times during his lifetime. His chief principle, expressed in the Latin phrase *similia similibus curentur* ("let likes be treated by likes"), derived from his belief that substances capable of producing specific symptoms in a healthy person could be used to cure those same symptoms in a person who was ill.[80] Although practitioners of mainstream medicine sometimes implemented a similar approach to Hahnemann's, as in the use of vaccines, these "old school" physicians (as they were known to homeopaths) more frequently combated disease "allopathically"—that is, with treatments that opposed or suppressed symptoms rather than imitating them. As Philip Nicholls explains in his well-balanced account of homeopathy's place in British medicine during the nineteenth century, allopathic physicians tended to think of disease in terms of warfare and thus to employ what were considered heroic measures against it, such as copious bloodletting, violent purging, and heavy dosing with caustic substances, such as mercury to treat syphilis.[81] Hahnemann vehemently objected to such extreme means, preferring instead gentle treatments using the lowest doses of medicines possible to effect a cure. Moreover, Hahnemann and his followers advocated treating their patients holistically, taking into consideration their complete psychological as well as their physical condition.

The greatest challenge to homeopathy's credibility, though, lay in Hahnemann's method of producing his medications, which was by a process of diluting the active (and often poisonous) ingredient to such an infinitesimal degree that none of the original material remained—a process referred to as "potentizing." Based on his idea that disease was caused by some derangement of a person's "vital force," the mysterious energy that gave life to one's entire being, Hahnemann argued that the cure must partake of the spirit as well; thus, potentizing is a "spiritual (a dynamic, virtual) action"—releasing the spirit of the material substance as it dilutes it.[82] As one could expect, Hahnemann's opponents accused him of quackery, calling his medicines nothing more than placebos; but even if ineffective, at least they were safe—and as Nicholls no doubt correctly claims, homeopathy's counter to the heroic therapy of the day was one of its most important contributions to medical science. Even so, Hahnemann's pharmacological procedures incurred the wrath not only of mainstream physicians but also of apothecaries, who were used to dispensing drugs independently of doctors, and whose drugs were not acceptable to homeopaths in any case.[83]

During his lifetime Hahnemann painstakingly recorded the effects of ninety-nine substances, reporting the symptoms they produced in himself and his colleagues in great detail, and providing in his *Materia medica pura* (the first volume published in 1811) an amazing repository of firsthand information as well as an exhaustive record of previous medical results regarding the actions of mostly plant, some mineral, and a few animal-based drugs. These extensive "provings" (Hahnemann's term for testing on healthy subjects) were essential because of his further tenet, as he says in the preamble to the *Materia medica pura*, that "every case is an individuality, differing from all others" (hence, the emphasis on holistic diagnoses and remedies).[84] The provings, then, are an invaluable resource for the history of pharmacological medicine. And not surprisingly, given the nightshade family's known history as potent poisons, they are well represented in the homeopathic pharmacy.

66 VICTORIAN NIGHTSHADES

Belladonna and dulcamara appear among Hahnemann's original twelve provings published in the first volume; and as we shall see, belladonna in particular found a home in homeopathy, as did all the known poisons like aconite and arsenic, "potentized" to the point that their venom was transubstantiated into healing power.

From his first experiments Hahnemann determined dulcamara to be "a very powerful plant" that showed promise as an "antipsoric," a substance that could cure itching and eruptions of the skin, symptoms that accompany a wide variety of pathological conditions and thus figure prominently in homeopathic therapy. Hahnemann then identified what he considered to be dulcamara's proven applicability: "It will moreover be found specific for some epidemic fevers, as also for various acute diseases the result of a chill." The process for potentizing dulcamara was given as follows:

> The juice expressed from the young stalks and leaves of this shrub-like plant before its flowering time, mixed with equal parts of spirits of wine. Two drops of the clear fluid lying over the sediment are added to 98 drops of spirits of wine, the phial shaken with two strokes of the arm, and in this way diluted through 29 phials (filled two thirds full with 100 drops of spirits of wine), and each potentized with two succussions [shakings] up to the decillion-fold development of power; one or two smallest globules [usually miniscule lumps of sugar] moistened with this serve for a dose.[85]

Needless to say, achieving the desired potency of the medicine required considerable effort and skill; consequently, as homeopathy gained followers, it spawned its own druggists, not only apart from the traditional apothecaries but also from the homeopathic physicians. By mid-century in England there were firms like Leath and Ross or Headland and Company selling homeopathic medicines, elaborate wooden cases for their storage, and books of "domestic medicine" to be used by novices in treating their own conditions. For example, the prolific homeopathic doctor E. Harris Ruddock first published in 1858 (according to the preface) *The Stepping-Stone to Homeopathy and Health*, a book designed for domestic use, whose companion volume, *The Homeopathic Vade Mecum of Modern Medicine and Surgery. For the Use of Junior Practitioners, Students, Clergymen, Missionaries, Heads of Families, etc.*, served as a more complete reference for do-it-yourself treatment.[86] In short, by mid-century, homeopathy had created a flourishing market for products designed for self-medication, the "every man his own physician" approach touted from the days of Nicholas Culpeper and subject to as much derision.[87]

Via homeopathy, bittersweet as "dulcamara" gained notoriety, no doubt for

DULCAMARA 67

its alleged efficacy as treatment in so many conditions. According to Ruddock's *Stepping-Stone*, dulcamara was indicated for "various affections, such as cold in the head, nausea, catarrh of the bladder, mucous diarrhoea, etc., resulting from exposure to *damp* or a thorough wetting; itching and stinging eruptions on the skin, and other conditions following a cold. If taken immediately after exposure to damp, *Dulcamara* will often entirely prevent the ordinary consequences of a cold."[88] It was this close identification with homeopathy that was no doubt responsible for dulcamara's fame in an entirely different quarter of Victorian culture: as the comic theatrical personification of a con artist purveying love potions, giving rise to its name becoming a satirical term for charlatanism in any form.

Dulcamara's association with quackery officially begins in Milan, Italy, on May 12, 1832, with the premiere performance of Gaetano Donizetti's *L'elisir d'amore (The Elixir of Love)*, his brilliant comic opera in which a fake medicine serves as a catalyst for a tender romance between a simple Basque peasant and a wealthy landlady during the early years of the nineteenth century.[89] As the story goes, Nemorino overhears his would-be love Adina reading the legend of Tristan's winning Isolde by his drinking a magical potion, which instantaneously arouses Isolde's enduring passion. As Nemorino is wishing for just such an elixir, a garrulous itinerant con artist called Dr. Dulcamara arrives in the village to sell his nostrums, which allegedly cure everything from toothaches and impotence to bad complexions and liver disorders. When Nemorino inquires if the doctor has the love potion of Queen Isolde, Dulcamara sells him a bottle of Bordeaux, the consuming of which helps Nemorino—inadvertently—to accomplish his desire, overcome a rival, and win Adina's heart. The opera concludes happily not only for Nemorino, but for the doctor as well: instead of being discovered as the charlatan he is, the irrepressible Dulcamara leaves the village hailed as a hero.

Although *L'elisir d'amore* is closely based on Daniel Auber's *Le Philtre*, a French opera with a libretto by Eugene Scribe that had premiered in Paris not quite a year earlier, Donizetti's librettist Felice Romani changed the characters' names, and thus is responsible for the allusions they suggest.[90] And certainly, along with the good doctor's instructions to Nemorino for taking the potion (as translated into English, "Very gently to begin it, tip the bottle, lightly shake it[,] . . . then uncork it, but uncap it so the vapor won't escape it"), the name "Dulcamara" recalls the homeopathic method of potentizing bittersweet.[91] Moreover, in both homeopathy and the opera, the only active ingredient in the elixir is wine; it is not exactly a placebo, but its success as a medicine works, we might say, through a spiritualizing process. In any case, from the time that *L'elisir d'amore* first appeared in England—at London's

Lyceum Theatre on December 10, 1836—"Dulcamara" became a recognized synonym for a charlatan.[92]

To London's opera audiences from June 1839, the embodiment of Dulcamara was the imposing figure of the Italian bass Luigi Lablache, who already had an international reputation as a great operatic performer; he had even given voice lessons to the Princess Victoria the year before she became queen, and from then on counted her among his most loyal fans. According to his biographer Clarissa Lablache Cheer, Donizetti had actually composed the role of Dulcamara with Lablache in mind, and when the latter performed the grandiloquent doctor, "audiences usually split their sides at the sight of him, wearing an old fashioned red wig and squeezing his huge form into out-of-date eighteen-century clothing."[93] Lablache performed *L'elisir* in London at least a dozen times over the next sixteen years, thereby defining the role onstage and in familiar advertisements: the British Museum now houses a celebrated nineteenth-century engraving of him as Dr. Dulcamara holding a phial of his eponymous elixir (see fig. 7).

FIGURE 7. Sigr. Lablache as Dr. Dulcamara in *L'elisir d'amore*. (Courtesy of the British Museum)

Thus it was that the name and figure of the quack doctor entered Victorian popular culture. In the 1850s Dulcamara made rather frequent appearances in Dickens's weekly *Household Words* with satirical reference to a variety of charlatans. The most amusing is Henry Morley's description of a Boston "Rapper" newspaper, the *New England Spiritualist*, advertising "Purifying Soap, Nerve-Soothing Elixir, and Healing Ointment which have such virtues as only Doctor Dulcamara knows how to recapitulate, with the additional recommendation that they are prepared from Spirit directions—heaven-sent potions."[94] The most famous reference, however, was the lead story in the December 18, 1858, issue titled "Dr. Dulcamara, M.P.," written by Wilkie Collins, who took potshots at the Right Honorable Sidney Herbert for a speech advocating various nostrums, but especially for this august statesman's promotion of Charlotte Yonge's embarrassingly popular and excessively pious novel *The Heir of Redclyffe*, published in 1853.[95] Collins's article has since achieved notoriety, because Dickens implies in one of his letters that at least one draft of it was too scathing to print.[96]

In 1859 *Bentley's Miscellany* used Dulcamara to make similar sport of another politician, the great Quaker advocate of humanitarian causes and parliamentary reform John Bright, describing him in "A Bright View of Reform" thusly: "The dean of the faculty, Mr. Bright, has positively given up his private practice to amend the health of the country, and is wandering about at his own sweet will, administering reform pills like a political Doctor Dulcamara, which would infallibly cure all the diseases under the sun."[97] In 1866 the satirical weekly *Punch* followed suit with a political cartoon by illustrator John Tenniel titled "Dr. Dulcamara in Dublin," satirizing Bright's championship of Ireland by picturing him selling his elixir labeled "Radical Reform" to an Irish crowd as the remedy for its English-bred "disease"—the illness caused by the enforced supremacy of the Established Church in Ireland and the ownership of large Irish estates by the English nobility (see fig. 8).[98]

Crowning all these references, however, was left to William S. Gilbert, the wordsmith for the soon-to-be inimitable pair Gilbert and Sullivan of comic-opera fame, who in December of the same year had his first theatrical success with a burlesque of Donizetti's *L'elisir d'amore* entitled *Dulcamara, or The Little Duck and the Great Quack*, which held the stage for four months and helped to spread the doctor's fame to less sophisticated audiences.[99] Gilbert's Dulcamara, however, peddles a face lotion called "Madame Rachel's 'Beautiful Forever,'" a reminder of bittersweet's long history of use to treat skin complaints, but also a satirical jab at an infamous product actually on the market produced by the con-artist cosmetician Sarah Rachel Russell.[100] And finally, in an 1893 cartoon entitled "Dr. Dulcamara Up to Date; or Wanted, a

DR. DULCAMARA IN DUBLIN.

FIGURE 8. "Dr. Dulcamara in Dublin," John Tenniel, *Punch*, vol. 51 (1866): 193. (Courtesy of HathiTrust; contributed by the University of California)

Quack-Quelcher," *Punch* attacked the lack of legislation to control lay practitioners and purveyors of patent medicines with another depiction of the doctor as a rotund gentleman in eighteenth-century garb—with Mr. Punch in the background complaining about Dulcamara's performance to a policeman (see fig. 9).[101] In short, to quote a translated version of the doctor's patter from *L'elisir*, "Dulcamara, whose skill and mystic aura, healing power and graces, throughout the world are famous ... and ... and ... in other places"[102] had risen to international celebrity as the iconic charlatan: to opera and popular theater audiences, he was a delightfully bombastic and irreverent presence; but onstage or off, the representation of a truly "suspect" species.

Solanum dulcamara, the lovely but oft-misidentified and too-frequently

Figure 9. "Dr. Dulcamara Up to Date; or Wanted, a Quack-Quelcher," Linley Sambourne, *Punch*, vol. 105 (1893): 218. (Courtesy of the Wellcome Collection)

misunderstood plant that was responsible for giving the doctor his name, was once again almost lost in the shuffle. Amusingly, one of its few apologists turned out to be Wilkie Collins's nemesis Charlotte Yonge, who in *The Herb of the Field*, a book written for children about indigenous plants published the same year as *The Heir of Redclyffe*, noted that woody nightshade's berries and flowers were "very pretty"—even though the former "would make you very sick and giddy for some days if you were to eat them, but would probably not kill you unless you were very weakly."[103] The more typical reactions to the plant, as we have seen, were confusion, indifference, or contempt. The following farcical letter to the editor of *Punch* in its June 5, 1869, issue about a real tragedy provides an accurate statement of bittersweet's perennial dilemma.

72   VICTORIAN NIGHTSHADES

It also addresses most of the themes covered in this account of the plant's Victorian history and foreshadows the story of its infamous relative *Atropa belladonna* to come:

NIGHTSHADE AND NIGHTSHADE.

Mr. Punch,

The Pall Mall Gazette quotes from the *Liverpool Mercury* an account of a poisoning by the undersigned, attributed to another. According to this narrative, a man pulled a root out of the ground, mistook it for a carrot, ate a piece of it, was presently seized with convulsions, and died within ten minutes. "The plant proved to be Deadly Nightshade, Solanum Dulcamara." Sir, the Solanum Dulcamara is not the Deadly Nightshade. It is the bittersweet, or woody nightshade, no more poison than the potato — a mere simple. Your friend, Dr. Dulcamara, derived his name from that member of the *Solanaceae*. It promotes the functions of the skin, liver, and kidney, as one of your young men can tell you from personal experience; and your fair readers who want to be fairer may like to know, it makes a cosmetic potion better than anything advertised, to remove tan, pimples, freckles, discolorations, bubukles, and whelks, and knobs, and flames of fire. I form a cosmetic lotion, and I dilate the pupil. Therefore a preparation of me is used by silly women to give what they call expression to their eyes. Solanum Dulcamara is only a distant relation of mine, and has none of my powerful properties. I am the deadly nightshade. Ha! Ha! I represent Medea in the Vegetable Kingdom — see her picture in the Royal Academy Exhibition, by Mr. Sandys, — and agreeably to my fatally killing qualities, on which I do not scruple to own that I pride myself, am botanically named your

Atropa Belladonna.

P.S. At your service, if you want to be returned *felo de se*. Am to be found among the ruins of Netley Abbey. S. D. bears red berries in clusters. Mine are black, shiny, and single.[104]

4

Belladonna

The Deadly Nightshade

> Etymologists declare that the name of belladonna, which has been given to the deadly nightshade (*Atropa belladonna*), was so given because those to whom it was administered fancied they saw beautiful females before them. There is no doubt that it produces illusions of a singular character, and cases of impulsive insanity have resulted from its use in repeated doses.
>
> —Mordecai Cooke, *The Seven Sisters of Sleep* (1860)

Dwale

The 1869 fatality that prompted the indignant letter from "Atropa Belladonna" to *Punch* quoted at the end of the last chapter not only records its ongoing confusion with *Solanum dulcamara*; the letter was a grim reminder of how the misidentification of belladonna—the most poisonous species of the Old World Solanaceae, the true deadly nightshade, and one of the most toxic plants growing wild on British soil—could indeed lead to tragedy. According to the Pharmacy Act of 1868, of the six vegetable poisons in its list of fifteen controlled substances, belladonna was one of only three that could be found growing wild in the British Isles; the other two, aconite (*Aconitum napellus*) and poppy (*Papaver somniferum*), were naturalized in places, but their presence either outside or in cultivation wasn't much cause for alarm.[1] In fact, they were grown not only for medicine but for ornament, both species having very attractive, showy flowers and neither posing much of a risk for poisoning either by mistake or design: that is, they don't look edible, and opium from the latter requires some knowledge and effort to obtain. Belladonna, however, is a very different story.

As even one of its admirers, the late Wiltshire naturalist Heather Tanner, frankly puts it, "Deadly nightshade looks poisonous"; thus, it certainly does not have the appeal of conventionally ornamental plants.[2] Nevertheless, belladonna is unusually striking in appearance: its stout, erect stem forks into long branches with large oval leaves, each one matched with a smaller leaf on its stem, between which leaf-pairs hang single bell-shaped, dusky-purplish-brown flowers. The plant's dark, subtle coloring and its height—it can even reach five or so feet—give it a stately, brooding presence that most illustrations, like the one in *English Botany* by the famous illustrator James Sowerby (see fig. 10), simply don't convey.[3] Indeed, the plant's somber beauty seems lost on most commentators, who focus instead on the infamous plump and succulent, but toxic, shiny black berries that appear in the fall, understandably the usual culprits in accidental poisonings caused by the living plant. Truly,

FIGURE 10. *Atropa belladonna,* James Sowerby, in Syme, *English Botany,* vol. 6, plate 934. (Image from the Biodiversity Heritage Library; contributed by New York Botanical Garden, LuEsther T. Mertz Library)

the stars must have been lined up against its victim of 1869, for the odds first of fixing upon a species rarely found in the wild like belladonna, and second, choosing to eat its root, must have been incredibly small. The most toxic part of the plant,[4] the root is, ironically but not surprisingly, the part most valued in medicine: it is the source of the alkaloid atropia (now known as atropine) and the basis for a popular analgesic liniment introduced in 1860 by Peter Squire, "Chemist-in-ordinary to the Queen" and one of the leading pharmacists of the period.[5] The berries, which are mostly responsible both for belladonna's bad reputation and. hence, for its rarity, are actually the least toxic, and quite variable in potency; thus, they are not "officinal"—that is, not considered of use pharmaceutically in the preparation of drugs, nor even worthy of mention in the newly revised *British Pharmacopoeia* that appeared in 1867. But what the berries lack in potency they have more than made up for in notoriety.

Since at least Elizabethan times, belladonna had been treated as a threat to human society because of those damnably attractive black berries. Gerard advanced their evil reputation, opening his comments with the information that "this kinde of Nightshade causeth sleep, troubleth the minde, bringeth madnesse if a few of the berries be inwardly taken, but if moe be given they also kill and bring present death." Gerard evidently loathed "*Solanum Lethale*"—commonly known in England at the time as "Dwale, or sleeping Nightshade"—famously issuing the horticultural edict quoted in chapter 1: to "banish therefore these pernicious plants" out of gardens and frequented places because they are irresistible to children and pregnant women, who "long and lust after things most vile and filthie; and much more after a berry of a bright shining blacke colour, and of such great beautie, as it were able to allure any such to eate thereof." Gerard's injunction was prompted by a recent incident on the Isle of Ely whereby two boys who ate the plant's "pleasant & beautiful fruite" died in less than eight hours, while a third survived thanks to an emetic of honey and water. Gerard certainly knew about belladonna's medicinal virtues, conceding, "The leaves hereof laid unto the temples cause sleepe, especially if they be imbibed or moistened in wine vinegar. It easeth the intollerable paines of the head-ache proceeding of heate in furious agues, causing rest being applied as aforesaid."[6] But this acknowledgment of the plant's soporific and analgesic benefits comes like an afterthought to his condemnation, whose tone insinuates that deadly nightshade berries "allure" their victims with criminal intent. No plant in England had ever been endowed with such a willfully wicked character, one proud to commit first-degree murder like the *Atropa belladonna* who allegedly authored the *Punch* letter; and Gerard's accusations launched a program of extermination that became part of the succeeding herbal and horticultural tradition.

Thus Nicholas Culpeper's long-lived and influential *Complete Herbal and English Physician* (1653) comments tersely about deadly nightshade (assigning it a temperature according to the old theory of "humors"), "It is of a cold nature; in some it causes sleep; in others, madness, and shortly after, death." Next, Culpeper repeats Gerard's counsel: "This plant should not be suffered to grow in any places where children are, as many have been killed by eating the berries." Culpeper then concludes his otherwise rather brief remarks by narrating an eleventh-century "remarkable instance of the direful effects of this plant" that would become a Victorian staple of *Atropa belladonna* lore.[7]

The story, taken from George Buchanan's *History of Scotland* (1582), relates how the Scots defeated an invading army of Danes led by King Sweno by serving them a drink spiked with "a quantity of the juice of these berries" as part of a feigned truce. Culpeper explains, "This so intoxicated the Danes, that the Scots fell upon them in their sleep, and killed the greatest part of them; so that there were scarcely men enough left to carry off their king."[8] In the original version Buchanan's description of what he calls "*Sleepy Nightshade*" positively identifies it as *Atropa belladonna*, including that its "Berries are great, and of a black Colour when they are ripe"; and he suggests that the Scots knew how to calculate its effects rather precisely, saying that "the Vertue of the Fruit, Root, and especially of the Seed is *Soporiferous*, and will make Men mad if they be taken in great Quantities. With this Herb all the Provision was infected, and they that carryed it, to prevent Suspicion of Fraud, tasted of it before, and invited the Danes to drink huge Draughts of it." Evidently the Scots had done some proving of belladonna in advance, since they didn't fear sampling the brew themselves, thereby piling one treachery upon another. Buchanan gives the grisly details of the rout that followed, telling how the obliging Danes, later found "fast asleep and full of wine," were either murdered while they slept or if they happened to awake, were slaughtered while "running up and down like Madmen." The "dead drunk" king nevertheless made it out alive by being "laid like a Log or Beast upon a Horse" that transported him to the waiting ships.[9]

The Scottish event likely reveals more about deadly nightshade's early reputation than Culpeper communicates, for it reinforces a long and interesting etymological connection between deception and stupefying drink on the one hand, and deadly nightshade as "dwale" on the other. The great Cambridge philologist Walter Skeat defines *dwale* as a word of Scandinavian origin meaning deadly nightshade, "so called because it causes stupefaction and dullness."[10] Skeat doesn't indicate when "dwale" came into being as a name for the plant, but it's suggestive that according to the *OED*, the word's earliest and now-obsolete English usage was as a noun signifying "error, delusion, deceit, fraud," appearing in a tenth-century Old English translation of Bede's *Ecclesiastical History of the English People*—and thus before the infamous episode in Scotland. Could it be, then, that the felonious, knockout "Sleepy Nightshade" inherited a previously pejorative term as a result of this occasion? According to Skeat, both the Danish word *dvale* and Swedish *dwala* mean "trance." There's "a soporific dwale-drink"—called *dvale-drik*—known in Denmark; and by the fourteenth century "dwale" usually referred to a sleep-inducing beverage in English literature, even making a brief appearance as such in Chaucer's *Canterbury Tales*.[11] In the first English-to-Latin dictionary, dwale is specifically identified as an herb with "*Morella somnifera*" or "*Morella mortifera*," both of which are old Latin names for deadly nightshade.[12] In short, the plant's old name "dwale" may have resulted from a conflation of terms to indicate both cheating and drugging, thereby increasing its seriously dangerous connotations of a mind-bending sort; and like the disparaging epithet "deadly" itself, "dwale" may have been assigned to *Atropa belladonna* not only as a warning but perhaps as a curse. In any case, by the nineteenth century, many commentators reporting Buchanan's story would add that Shakespeare's reference in *Macbeth* to "the insane root that takes the reason prisoner" alludes to the same historical incident of Scottish treachery and thus to deadly nightshade, a theory reinforced by Buchanan's report that Macbeth was one of the generals of the victorious and deceitful Scots.[13]

With such dreadful reports coming from Gerard, Culpeper, and perhaps even Shakespeare, deadly nightshade's reputation could not help but plummet, and Culpeper's contemporary and rival William Coles seems to be one of the few writers of the day to defend its virtues. His comment in *The Art of Simpling* (1655) sounds very like an early Samuel Hahnemann: "And that Nightshade that carries death in its very name, prevents death by procuring sleep, if it be rightly applied in a Fever."[14] Presumably, though, Coles's method of "rightly" procuring belladonna's therapeutic benefits, like that mentioned by Gerard, had to do with the external application of its leaves and not with ingestion of any part of the plant.

By the eighteenth century the tradition of maligning belladonna for those treacherous, child-baiting berries had become canonical. The most dramatic and probably devastating example can be found in the 1735 edition of Philip Miller's influential *The Gardeners Dictionary*, the most important horticultural text of the century. The head gardener for the famous Company of Apothecaries' Physic Garden at Chelsea in London and the foremost authority on plants in England at the time, Miller himself grew deadly nightshade, but his entry on "belladona" was almost the kiss of death for the species in the wild. He begins, "This plant grows very common in many Parts of England about Farmers Yards, and in shady Lanes, but is never kept in Gardens, unless in those of Botanists; nor indeed should it be suffered to grow in any Places where Children resort, for it is a strong Poison"—and so on, recounting another recent incident of "several Children being kill'd with eating the Berries." Miller repeats a story told by his contemporary the botanist John Ray of a mendicant friar who drank "a Glass of Mallow-wine" adulterated with belladonna, with what would prove to be typical results: "In a short Time he became delirious, soon after was seized with a grinning Laughter, after that, several irregular Motions, and at last a real Madness succeeded, and such a Stupidity as those that are sottishly drunk have; which, after all, was cured by a Draught of Vinegar." And following an account of the Scottish legend taken verbatim from Culpeper, Miller concludes by sentencing deadly nightshade to extinction: "This Plant, being of so deadly Quality, should be extirpated wherever it grows wild, before the Berries are ripe, to prevent the dangerous Effects which may happen by their being eat. There are some Persons who give a Reward annually for destroying all the Plants which grow in their Parishes; and Her Grace the Duchess of Marlborough, constantly orders it to be rooted out from her Park at Woodstock, where it formerly grew in great plenty."[15] Although this recommendation was edited out of later editions of *The Gardeners Dictionary*, which even contain information about belladonna cultivation,[16] there is no question that Miller's advice had an effect, for later commentators no longer would call the plant "very common."

Thomas Martyn, the translator and editor of Rousseau's *Letters* echoes most of his herbalist predecessors—and even goes them one better: "The same poisonous effects follow from eating the young shoots of the spring boiled, as of the crude berries of autumn." Nevertheless, no doubt thanks to the program of eradication set in motion two centuries earlier, by 1785 Martyn could say that "*Deadly Nightshade* is rarely cultivated, and not common wild; it skulks in gloomy lanes, and uncultivated places, but is too frequent near villages in some countries."[17] Martyn's verb "skulk" is a sign of belladonna's outlaw status in England; and at the end of the eighteenth century Priscilla Wakefield's brief but disparaging remarks in *An Introduction to Botany*, quoted in the previous chapter, reinforce the image of belladonna as fugitive: "Woods, hedges, and gloomy lanes mostly conceal this dangerous plant; though it too frequently lurks near the husbandman's cottage, whose children are endangered by the tempting appearance of its bright shining black berries."[18] Wakefield's tone and the sinister verb "lurk" convey the same sense of a disreputable character as Martyn's description, further suggesting an escaped criminal and desperate species now on the lam.

Thus, deadly nightshade entered the Victorian era as a sort of botanical

Moll Flanders, a felonious highway rogue with no compunction whatever about drugging gullible Danes or murdering innocent children. The brief entry under its genus name "Atropa" in one of the horticultural bibles of the era, the wonderfully informative and very popular *Johnson's Gardeners' Dictionary*, first published in 1846 and continually (under slightly variant titles) into the 1900s, accurately records the prevailing opinion and outcast state of belladonna at mid-century. The plant initially appears in the 1852 edition with this ominous commentary: "We introduce this native weed for the purpose of warning country people from eating its berries, fatal accidents frequently occurring in consequence. The berries are at first green, but become black and juicy, of no horticultural value."[19] Obviously, belladonna was not a plant to be found in the formal park, the parterre, the greenhouse, or the cottage garden, nor was it likely to be found in the wild, "native weed" or not. But as in the 1869 incident, the plant's very rarity had lately proven to be an invitation to disaster, since there weren't many living specimens around to assist the uninformed with its identification.

Belladonna's presence in the *Gardeners' Dictionary* with its "warning" for "country people" was likely prompted by a sensational 1846 incident when just such a disaster had occurred, one that caused the plant's Victorian reputation to reach its nadir. In August that year an herb collector named James Hillard caused a shocking series of poisonings in London around Whitechapel, including two deaths, by selling the berries for making tarts. In the *Pharmaceutical Journal's* October 1846 issue, the article "Poisoning with the Berries of *Atropa Belladonna*, or Deadly Nightshade" told the story, which had been "published in all the newspapers." A thirty-four-year-old man and a three-year-old child were the fatalities, both having consumed the berries baked in pies. The *Journal* dutifully gave the progress of the symptoms, including delirium "which increased to a state of absolute madness," reporting that the man ate thirty berries and died in eighteen and a half hours, whereas the child consumed fewer and died in twenty-nine hours.[20] The toxicologist Alfred Swaine Taylor would report at length on a third victim, a fourteen-year-old boy, who was admitted under his care at Guy's Hospital, and after experiencing the typical symptoms of delirium over several days, at last recovered.[21] The purveyor Hillard was tried for "willful murder" because he had been in the herb business for some time and should have known what he was selling; but the record of the case in *The Proceedings of the Old Bailey* reveals an almost laughable ignorance about the fruit's identity—declared variously to be "nettle-," "wortle-," or "hoccle"-berries—on the part of practically everyone involved including Hillard, who, perhaps like a wily Scot, insisted he had eaten a dozen of the berries himself.[22] He was ultimately convicted of man-

Figure 11. "Atropa Belladonna, or Deadly Nightshade," *Pharmaceutical Journal*, vol. 6, no. 4 (October 1846): 176. (Courtesy of HathiTrust; contributed by the University of California)

slaughter and jailed for six months, but the incident became a *cause célèbre* in botanical and horticultural circles concerning the serious consequences that can result from not knowing how to recognize *Atropa belladonna* or its berries. Thus the *Pharmaceutical Journal*, the official publication of Great Britain's prestigious and powerful Pharmaceutical Society, included a woodcut of the plant to assist with identification (see fig. 11). It also concluded its account by reprinting a longer article calling for education about deadly nightshade from the September 12, 1846, *Gardeners' Chronicle*, written in response to the same episode at the height of its notoriety, while Hillard was "lying in prison" awaiting his trial.[23]

This second article undoubtedly captured the attention of a huge readership, for the *Gardeners' Chronicle and Agricultural Gazette* was (appropriately) as outstanding in its field as the *Pharmaceutical Journal* was in matters related to drugs. Founded by Joseph Paxton and John Lindley in 1841,[24] the periodical served as the leading British forum for all things horticultural, with articles by the likes of Charles Darwin and Joseph Hooker and with "The Horticultural Part Edited by Professor Lindley," as the cover proudly announced, thereby establishing the weekly's impeccable credentials. Although Lindley did not write the article in question, he may well have provided the accompanying magnificent drawing of belladonna berries that is its dramatic centerpiece (see fig. 12), since the illustration would appear in his *Medical and Oeconomical Botany* published three years later.[25]

Entitled "The Deadly Nightshade," the *Gardeners' Chronicle* article was one of an occasional series on "Familiar Botany" authored by "R.E.," whose

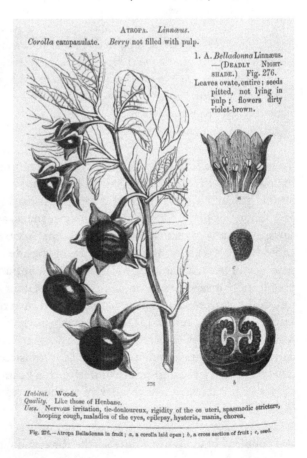

FIGURE 12. *Atropa belladonna*, John Lindley, *Medical and Oeconomical Botany*. (Image from the Biodiversity Heritage Library; contributed by Missouri Botanical Garden, Peter H. Raven Library)

previous offerings manifested a wide-ranging knowledge of plants and a penchant for poetic turns of phrase. This time, however, R.E. outdid himself. Although the article fulfills its purpose of promoting education by providing a wealth of information about the appearance, lore, and poisonous effects of deadly nightshade, all presumably provided in the service of preventing future catastrophes, its melodramatic excess certainly did not help the plant's reputation or foster its survival. Following an epigraph quoting the "insane root" passage from *Macbeth*, the opening paragraph, unquestionably crafted to capture all the excitement of the current London incident, is a masterpiece of disparaging botanical portraiture:

> In ruined and desolate places, in the skirts of woods, among the offal of a garden, there grows a "wicked weed" which our ancestors called Dwale. Its very name is portentous of misery and woe. Unlike some of those plants that conceal their venom under a fair aspect, or disguise it by a fragrant odour, this is fetid in its leaves, and repulsive in its flowers, which are dull pale chocolate coloured bells, with a lurid yellow bottom. There is not a point of beauty about the thing till it bears its fruit, but then it becomes only too attractive. No cattle will touch it; not a fly or grub finds a resting place or a pasture among its leaves, and it may even be said to be shunned by its own species, for it grows year after year singly, in the same place, without a companion near it.[26]

There is no question that the author has more interest in sensational effect than accuracy here, since the general tone and parts of the description are frankly subjective (that the flowers are "repulsive," for example, is a matter of opinion) or blatantly erroneous (insects, snails, and mammals, including cows, eat deadly nightshade—some with impunity; and members of its own and other species have no problem growing beside it). Moreover, the plant's presence in wasted habitats, while correct, is not a matter of some sinister predilection as seems to be implied, but the result of a long history of misfortune and massacre: that is, the plants found in such funereal locations are the vestiges of former cultivation, the surviving stock of a disinherited race. And while the beauty of its berries cannot be denied even by deadly nightshade's detractors, the phrase "only too attractive" exquisitely conveys the insidious, "wicked" quality the author attributes to this beauty. The only missed opportunity for abuse here is the lack of further explanation as to why the old name "Dwale," is "portentous of misery and woe," but the remark may allude to an alternate etymological tradition that appeared at mid-century in several popular botanies. Most notably, the Reverend Charles Alexander Johns's

BELLADONNA 83

well-loved *Flowers of the Field*, first published by the Society for Promoting Christian Knowledge in 1853, concludes its entry on deadly nightshade with the information that its "English name *Dwale* is derived from a French word *deuil* which signifies 'mourning'"—a claim that incurred the wrath of Skeat as "utterly wrong" (and interestingly, got edited out of later editions).[27]

Despite the gloom that pervades the introduction to the *Gardeners' Chronicle* article, the tone occasionally lightens to playfulness, as when the berries are called "a fruitful source of fatal accidents," and the Scottish affair is discharged waggishly: "The old chroniclers tell of a legion of Danes having been feasted by the Scotch, who finished them off by a mess of Dwale, from which they never woke." Moreover, following his recital of numerous instances of accidental and intentional poisonings, R.E. even acknowledges that "some constitutions are able to resist this poison better than others," and that "it seems doubtful indeed, whether any effect at all is produced by small quantities upon some people." But these anomalies are no doubt mentioned not out of dispassion, but because of their possible relevance to the case at hand: "These facts give some colour to the assertion of the man now awaiting his trial, that he did not know the berries to be poisonous, because he had eaten them himself."[28] In short, the thrust of the article is to warn the public of belladonna's deleterious effects, and R.E. never loses sight of this toxicological focus.

The article's conclusion returns to the occasion that prompted it in the first place and ends with its thesis, one that seems to recommend treating belladonna as a vegetable outlaw and its berries as contraband:

> But how deplorable it is that the population of a city like London should be so profoundly ignorant as not to know these berries when offered for sale. We hear of no policeman stopping Hillard's trade; indeed, it was at first supposed that they were Sloes that he was selling; even the reporters in the police-courts seem to have known no better. Would it not be as well if as much botany were introduced into our national schools as would prevent such fatal consequences as these? A very little instruction would render such instances impossible.[29]

Although its call for botanical education is admirable, the article may well have done more harm than good by broadcasting belladonna's status as a plant pariah without any redeeming qualities. Its overwrought tone certainly did not escape the notice of one *Chronicle* reader self-identified as J. Wighton, who in a later issue pointed out R.E.'s factual errors and guardedly defended belladonna, first in reference to its berries—"Though the latter are very poisonous, and though the plant contains much, still I do not think it is so bad

84   VICTORIAN NIGHTSHADES

as has been lately represented"—and then in telling the story of the plant's current sad plight: "The Belladonna, in general, is found in thickets, where it happens to escape the notice of the labourer, who is sure to destroy it, consequently it is very rare; indeed in some parts of the country it is hardly known. But none need mistake it who has seen the drawing of it in a late number, with its enticing but deadly berries."[30] The centuries-old assault on belladonna had obviously taken its toll, for this is truly a tragic tale of a plant forced into hiding merely to survive.

In any case, the fifties saw a rash of botanical publications written to educate a general audience, which, in response to the London fiasco, made a point of describing belladonna's berries in such a way as to avoid misidentification. Such an article even appeared in an 1850 issue of the *Lady's Newspaper*, a weekly devoted to current affairs and women's concerns, sandwiched between national news and a column entitled "The Paris and London Fashions." The article "Belladonna, or Deadly Nightshade" launched a series on English poisonous plants in part inspired by the three deaths and "dreadful sufferings" of others who had eaten berries of the "terrible Belladonna" sold in London in 1846 for "culinary purposes." The author, Georgina Ross, after rehearsing some of the usual lore including the Scottish story (with the obligatory mention of Shakespeare's "insane root"), carefully noted how the ripe berry was "encircled by the five little pointed green leaflets of the calyx," and minutely described the rest of the plant, which was illustrated on the following page.[31] Clearly, the climate of popular opinion in the late forties and early fifties with regard to belladonna could not have been more negative; and it is no wonder that in his essay celebrating the vegetable kingdom written for the Great Exhibition *Art-Journal* catalogue, Professor Forbes accuses the nightshade family of harboring "deadly enemies," since *Atropa belladonna* had proven to be not just a legendary but a contemporary culprit.

Another influential attack presented as a public service in the interest of botanical literacy came in Johns's aforementioned *Flowers of the Field*, which notes that "the berries are black, and as large as cherries, which they somewhat resemble in appearance, but may be readily distinguished by the calyx at the base." But Johns also helped to spread deadly nightshade's ill fame, commenting, "This noxious plant, the most dangerous growing in Britain on account of its active poisonous properties and the attractive appearance of its berries, is fortunately of rare occurrence, growing principally in old quarries and among ruins."[32] And the moralizing novelist Charlotte Yonge rose to the occasion with *The Herb of the Field*, which took book form in 1853 from essays "Reprinted from 'Chapters on Flowers' in the *Magazine for the Young*," as the title page acknowledges.[33] Her comments first about henbane and then about deadly nightshade exemplify how both species had become unusual sights and the latter an icon for evil. After describing the appearance of henbane's flower, Yonge instructs her readers, "If you find any keep the little ones from touching it, but don't pull it up, for it is so rare that botanists think a specimen a great prize, and lament if they find it gone from the spot where it once was known."[34] She continues, "Even more rare than the henbane, and still more poisonous, is the dwale, or deadly nightshade";[35] but rather than advising readers not to destroy the plant if spotted, she uses child poisoning by nightshade as the occasion for moralizing about neglectful adults, greedy children, and the nature of sin, ultimately leaving the plant to bear the burden of responsibility: "Poison fruits seem to be placed in this world in order to put us in mind of temptation and sin, which allure us at first and then destroy us. We may almost feel sure that the earth brought forth no poison before sin entered into the world, and death by sin."[36]

Given such a preface, Yonge's comment about actually seeing a specimen of the plant comes almost as a non sequitur, but she does try to convey a sense of belladonna's distinctiveness: "The dwale is very uncommon. I have only seen a plant of it. It was growing on an old bridge; it was very tall and branching, reaching some way above the parapet, with a quantity of light green downy leaves, and a profusion of dark, dull, reddish purple, bell-shaped blossoms, such a plant as no one could ever mistake once met with a description of it."[37] Despite the uniqueness Yonge attributes to belladonna's appearance, it would probably be hard to recognize the plant from her portrayal; and the fact that an enthusiastic botanizer like herself had seen only a single living specimen is further evidence that belladonna's reputation had taken on a horrific life of its own quite separate from its real existence.

Even that great Victorian popularizer of indigenous plants Anne Pratt could be alarmist when presenting *Atropa belladonna*. In her monumental *Flowering Plants, Grasses, Sedges, and Ferns of Great Britain*, the first version published in 1855 and likely the most invaluable repository of ethnobotanical information written for a general readership available at the time, Pratt gives "Dwale, or Deadly Nightshade" plenty of attention, citing a wide variety of sources, most of which I've already mentioned. But her portrayal not only echoes the ominous tone inherited from deadly nightshade's evil late-sixteenth-century and lurid eighteenth-century past; it also reflects the prejudice against the plant that had intensified after the recent incident in London. Beginning with "This is a rare plant, and, as its name imports, is so poisonous that we cannot wish it were more frequent," Pratt describes the flowers as "drooping bells of a dark lurid purple hue, which have a faint but unpleasant odour," and notes, "The whole herb has a dull gloomy appearance."

86 VICTORIAN NIGHTSHADES

Her comments indicate a familiarity with Johns's botany, since she repeats his etymology and gives the same information about how to identify the berries, which she follows with reference to Hillard's story:

> Not one of our British plants is so deadly as this, for its black shining juicy fruits, like small cherries, are highly poisonous, and produce fatal effects even if a small portion be taken. The calyx attached to these berries readily distinguishes them from cherries, but fatal accidents have occurred in their use by the ignorant. Even within the last few years a man was prosecuted for selling these berries in a basket about London, and though it appeared that he was unacquainted with the dangerous nature of the fruits, yet several persons suffered in consequence. Children have sometimes died through eating these sweet berries, and doubtless accidents would be more frequent but for the rareness of the plant, which has probably been in a measure extirpated by botanists and herbalists of former years.[38]

The last sentence once again locates belladonna's unforgivable sin, its perpetual appeal—and thus its danger—to children, a crime so heinous that eradication had formerly been considered justified even "by botanists and herbalists," as we have seen. Although "extirpation" was no longer part of the official Victorian botanical agenda, the program of national education promoted at mid-century was no doubt a double-edged sword: as more people learned to recognize the plant, more plants came to be destroyed.

DENIZEN

In short, by the mid-nineteenth century belladonna had become such a rarity that it was much better known for its widely broadcast, wicked reputation than for its presence as a living species. The rarity is obvious when reading the mid-century botanical journal the *Phytologist*, in which contributors always report their locating *Atropa belladonna* as a noteworthy event. The longest and most memorable comments come from one of the great Victorian botanizers in the south of England, William Arnold Bromfield, whose records of wild Hampshire plants appeared in several numbers during 1849. In general terms, Bromfield or his several Hampshire informants reported finding belladonna locally on the margins of human habitations: "In woods, thickets and waste shady places, along fences, amongst ruins, and on the sea-beach, but not common in the county, and extremely rare in the Isle of Wight, if not extinct there." A resident of the latter, Bromfield himself had never seen the plant on the island, but one of his correspondents, Mrs. Charles (Lady) Brenton, had provided evidence of its existence several years earlier near the village of Knighton—where it was "now quite extinct"—via "a coloured sketch made by her from a living specimen at the time."[39]

On the mainland, the "ruins" Bromfield specifically refers to certainly included former monasteries like Netley Abbey, which was presumably the home of *Punch*'s belladonna scribe since it is the site named in the fake letter's postscript.[40] Bromfield doesn't say anything about the abbey, but he located specimens nearby "in various places along the shore between Southampton and Netley, on the shingly beach, and under palings and banks."[41] Nine years later, an article on the plants of Netley Abbey in *The Phytologist* would verify belladonna's existence as part of the "coarse vegetation" that had taken over the grounds: "Trees of great girth and height have grown on the rubbish which covers the floors of the sacred edifices; deadly nightshade, nettles, and brambles, with other rank and bad-smelling plants, fill the area which was formerly trodden by men engaged in the worship and service of God, and whose dust is now concealed by the mouldering ruins and coarse vegetation."[42] Although the author evidently intends to suggest the ironic contrast between the unholy plants and the hallowed previous residents, the "deadly nightshade" probably comes from stock cultivated by the monks for medicine, as it did farther north at its most famous monastic site described in chapter 2, the Vale of Nightshade's Furness Abbey. A few pages later the *Phytologist* article comments, less dramatically but more specifically, "In the ruins of Netley Abbey the Atropa belladonna (Deadly Nightshade) grows, but very sparingly," which is considered "a fortunate circumstance" for the usual reason: "Its berries, tempting to the eye, are most dangerous; it is one of our most virulent of poisonous plants."[43]

Besides the ruins, the other habitats that Bromfield names for belladonna also reflect human influence, if perhaps in a less direct manner. This is certainly the case with Longwood Warren near Winchester, where the plant could be found "in very great abundance," a circumstance first reported, he says, "by my very zealous friend Miss G. E. Kilderbee!!!"[44] The triple exclamation marks are more than a sign of Bromfield's enthusiasm: they are a code indicating that he had verified "both plant and station" himself and could thus describe this location minutely, which he proceeds to do.[45] Following a sweeping overview of the warren's extensive terrain, which "swarms with rabbits noted in the market for their superior flavor,"[46] Bromfield offers an account of the vegetation so unforgettably overcharged that it would appear later in yet another British plant book written for a general audience, Edwin Lees's *The Botanical Looker-Out among the Wild Flowers of England and Wales*, in which it no doubt appears for its powers of titillation as a habitat worthy of Gothic fiction: "All the fetid, acrid, venomous and unsightly plants that Britain produces seem congregated on this blighted spot, a witch's garden of malevolent and deadly herbs, ready for gathering into her cauldron, which for aught I know may be nightly simmering and seething in this lone spot, as fitting a rendezvous for the powers of darkness on Hallowmaseve, as their favorite Blocksberg in the Hartz forest, for a Walpurgisnacht commemoration."

Although Bromfield needs to revise his witch's calendar (Walpurgisnacht is May eve, April 30, not Halloween, October 31), the "noxious brood" of plants he records finding certainly includes several traditionally linked to witchcraft, among them, of course, the members of the Solanaceae family: "the deadly but alluring dwale, the fat dull henbane," and "the rampant nightshade" or bittersweet (*Solanum dulcamara*), which as we have seen and Bromfield correctly acknowledges, really ought to be considered "free itself from the stain of bloodguiltiness."[47] But of more interest here is not the nightshade link to witches (of which more to come), but to warrens. Both bittersweet and belladonna are, according to the historical ecologist Martin Ingrouille, among the plants "particularly associated with rabbit burrows" and left to grow untouched because these nightshades are "unpalatable."[48] However, in a footnote Bromfield indicates otherwise, commenting, "Miss L. Legge, of Hinton Ampner, informs me that these animals devour the leaves of the Atropa with avidity, and strip the plants of their foliage as high as they can reach up the stem to browse upon it."[49]

In any case, rabbits and belladonna seem to have a parallel history in England, with both having been brought from southern Europe by the Romans in the first century CE, and more fully established via domestication and cultivation, respectively, in the twelfth century following the Norman Conquest, especially at the monasteries.[50] The rabbits most likely spread the belladonna seeds in their droppings; but for whatever reason, the two species seem to have stuck together, sharing habitats in England since their arrival. Moreover, most Victorian commentators agreed with Miss Legge that rabbits eat belladonna with impunity, and their flesh evidently can become poisonous if then consumed by humans—the "superior flavor" of the Longwood rabbits notwithstanding. Indeed, the possibility of belladonna poisoning by baked bunny meat would become a national concern and debate in 1865 when a Dartmoor family got sick from eating a rabbit pie, presumably implanted with atropia for murderous intentions by one Dr. Charles Sprague, who was finally acquitted, in part because the rabbit itself might have brought the belladonna to the table. The September 1, 1865, *Pharmaceutical Journal* article "Alleged Poisoning by Atropine" gave all the details.[51]

The rabbits were not enough to relieve Longwood Warren from being "dead, dreary and baleful," but these dour attributes had helped to ensure belladonna's survival.[52] Bromfield's last words on the species give the customary reason for belladonna's being suffered to grow unharmed on these out-of-the-way, "thriftless" downs: "Were Longwood Warren less secluded and nearer to Winchester, the extirpation of this virulent plant would be a matter of public necessity, for the avoiding of accidents to children or ignorant persons."[53]

Theatricalities aside, Bromfield and his informants' excitement over finding a rare, if "virulent," plant indicates a new attitude of respectful restraint and tolerance emerging among naturalists toward all species, an attitude that would ultimately prevent *Atropa belladonna*'s demise in the wild in England. The growing interest in the geographical distribution of flora and fauna would culminate in the 1859 publication of Darwin's *On the Origin of Species,* a work that greatly advanced an appreciation for all forms of life as part of nature's intricate organization and subject to its laws, thereby gradually diminishing the rhetoric of evil that had long been directed toward poisonous plants and animals. As plant geography emerged as a discipline, belladonna's spotty British representation became a matter of intense curiosity, with botanists wondering how it was still managing to hold its ground, so to speak, and more important, why it was growing in England in the first place.

The first botanist to attempt a systematic and comprehensive account of the distribution of British plants was Hewett Cottrell Watson, whose authoritative *Cybele Britannica,* published in four volumes from 1847 to 1859, was a landmark for British botany—and a major influence not only on Bromfield and most of the botanists of the day, but most importantly on Charles Darwin and his developing evolutionary theory. Named for the ancient goddess "who was supposed to preside over the productions of the earth," Watson's *Cybele Britannica* "intended to show [plants'] relations to the earth, as local productions of the ground and climate."[54] In other words, Watson was particularly concerned to determine what plants were originally and thus, it was assumed, naturally and legitimately "at home" on British soil, having gotten there without human involvement (what was less known at the time was how plants could have gotten there otherwise). To gauge what Watson calls their "civil claims," he devised "a scale of terms" beginning with the conventional "Native" for species considered to be aboriginal, or not having been introduced by human agency, and moving toward increasing foreignness or unaccountable, but questionable, origin through four more categories: "Denizen," "Colonist," "Alien," and "Incognita" (a sixth category, "Hibernian," designated plants that were "not found in Britain proper" at all, but were evidently native to Ireland).[55] Because of its long tenure in England and in deference to previous claims, Watson labeled belladonna a "Denizen," a somewhat purgatorial category whose usage in this context was evidently adapted from English common law, which allowed certain foreigners legal rights to citizenship except inheriting and office-holding, thereby denying full claim to nationality: "At present maintaining its habitats, as if a native, without the aid of man, yet liable to some suspicion of having been originally introduced."[56] Certainly, "without the aid of man" is an operative part of the definition here, since "man" had for centuries proven to be belladonna's sworn enemy—a circumstance that lent plausibility to its having "native" status.

In 1849 Watson estimated that belladonna could be found in only twenty British counties, fewer than a quarter of the total number, with a local habitat categorized as "Viatical," a term indicating "plants of the road-sides, rubbish heaps, and frequented places," sites that reinforce his "suspicion of its having been originally introduced" since these locations are disturbed areas, with soils that have been significantly altered by the course of civilization. However, Watson's extended comment about belladonna registers doubt about the accuracy of his judgment in light of opposition from leading authorities: "Admitted a native by Hooker, Henslow, Babington, etc. Various botanists, however, record its local habitats as liable to suspicions; and some deem it a plant originally introduced by monks. On the whole, the testimony seems in favour of holding it as a true native, on the chalk and limestone tracts, even although many of its present localities, about old castles and ruins of religious buildings, may have been of artificial origin in the past."[57]

The opinions of William Jackson Hooker, director of Kew and editor of *Curtis's Botanical Magazine;* John Stevens Henslow, professor of botany at Cambridge, author of *The Catalogue of British Plants,* and mentor of Charles Darwin; and Charles Cardale Babington, author of the *Manual of British Botany,* could not be ignored; but at the time Watson was writing, there was no hard evidence available to disprove belladonna's "native" status. That belladonna came in with the Romans would not be verified until the early twentieth century when the paleobotanist Clement Reid identified seeds found at the bottom of wells at Caerwent and Silchester while excavating Roman sites.[58] By 1995 Martin Ingrouille in his *Historical Ecology of British Flora* could say that "*Atropa belladonna* was very common in the Roman period,"[59] a circumstance that probably accounts for its greater distribution "on the chalk and limestone tracts" of southern England, where the Roman occupation was most extensive. If Buchanan's and Holinshed's histories can be trusted, deadly nightshade was abundant in Scotland in the early eleventh century—with the name "dwale" marking its pre–Norman Conquest arrival. There was little question, however, about belladonna's monastic connections, since its continued existence "about old castles and ruins of religious buildings" had long been a matter of record.

Toward the end of the Victorian era, the increased understanding about the migratory patterns, adaptive habits, and indeed, the evolution of species led to a more nuanced idea of nationality, one that recognized how tenuous were the claims of any species to be called "native." In a wonderful essay entitled "British and Foreign" written for *Cornhill Magazine* in 1889, the prolific science writer, novelist, and ardent popularizer of Darwinian theory Grant Allen explained the state of affairs quite succinctly: "Strictly speaking, there is nothing really and truly British; everybody and everything is a naturalized alien. Viewed as Britons, we all of us, human and animal, differ from one another simply in the length of time we and our ancestors have continually inhabited this favoured and foggy isle of Britain."[60] Allen covers the entire geological and natural history of England, using specific examples of species to show how arbitrary all claims of primal citizenship must be; and also (with a tip of his hat to Darwin's co-evolutionary theorist Alfred Russel Wallace) how global patterns of distribution have revealed that human agency alone cannot account for many species' introductions—oceans, birds, and shifting land masses can be just as effective in moving organisms from one location to another. Thus, Allen accepts belladonna as a "British species," whose presence must nevertheless be considered "very suspicious"—for reasons other than Watson's:

> In other cases, the circumstances under which a particular plant appears in England are often very suspicious. Take the instance of the belladonna or deadly nightshade, an extremely rare British species, found only in the immediate neighborhood of old castles and monastic buildings. Belladonna, of course, is a deadly poison, and was much used in the half-magical, half-criminal sorceries of the Middle Ages. Did you wish to remove a troublesome rival or an elder brother, you treated him to a dose of deadly nightshade. Yet why should it, in company of many other poisonous exotics, be found so frequently around the ruins of monasteries? Did the holy fathers—but no, the thought is too irreverent. Let us keep our illusions, and forget the friar and the apothecary in "Romeo and Juliet."[61]

Allen is of course being facetious about the monks' possibly criminal motives for bringing belladonna to Britain, but for once the blame for its wickedness has shifted from plant to perpetrator, at last relieving the former of moral responsibility. And unlike many of the more recent arrivals, it was not invasive like the "pretty blue veronica" that had lately become "one of the commonest

92   VICTORIAN NIGHTSHADES

and most troublesome weeds throughout the whole country."[62] Rather, as Allen explains using two other old but very rare species in Britain as similar cases, "belladonna has never fairly taken root in English soil. It remains, like the Roman snail and the Portuguese slug, a mere casual straggler about its ancient haunts."[63] Even with legitimate long-term claims to denizenship, there would always be something inherently "Not English!" about belladonna, as Dickens's smugly chauvinistic Mr. Podsnap would say, no matter how long it had been in the country.[64] Appropriately, rather than the Scandinavian "dwale" or German "nightshade," both terms with a longer history in Britain and close linguistic ties to Old English, the Latinate "belladonna," which evokes images of wild-eyed beautiful women, won out as the common name for the plant, the medicine, and of course, the poison.

Vegetable Neurotics

The fatal poisoning by belladonna that caused a stir in *Punch* and the popular press in 1869 also made its way that year into the *Pharmaceutical Journal*,[65] whose perennially keen professional interest in poisonings had recently hit fever pitch. Since the beginning of the century, advances in chemistry had resulted in the discovery of miraculous new alkaloids like morphine and compounds like chloroform that could effectively kill pain and provide anesthesia during surgery, but their unrestricted availability in the form of over-the-counters like laudanum or chlorodyne (a patent cocktail featuring chloroform and cannabis), as well as the easy public access to a host of other poisonous substances, sold as, among many other things, painkillers, cough medicines, infant pacifiers, tonics, salves, pesticides, vermifuges, dyes, cosmetics, and cleaning agents, had led to widespread abuse, innumerable accidents, and frequent fatalities. In fact, "a case of poisoning" had become a headline cliché, and many, like the infamous Bradford sweets incident of 1858, in which peppermints laced with arsenic killed twenty people and sickened some two hundred others, were inadvertently caused by the druggists themselves.[66] The July issue of the *Journal* had, in fact, opened with an article entitled "Prevention of the Misuse of Poisons," which called attention to two other recent fatalities, both resulting from pharmaceutical error.[67] The risks that poisons presented, though, merely served to underscore the fact that they had become extremely valuable commodities, embraced by everyone to be a necessary evil and by many as one of the great consolations of Victorian life. Poisons, after all, put the "toxic" in "intoxication," which was a hallowed and ubiquitous Victorian condition, cutting across age, gender, and class lines to reach such a peak by the sixties that drug addiction was rampant and overdosing a seri-

BELLADONNA 93

ous problem even "in the best-regulated families," as Dickens's Mr. Micawber would say.[68] Verily, the High Victorian period was at hand.

In an effort to curb the misuse of hazardous substances, raise the standards of the profession, as well as to consolidate its own interests, the Pharmaceutical Society had finally succeeded in pressuring Parliament to take action, a move resulting in the passage of the Pharmacy Act of 1868, which required that fifteen specifically named notorious and widely available toxins be labeled "poison" (with opium being the worst offender even though its omnipresence had almost gotten it omitted from the list), and sold only by chemists and druggists registered with the Society.[69] One of the fifteen was "belladonna and its preparations," which referred to the so-called "galenicals" or compounds made from its leaves, stalks, and root into extracts, tinctures, liniments, or plasters—the latter two mostly for pain relief via external application—or as "solution of atropia," a liquor composed of the alkaloid derived from the root mixed with alcohol and water.[70] Prescribed for a whole spectrum of ailments and other uses ranging from sedation to "dilatation" of the pupils, *Atropa belladonna* had in the 1860s come into its own as one of Britain's essential medicines—so much so that by 1882 in an article in the *Lancet*, "Some Medical and Surgical Uses of Belladonna or Its Alkaloid," Dr. J. H. Whelan would place it second in importance only to opium, which the medical community and the general populace alike understood to be the number one painkiller: "But while opium may be called the *prima donna* of drugs, belladonna vies with it, and may be called *bella donna* on its own merits and justice." Whelan praises belladonna in the form of atropine as a potential lifesaver when used before chloroform inhalation; predicts that with its use "cases of 'death from shock' ought to fade from the death register of surgical practice"; and notes that it is even "serviceable in annoying nocturnal emissions," among other applications.[71] Nevertheless, in the wrong hands a bottle of belladonna tincture or solution of atropia could be just as dangerous as the living plant—and far more accessible, serving only to increase deadly nightshade's long-lived reputation as a vegetable poison of mythic proportions. As purveyors and preparers of drugs made from raw plant material, the pharmacists had good reason to concern themselves with fatalities caused by belladonna in whatever form it presented itself.

The article in the July 1869 *Pharmaceutical Journal*, "Poisoning by Belladonna," relates in detail the full story of three miners scouting for a vein of lead on the Isle of Man who got sick after eating what looked like an edible root pulled up from a shrubby plant. Shortly after partaking, the miner who erred in the identification and ate the most, one Thomas Christian, "was seized with violent convulsions," whereas the other two, who had merely

94 VICTORIAN NIGHTSHADES

taken a taste, "also began to feel similar symptoms," and went for help in the way of an emetic or antidote. A fourth, wisely abstinent miner stayed with Christian, who purportedly died "in less than ten minutes." The article concludes with reference to the *Liverpool Mercury*'s erroneous claim that the plant in question was *Solanum dulcamara* and credits "Dr. Dudgeon," quoted in an article from the *Times*, with the correction, adding that the doctor "believes there is no case on record where death has ensued so speedily from *Atropa belladonna* as in this instance."[72]

As members of the Pharmaceutical Society and druggists trained in materia medica, the readers of the *Journal* no doubt applauded the correction regarding the plant's identity, and most would have accepted the doctor's other opinions in this case; for anyone conversant with contemporary medicine recognized that Robert Ellis Dudgeon, co-founder of London's Hahnemann Hospital and a well-respected, articulate homeopath, certainly knew his belladonna. Not only was Dudgeon an ardent proponent of belladonna's use and personally well-versed in its therapeutic effects, if in the small doses usually prescribed in homeopathic practice; as a translator of all Samuel Hahnemann's seminal works over the course of the nineteenth century, author of his own *Lectures on the Theory and Practice of Homeopathy* (1854), and one of the editors of the *British Journal of Homeopathy* since 1846,[73] Dudgeon was intimately acquainted with the record of Hahnemann's "proving" of belladonna—the extensive experiments the master and his colleagues had performed on themselves to determine the plant's physiological operation on healthy adults, presumably like the Manx miner. The individual symptoms Hahnemann listed in his *Materia medica pura* recording the personal provings as well as many reports from "old-school authorities" go on for over fifty pages and suggest both the variety and idiosyncratic nature of belladonna's effects, which include everything from vertigo, headaches, blurred vision, difficulty speaking or swallowing, to fever, rash, muscle spasms, deep sleep, troubled sleep, or insomnia, many curious episodes of delirium, as well as convulsions, coma, and death.[74] (Today, a mnemonic used to aid in detecting an overdose of atropine has conveniently generalized these symptoms as "hot as a hare, blind as a bat, dry as a bone, red as a beet, and mad as a hen.")[75] For Hahnemann the sheer length of the list was no cause for alarm, however, but for rejoicing: in keeping with the homeopathic rule of "like cures like," more symptoms meant more remedial applications. Thus Hahnemann hailed belladonna as a "polychrest," or all-purpose drug, and promoted it as one of the most indispensable medicines in the homeopathic pharmacy. His defense of its usefulness in the *Materia medica pura* is spectacularly outspoken: "Those small-souled persons who cry out against its poisonous character must let a number of patients die for want of belladonna, and their hackneyed phrase, that we have well-tried mild remedies for these diseases, only serves to prove their ignorance, for no medicine can be a substitute for another."[76] If Hahnemann's hostility toward his fellow practitioners is apparent here, it is no doubt because his promotion of belladonna resulted in probably the most ill-advised and acrimonious episode of his career.

In the introduction to the *Lectures,* Dudgeon tells the story of how Hahnemann cured six children and prevented six others and possibly more from succumbing to scarlet fever during an epidemic in 1799 in the town of Königslutter, Germany. Following his first principle, the law of similarity, Hahnemann reasoned that the rash that often erupted in healthy people from an overdose of belladonna was a recommendation for its use to treat the similar symptoms in the appropriately named "scarlet fever," a virulent, infectious, and often fatal children's disease that was one of the medical horrors of the day. Unfortunately for his reputation, however, Hahnemann got the idea of selling subscriptions for a pamphlet that, when published, would reveal the name of his newly discovered prophylactic—a move that further incensed the local physicians and apothecaries whom he had already offended with his iconoclastic practices. He eventually published his results without subscription, but the ire he had provoked among his allopathic colleagues succeeded in driving him out of town.[77]

Nevertheless, Hahnemann's championing of belladonna as both a cure and a preventative of scarlet fever brought it into universal celebrity among practitioners of every persuasion during the nineteenth century; famously, Beth in Louisa May Alcott's classic American girls' novel of 1869, *Little Women,* takes belladonna (to no avail) after being exposed to a family of sick children.[78] Although belladonna would ultimately prove ineffectual as a prophylactic, its use for this purpose at mid-century and beyond enjoyed so much popularity that Dudgeon compared its success to the vaccine for smallpox pioneered by Edward Jenner at the turn of the century.[79] In short, the Pharmaceutical Society—aligned as it was both professionally and philosophically with the regular medical establishment—might begrudge Dudgeon his authority and decry his methods, but it surely would respect his judgment about belladonna's operation on the healthy human frame.

Besides, the *Pharmaceutical Journal* had just the year before endorsed, at least tacitly, the homeopathic practice of proving poisons, with opium and belladonna being the two major drugs in question. The April 1868 issue had reported on the annual Gulstonian Lectures given at the Royal College of Physicians by one of the young regulars, Dr. John Harley of the London Fever Hospital, whose lectures entitled "The Physiological Action and Therapeutic

Uses of Conium, Belladonna, and Hyoscyamus Alone and in Combination with Opium" were particularly timely, given the restrictions soon to be placed on belladonna and opium when the Pharmacy Act passed three months later.[80] Harley's investigations had as their stated objective "to ascertain, clearly and definitely, the action of the drugs employed on the healthy body in medicinal doses, from the smallest to the largest"—an objective identical to Hahnemann's practice of proving. Moreover, Harley's approach, including his tests on animals, perfectly complemented a growing body of research in the developing field of toxicology, since the line between medicinal doses and overdoses of the four designated poisons had never been definitively drawn. Harley's groundbreaking study published the following year immediately became a Victorian medical classic—with a most Dickensian name: *The Old Vegetable Neurotics: Hemlock, Opium, Belladonna and Henbane, Their Physiological Action and Therapeutical Use Alone and in Combination.*[81] Through exhaustively recorded experiments on human subjects, himself included, as well as on horses, dogs, and other animals, Harley provided the most detailed and reliable information to date concerning the effects produced by four drugs well known for their "neurotic" proclivities—that is, their ability to act on the nervous system, as the OED explains, citing Harley's title as the single example for this now-obsolete usage. Harley's observations also reveal that because of belladonna's particularly alarming capacity for producing delirium, it lays special claim to the term "neurotic" in its more familiar usage as well.

The Old Vegetable Neurotics begins as a call to arms for the medical profession to get smart about prescribing the powerful and potentially dangerous narcotics named in the title. Hemlock, opium, belladonna, and henbane had all been known since antiquity to possess formidable sedative and lethal properties (hemlock was of course the Greek executioner's drug, whose paralyzing effects Plato describes in his account of the death of Socrates); but using these plants for their therapeutic benefits had always been risky. Even since the beginning of the century when isolating vegetable alkaloids became possible, thereby offering some degree of control over potency, dosage was still a matter of guesswork. Although Harley acknowledges from the start the extreme difficulty of precisely predicting the action of all drugs in any given case, considering the many variables at play, including "individual peculiarity," he nevertheless rails at his fellow physicians for their "want of intelligence" and irresponsible approach to medication. He complains in the preface, "In the present day, the use of secret nostrums is openly sanctioned and adopted. Patients are allowed to drug themselves to death with anodynes and narcotics. The profession includes numbers of men who, if they have faith in their practice, evince an ignorance discreditable to an Anglo-Saxon

Leech, and who, if they have not, are the basest of charlatans."[82] Harley set out to correct this problem through meticulously recorded trials designed to gauge the effects of increasing dosages of the four drugs and their alkaloids, "alone and in combination" with each other but particularly with opium, since the latter was the chief anodyne and narcotic that doctors prescribed. It was accepted as medical fact that opium's ability to relieve pain was unparalleled by any other drug; but it is a palliative, not a cure—except for one problem sardonically identified by who else but Samuel Hahnemann: "No medicine in the world suppresses the complaining of patients more rapidly than opium and misled by this, physicians have made immense use (abuse) of it, and have done enormous and wide-spread mischief with it."[83] Harley couldn't have agreed more, and his research goes a long way toward establishing how the three other "neurotics" not only serve as opium's handmaid, but in many cases are its superior.

In a comment that hints at the current critical state of affairs related to narcotic poisoning, Harley explains at the outset that a central goal of his research has been to address the "question of the antagonism" between opium and belladonna, because of late it "has assumed a very serious and important phase."[84] He was asking, in short, if, as several clinicians had proposed, belladonna could be an antidote to opium poisoning; and with tables of evidence to support his answer, it was a qualified "no."[85] In fact, in most doses belladonna tended to increase opium's effects—a result that did, however, offer some therapeutic advantages.[86] But what truly impressed Harley about belladonna was its extraordinary effect on the heart. In his words, "Belladonna must be regarded *first* as a direct and powerful stimulant to the sympathetic nervous system, or in other words, to the heart and bloodvessels."[87] From this singular capacity as a "vasculo-cardiac stimulant" proceed all of belladonna's other therapeutic benefits, which Harley identifies as the following: it is a diuretic, an oxidizing agent, an anodyne with hypnotic properties, and an antispasmodic. Harley then enlarges upon his stunning discovery:

> First as a *vasculo-cardiac stimulant*. It is remarkable that this, the primary and essential result of belladonna, should have been so long overlooked. So simple and immediate is the influence of this plant in exciting the action of the heart and so powerful in sustaining the force and rapidity of the circulation, that none but its own natural allies datura and hyoscyamus at all approach it; and in the directness and simplicity of its action it is superior to either of these. Simply then as a general diffusible stimulant, belladonna surpasses all other drugs, whether derived from the animal, vegetable, or mineral kingdom.[88]

As this dramatic statement reveals, Harley had successfully located, primarily in belladonna and secondarily in its solanaceous "allies," datura and hyoscyamus (or in common parlance, thornapple and henbane), an effect that would secure for the former a permanent place in modern therapeutics. Harley advocated belladonna's use for all medical "conditions and diseases in which there is depression of the sympathetic nerve-force," which included a number of acute and chronic illnesses as well as medical emergencies like the "failure of the heart's action from chloroform or other cardiac paralysers"—thus anticipating its importance in anesthesia for the decade and, indeed, the centuries to come.[89] Dr. Whelan's 1882 *Lancet* article mentioned earlier recommends injecting atropine subcutaneously before using chloroform;[90] and while the latter has been replaced by relatively safer anesthetics, belladonna is still used in cases of cardiac arrest during surgery. As Dr. Robert S. Holzman proclaims in a 1998 article in the journal *Anesthesiology*, "Hardly a practitioner of anesthesia begins administration of an anesthetic agent without the ready availability of atropine."[91]

Like Hahnemann before him, Harley was convinced that belladonna had not as yet achieved its full potential as a medicine, which he considers to have been largely "confined to its anti-spasmodic and anodyne properties";[92] and he reports case after case—from pneumonia and typhus to acute and chronic nephritis to neuralgia—in which his use of belladonna to treat patients seems to have been not only palliative, but remedial.[93] However, his recommendations came with a caveat: establishing for belladonna what is now known as the "therapeutic window"—the safe range of dosage for a drug, or the difference between effective medicine and egregious error, we might say—was a matter of crucial importance and obviously fraught with difficulty. For example, in the case of the cardiovascular stimulant, "In using it as such, we must bear in mind the fact which has so often forced itself on the attention during the course of our enquiry, that the stimulant action of the drug is soon superseded when the dose is excessive by depressant effects. In other words, the power of the drug to exhaust is in direct proportion to its power to stimulate." Moreover, both the dose itself and the margin for error were quite small: "Given with a view of exciting or sustaining the heart's action, the dose [of subcutaneously injected solution of atropia] will range from the 1/100 to the 1/60 of a grain, and it should never exceed the 1/40."[94] Finally, Harley's general directions for dosage came with this rather ominous warning: "In the medicinal use of the drug its deliriant effects should rarely or never be induced."[95] Indeed, such symptoms meant that someone had just shattered the therapeutic window, as it were, and matters would soon be getting out of hand.

BELLADONNA 99

Of course, all four of the old vegetable neurotics had long been known for their delirium-causing capability—in fact, trying to counter opium's frequent and unpredictable deliriant effects was one of the chief concerns of Harley's research. But delirium à la belladonna was a thing of legend, and Harley's rather tight-lipped comparison of opium and belladonna with regard to this feature of their "neuroticism"—their effect on the nervous system, which includes the brain—conceals volumes in its reserve: "The cerebral action of excessive doses of the two drugs differs chiefly, if not entirely, in the intensity of the soporific effects. Both cause delirium, but sleep converts that produced by opium into a dream, while the insomnia which accompanies the belladonna action allows of its active manifestation."[96] In other words, too much belladonna can make people crazy in a most visible, bizarre, wildly outrageous, and often frightfully embarrassing manner. For example, in his *Outlines of Botany* (1835), Professor Gilbert Burnett of King's College, London, reports an early episode from Plutarch in which Marc Antony's soldiers, "distressed for provisions" during their retreat from the Parthians, "challenged the wonder of observers" with their behavior upon consuming an herb that "brought on madness and death," which was apparently deadly nightshade: "He that had eaten of it immediately lost all memory and knowledge, but at the same time would busy himself in moving every stone which he met with, as if he was engaged on some very important pursuit."[97] Moreover, two of Harley's most important sources cite another, more recent episode of full-blown mass delirium caused by the consumption of belladonna berries.[98]

The story was reported in 1813 by one M. Gaultier de Claubry, who watched more than 150 French soldiers encamped near Dresden during the Napoleonic Wars succumb to the symptoms of belladonna poisoning. This version, from the first great work in pharmacology, Jonathan Pereira's *Elements of Materia Medica and Therapeutics*, is an English translation of de Claubry's original account:

> Dilatation and immobility of the pupil; almost complete insensibility of the eye to the presence of external objects or at least confused vision; injection of the conjunctiva with a bluish blood; protrusion of the eye, which in some appeared as if it were dull, and in others ardent and furious; dryness of the lips, tongue, palate, and throat; deglutition [swallowing] difficult or even impossible; nausea not followed by vomiting; feeling of weakness, lipothymia, syncope [giddiness, fainting]; difficulty or impossibility of standing, frequent bending forward of the trunk; continual motion of the hands and fingers; gay delirium, with a vacant smile; aphonia [loss of voice] or confused sounds, uttered with pain; probably ineffectual desires of going

to stool; gradual restoration to health and reason, without any recollection of the preceding state.[99]

Here is a scene that staggers the imagination—or "challenges the wonder," as Professor Burnett would say. The sight of 150 men with dilated and discolored eyes bulging and staring blankly or wildly while alternately dry-heaving, stumbling, falling, bobbing up and down, gesticulating, groaning, grunting, and best of all, experiencing "gay delirium, with a vacant smile" must have been both horrifying and hilarious, a spectacle for which Harley's description "active manifestation" hardly does justice. All of these symptoms had appeared somewhere in Hahnemann's list from his provings, but there they showed up in isolated cases, not as part of one maniacal group event. This, however, was a textbook case in belladonna poisoning. The dilated pupils are a first clue to the type of poison consumed, since they are belladonna's trademark, a symptom that occurs even in low doses or with topical application. Accordingly, belladonna is indispensable in ophthalmology for examining the eye and treating certain ocular disorders; but when ingested it can cause great disturbances in vision—exceeded only, it seems, by the fantastical inner sights to which the outside observer of necessity cannot be privy. Belladonna's ability to produce phantasms—assuming identification of the plant in the Greek text is accurate—had been known since the time of Theophrastus, the "Father of Botany"; and according to T. R. Forbes's "Note on Belladonna" (1977), it was Theophrastus's seventeenth-century translator, the botanist Bodaeus, who proposed that the name originated from the plant's curious ability "to arouse sexual fantasies of beautiful women," attributing the claim to Pliny.[100] Such imaginings would certainly explain the soldiers' "gay delirium," a frequently reported symptom that often provokes more than "a vacant smile"—as Pereira himself witnessed. His *Elements of Materia Medica and Therapeutics* not only includes a reference to the "remarkable and fatal effects on the Roman soldiers" mentioned above as well as the account of the French soldiers; it also contains the following description of the delirium caused by belladonna poisoning (with two fatalities) in seven patients he treated during his practice at the London Hospital:

> The delirium was of the cheerful or wild sort, amounting in some cases to actual frenzy. In some of the patients it subsided into a kind of sleep attended with pleasant dreams, which provoked laughter. The delirium was attended with phantasms; and in this respect resembled that caused by alcohol; ... but the mind did not run on cats, rats, and mice, as in the case of drunkards. Sometimes the phantasms appeared to be in the air, and various

BELLADONNA

attempts were made to catch or chase them with the hands; at other times they were supposed to be on the bed. One patient (a woman) fancied the sheets were covered with cucumbers.[101]

Pereira's personal observations give a better glimpse into what goes on inside the head of a person made delirious by belladonna; and whatever these visions are, some at least appear to be highly entertaining—at least up to a point. As Hahnemann recorded, in several cases these hallucinations must have been frightening, since they provoked a violent response, such as an inclination to bite.[102] More typically, the images seem to be attractive like the airborne fantasies here, which are such a common symptom that they gained a specific name in mid-nineteenth-century medicine. In a lecture on "The Nightshades" given in 1870, Dr. F. T. Griffiths noted that "amongst the symptoms of intoxication by belladonna is that which is termed *carphologie*, which means a seeking for little objects; the affected person imagines he sees insects everywhere around him, small birds continually flying before him, and he madly excites himself in their useless pursuit."[103] The hand and finger movements of the French soldiers and the Roman soldiers' searching under stones, then, would seem to result from illusions of this type, whereas the delirious woman's vision of cucumbers on her bed may require some other, possibly more salacious explanation, such as the female alternative to the male sexual fantasy. In any case, Pereira reports that Linnaeus categorized belladonna as "*a phantastic*," and with excellent reason, as these examples show.[104]

The term applies equally well to a solanaceous relative naturalized in Britain, the cosmopolitan *Datura stramonium* or thornapple, whose intoxicating properties, which are virtually identical to belladonna's, led to another famous large-scale poisoning in America so strikingly similar to the nineteenth-century French-soldier episode on the Continent as to be its seventeenth-century counterpart. In 1676 British soldiers sent to put down a rebellion of colonists near Jamestown, Virginia, were made ridiculously delirious for days by a salad of thornapple leaves. In his contemporary *History of Virginia* (1705) Robert Beverley, a planter who offers his credentials as "a native and inhabitant of the place," says that the soldiers ate "plentifully" of the "Jamestown weed,"

the effect of which was a very pleasant comedy; for they turned natural fools upon it for several days: one would blow up a feather in the air; another would dart straws at it with much fury; and another stark naked was sitting up in a corner, like a monkey, grinning and making mows at them; a fourth would fondly kiss and paw his companions, and snear in their faces, with a countenance more antic than any in a Dutch droll. In this frantic condition they were confined, lest they should in their folly destroy themselves; though it was observed that all their actions were full of innocence and good nature. Indeed, they were not very cleanly, for they would have wallowed in their own excrements if they had not been prevented. A thousand such simple tricks they played, and after eleven days returned to themselves again, not remembering anything that had passed.[105]

Obviously, the "gay delirium" and the carphological behavior were operating here in full swing; and like the later episode, this one was extremely protracted. Fortunately, both mass intoxications reached a similar conclusion for the affected soldiers—there evidently were no fatalities, nor any memories of what had taken place. The Jamestown episode itself will not be forgotten, however, for thornapple's common American name "jimsonweed" was given in its honor.

The third delirium-producing nightshade growing wild in Britain is the last of Harley's old vegetable neurotics, *Hyoscyamus niger* or henbane, a plant known for millennia to pack a formidable psychoactive punch and the member of the Solanaceae family that is the closest rival to opium in its narcotic effects, as Harley acknowledges.[106] Moreover, the two together compose "the most powerful hypnotic and narcotizing combination that can be formed."[107] In his *Elements of Materia Medica and Therapeutics*, Pereira had already noted that in small doses henbane often had more of a calming influence than opium, to which it was a preferred substitute under certain circumstances and with certain constitutions.[108] Consequently, henbane enjoyed a degree of popularity in the nineteenth century as an effective tranquillizer and evidently as a recreational drug. Charles Dickens was certainly aware of its virtues (and vices). In 1838 Dickens wrote his wife, Kate, that on taking henbane to relieve "such an ecstasy of pain" in his side that he couldn't sleep, "the effect was most delicious"—"exhilarating me to the most extraordinary degree yet keeping me sleepy."[109] Later in *Our Mutual Friend*, Dickens describes speculators intoxicated by "O mighty Shares!" as (metaphorically) "under the influence of henbane or opium."[110] And whereas its solanaceous cousin belladonna frequently produced trips of the raw-edged, rocky sort we have seen, henbane's rides tended to be smooth, as Harley explains: "The difference between the two drugs may be summed up in these few words. Compared with belladonna, the influence of henbane on the cerebrum and motor centres is greater, while its stimulant action on the sympathetic is less."[111] There was, however, an exception to henbane's operation as a kinder, gentler nightshade: Harley cautions against its use to treat weak, elderly patients, because they appear to be much more susceptible to its hallucinatory effects. As evidence, Harley describes in graphic detail a forty-eight-hour episode of energetic delirium experienced by a gentleman in his mid-seventies, for whom "there was complete insomnia, great mental vivacity such as he had not exhibited for years, perpetual talking, and occasional catching at surrounding objects." The old fellow evidently entertained his nurses all night long "recounting the adventures of a friend in the Peninsular campaigns," recalling memories of his own past in India and elsewhere, and moving in and out of coherent speech. All the next day he was "intent upon taking a journey," repeatedly calling for a carriage and trying to get up when he thought it had arrived. Clearly, his efforts at mental travel, at least, were quite successful, for after he had "wandered in the country," Harley says, "the next minute he introduced himself with a loud voice in a friend's house at Torquay, and, while engaged in imaginary conversation, suddenly raised the eyelids and looking across the empty space in the direction of the bare wall, said with much emphasis, 'That's a fine dahlia!'" Harley finally succeeded in putting the man's mania to an end with a dose of morphia (opium's alkaloid morphine), so that "four hours afterwards he fell asleep, and slept tranquilly and almost continuously for the next twenty-six hours."[112]

Harley's amusement at the henbane-induced insanity of the old man can easily be read between the lines of his narrative—and the other bouts of delirium described above certainly have their comical moments; but this is never to deny that the old vegetable neurotics unquestionably had a dark side— the one to blame for the name "nightshade" in the first place. For instance, the extremely toxic, hypnotic, and hallucinatory—that is to say, neurotic— powers of belladonna, thornapple, and henbane, along with their magical relative mandrake, are responsible for their infamy as the essential drugs in the witches' pharmacopoeia, and they were employed for other nontherapeutic purposes as well. These more-or-less clandestine applications—or so they would have appeared in the nineteenth century—will be taken up in the next chapter.

John Harley's research established him as the foremost authority on medical belladonna and henbane, championing these nightshades for a new generation of doctors and pharmacists. His findings were not only cited in future editions of *Dr. Pereira's Elements of Materia Medica and Therapeutics*, edited and abridged after Pereira's death in 1853 by Robert Bentley and Theophilus Redwood, two professors of materia medica and botany and pillars of the Pharmaceutical Society;[113] the results were also incorporated into the 1876 text of *Royle's Manual of Materia Medica and Therapeutics*, an important compendium of medical knowledge valued by students, which Harley himself edited.[114] As these publications make clear, John Harley quite literally wrote the book on the therapeutic use of the nightshades for the last quarter of the nineteenth century.

Atropa belladonna's increasing use as a drug beginning in the thirties following the isolation of atropine, and especially in the sixties and later following John Harley's advocacy as well as Peter Squire's development of an atropine-based liniment, meant that much larger supplies of the plant were in demand than ever before, and they had to come from somewhere. According to the various pharmacopoeias currently in use, the conventional wisdom was that wild belladonna plants were superior to cultivated ones, but their dearth in England meant that dried leaves and roots usually had to be imported, mostly from Germany, and these were often of inferior quality.[115] Not surprisingly, then, unlike their counterparts in the past, the modern-day herbalists—that is to say, the enterprising and poisonous-plant-loving Victorian pharmacists—became the great champions of belladonna's (and henbane's) cultivation on British soil. Belladonna had always been grown at physic gardens like the ones in Chelsea or at Guy's Hospital in London, but starting in the thirties, large-scale farming of medicinal plants began in earnest in several important locations. In a series of articles written for the *Pharmaceutical Journal* in 1850, Jonathan Pereira reported on medicinal plants grown around Mitcham in the county of Surrey, the most important area of England at the time devoted to raising herbs for the London market, where that year eight hundred acres were under cultivation. Although peppermint and lavender were the largest crops, poisonous species like aconite, belladonna, henbane, poppies, savine, foxglove, and hemlock were also grown.[116] According to Peter Squire, an average of six tons per year of belladonna and twenty of henbane were produced to supply the principal dealers in London during 1843–45.[117] The village of Market Deeping in Lincolnshire was the hub of another medicinal plant cultivation area whose principal crops also included peppermint, belladonna, and henbane. A *Gardeners' Chronicle* article in 1861 told the story of how the increasing demand for belladonna, henbane, and hemlock had led to the great expansion of their cultivation in the Herb Garden of W. Holland, since it was "quite impossible to get a supply" from wild plants nearby.[118]

The pharmaceutical firm of Stafford Allen and Sons, which grew large quantities of belladonna and henbane at their operations near Long Melford in Suffolk, produced in the early twentieth century a promotional booklet featuring brief histories of both plants accompanied by photographs, one showing belladonna being harvested in the fields.[119] Although the plants in the pictures are hard to identify, the extent of their cultivation is apparent. Finally, inspired by Pereira's articles on Mitcham, Robert Bentley in 1860 wrote a similar series on medicinal plants grown at Hitchin, Hertfordshire, for the *Pharmaceutical Journal* and reprinted in the *Gardeners' Chronicle*, the fourth of which included a detailed account of present-day belladonna cultivation. The two growers, Ransom and Perks, planted "on average" about four acres per year to supplement wild plant supplies, which were "by no means abundant in this country" and "becoming less so every year." By 1860, therefore, a major change in the source of belladonna for pharmaceutical use had occurred: "Nearly all the preparations of belladonna, in this country at least, are made from the cultivated plants." Moreover, the quality of the product could be better controlled by the speed at which the plant got to the laboratory, which was right on the premises: "From the herb growing in the field to its conversion into extract fit for use, only occupies from 12 to 16 hours."[120] Large-scale cultivation therefore replaced Ransom's former source of supply, which, as Eileen Wallace reports, had come from wild belladonna collected and sold to the firm by village women and children. The irony that the very ones thought to need protection from belladonna were once paid (a pittance) to harvest it need not be pointed out.[121]

By the turn of the century, family firms like Ransom and Sons, Peter Squire and Sons, and Stafford Allen and Sons had achieved substantial commercial success cultivating belladonna, henbane, and other medicinal plants, giving England an enviable international reputation for producing high-quality pharmaceuticals by way of scrupulous oversight from seed to drug, a method vastly superior to Continental practices, where peasant gatherers still collected most raw plant material from wild sources. In an article for the *American Journal of Pharmacy* in 1900, F. B. Kilmer describes the English drug farms, where "scores of acres of belladonna" grew, to be turned to medicine in laboratories adjacent to the fields, hidden from sight "by a thick hedge or a wall of solid mason work." Indeed, Kilmer continues, "the drug farms are especially well barricaded. Many of the fields are far back from the main roads and the lanes are guarded with signs of 'No Thoroughfare' or 'No Trespassers.'"[122] Thus, as the Victorian era came to a close, belladonna was being grown, to use Holmes's words, on an "agricultural rather than horticultural scale," necessitated no longer by the plant's scarcity, but by quality control and commercial expediency.[123] As in the days of the old monasteries, *Atropa belladonna*, maligned and condemned almost to extinction, once again found sanctuary, to be cloistered and coddled as a unique and extremely valuable medicinal species.

5

Victoria's Secrets

Sex, Drugs, and Belladonna

> Although my eyes got gradually accustomed to the darkness, I was almost on top of the outhouses before I saw the thick blur of the deadly nightshade. It was like a lady standing in her doorway looking out for someone. I was prepared to dread it, but not prepared for the tumult of emotions it aroused in me. In some way, it wanted me, I felt, just as I wanted it; and the fancy took me that it wanted me as an ingredient, and would have me.... There was no room for me inside, but if I went inside, into the unhallowed darkness where it lurked, that springing mass of vegetable force, I should learn its secret and it would learn mine. And in I went.
>
> —L. P. Hartley, *The Go-Between* (1953)

The Fatal Woman

In 1897 a novel by Caroline Ticknor, a scion of the Boston publishing dynasty Ticknor and Fields, satirized the evolution that had occurred in the meaning of deadly nightshade's species name over the course of the nineteenth century. *Miss Belladonna: A Child of To-Day* recounts the misadventures of its precocious eponymous narrator and her three younger siblings, all four of whom are "very handsome" she tells us, their physical attractiveness (and possibly her vanity) aligning with her name's original Italian meaning. But in the first chapter, Belladonna identifies herself further as a "homeopathic child" whose sisters Chamomilla and Ipecacuanha and brother Mercurius have similarly "medicinal names," given them by their father as "a kind of joke" played on his brother, who was a "'regular practitioner'": that is, "papa just did it to make

Uncle Jim mad, because he made such a fuss when father gave up castor oil and quinine for sugar pills."[1] In short, the heroine's name most directly and ironically derives from the highly poisonous plant commonly known as deadly nightshade; but except for complicity in the childish scrapes the novel recalls, this charming young Belladonna is quite innocuous—we could say as free from toxicity as the homeopathic medicine for which she is named.

To state the obvious, Ticknor's satire relies on the fact that at the turn of the twentieth century the use of "belladonna" to refer to a drug—even a homeo-pathic one—was at least as familiar as the word's reference to the "beautiful lady" signified by its Latinate etymological roots. But behind these divergent meanings lies the history of their connection, one ultimately grounded in the ancient and evidently cross-cultural tradition of equating women with plants. The popularity of floral names for girls is just one indication of this pervasive identification, for Roses, Lilies, Daisies, and Violets sprang up as numerously in Victorian nurseries as in Victorian gardens. These names recall the beauty and delicacy of flowers, their associations with enclosed domestic or culti-vated spaces, and we might also add with their sweet, fragile, serene, and even stationary existence. As the nineteenth-century language-of-flower craze reveals, the woman-flower connection had even gained momentum in the past hundred or so years, the surge following Linnaeus's classification system based on flowers in their capacity as plants' sexual parts. As Amy M. King ar-gues in *Bloom: The Botanical Vernacular in the English Novel*, the conflation of horticultural terms with human matrimonial ones led in polite parlance and parlors to the identification of a marriageable girl with a "bloom," a blossom-ing plant ready for reproduction, for sexual fulfillment and propagation.[2] (It seems germane here that "flowers" was also a term for menstruation, although it predates Linnaeus, in use at least from the seventeenth century.[3]) Of course, not all flowers have given their names or attributes to women: in nearly every case the extremely poisonous ones, species like aconite, henbane, and hem-lock, are not even anthropomorphized, but have retained an ungendered, es-sential planthood instead. However, in the flora of Victorian England there is one spectacular exception: *Atropa belladonna*, the real deadly nightshade, justifiably called by Ruskin the "queen" of the Solanaceae, came to life as a type of the sexy, marriage- and man-destroying, iconic Fatal Woman in the cul-tural imagination of the age, thanks to the combination of its ultra-feminine name with its virulently poisonous nature.[4] As the Victorian public became more familiar with deadly nightshade's scientific name through programs of botanical education, Belladonna literally took human form in the "botanical vernacular" as the spoiler in sentimental discourse: the beautiful but treacherous lady associated with unmentionable matters of a sexual nature—the privacies of the boudoir and the female body—a plant embodied as horticulture's whore.

Most scholars agree that the Italian physician and botanist Pietro Andrea Mattioli, in his discourses on the *Materia medica* of Dioscorides (1554), introduced Venice's popular feminine name for deadly nightshade to the rest of Europe. During the Renaissance, *herba bella donna* was intimately associated, it seems, with Venetian beauties—notably the city's actresses and courtesans—because of its use as a cosmetic; but as T. R. Forbes reports in his detailed inquiry into the name "Belladonna," there were various explanations for its specific cosmetic application. The most familiar of these has to do with belladonna's celebrated mydriatic effect: its ability to dilate the pupils and thus to give the user a wide-eyed, seductive appearance suggestive of sexual arousal.[5] It's possible that Titian's painting from the period, *Woman with a Mirror* (ca. 1515), which portrays a Venetian beauty at her toilette, suggests that she has just infused her eyes with belladonna, since the large oval mirror behind her serves as a visual pun for an enormous darkened pupil. But there were other beautifying possibilities: the juice could be employed as a wash to remove pimples or blemishes; to whiten the skin to an attractive pallor; or obversely, to heighten the woman's color by using it as rouge.[6] Whatever its application, though, belladonna's association with women for whom beauty was not just an advantage, but even a professional necessity, condemned the plant to a vegetable demimonde as the botanical representation of whoredom. In England Gerard was the first to report that *Solanum lethale*, then commonly known in English as "Dwale, or deadly Nightshade," was called "*Bella dona*" by the "Venetians and Italians"; and although he doesn't mention the plant's cosmetic uses, he hints at the negative sexual stereotyping when he describes the berry as being "of a bright shining blacke colour, and of such great beautie, as it were able to allure any such to eate thereof." In short, there is evidence that by the early modern period belladonna's reputation as a harlot of a plant associated, as Gerard says, with the vicious appeal of "things most vile and filthie," had reached England from the Continent; so it is not surprising that the Jacobean playwrights Beaumont and Fletcher called a "gentlewoman whore" a "night-shade" in *The Coxcomb*, a play produced not long after the first 1597 publication of Gerard's *Herbal*.[7]

Belladonna achieved full femme fatale status during the eighteenth century when, because of the plant's extremely poisonous nature, Linnaeus named the genus after the third of the Fates Atropos, the "unturning" one, who cut the thread of life for every human being. Added to the species name meaning "beautiful lady" provided by the Italians, the combination "Atropa belladonna" became one of the few scientific nomenclatures evidently registering irony as a description of a two-faced, murderous woman. Thomas Martyn, in his edition of Rousseau's *Letters on the Elements of Botany*, calls attention to this fact when he introduces deadly nightshade, commenting, "How the same plant should come to have the gentle appellation of Bella-donna and the tremendous name of Atropa, seems strange, till we know that it was used as a wash among the Italian ladies, to take off pimples, and other excrescencies from the skin; and are told of its dreadful effects as a poison."[8]

Possibly in deference to notions of Victorian gender propriety, the nineteenth-century botanies written for ladies by John Lindley and Jane Loudon refrain from explicit remarks about belladonna's association with tarnished sexuality. Nevertheless, Lindley (and probably Jane) knew the more salacious details of belladonna's story, for in his commentary for England's first comprehensive flora compiled by Jane's husband John, *Loudon's Encyclopedia of Plants* (1829), Lindley proclaims the plant's ill repute in patently sexual terms: "*Atropa*. A mythological name. Atropos was one of the Fates, and it was her especial duty to cut the thread of human life. The fruit of this genus is well adapted to fulfilling her office. A. belladonna (fine lady) has its specific name, according to some, from its being used as a wash among the ladies, to take off pimples or other excrescences of the skin; or, according to others, from its quality of representing phantasms of beautiful women to the disturbed imagination."[9]

There is no doubt from reading Lindley's description that the plant has now assumed the character of the euphemistically "fine" ladies who employ it as makeup to render themselves more desirable—he gleaned this much information from Thomas Martyn, whose cosmetic information Lindley repeats almost verbatim. But the lore about the plant producing "phantasms of beautiful women to the disturbed imagination" adds a new twist to the danger it poses by suggesting that belladonna puts men at greater risk than the here-unmentioned children who have historically been its most celebrated victims. Lindley doesn't offer the source for his comment, but it considerably increases the sexual connotations surrounding belladonna's image—as if these hallucinations are the ghostly female incarnation, the spiritual embodiment, of the plant itself. In any case, Lindley's portrayal confirms *Atropa belladonna's* reputation as a temptress of supernatural proportions, a treacherous female whose powers of seduction lead men irresistibly to their own destruction. Lindley doesn't mention belladonna's further connection to the attractions of Italian women, but he does note its profound action on the eyes, commenting that "the inspissated juice of the berries is used in the form of extract for anointing the eyelids in some opthalmic [*sic*] complaints," and that "its effect in dilating the pupil is quite remarkable"—a fact that accounts for the other, now more-familiar cosmetic application.[10]

By 1874 an article in the popular lay science journal *Hardwicke's Science-Gossip* would make explicit Belladonna's identity as a femme fatale, giving more sobering details about the "ancient belief that the nightshade is the form of a fatal enchantress, luring to destruction by her beauty";[11] and a decade later the folklorist Hilderic Friend would repeat the myth that deadly nightshade is in fact a supernatural temptress disguised as a plant. Friend's vast and popular *Flowers and Flower Lore* (1884) explains the genus and species names in a manner somewhat reminiscent of Lindley's; but he elaborates on the plant's beautifying functions while filling in details on the supernatural theme: "One of the names of this plant, Fair Lady, refers to an ancient belief that the Nightshade is the form of a fatal enchantress or witch, called Atropa; while the common name Belladonna refers to the custom of continental ladies employing it as a cosmetic, or for the purpose of making their eyes sparkle."[12] Obviously, in the more than fifty years that intervened between Lindley's and Friend's interpretations of *Atropa belladonna*'s name, the plant's mysterious and magical reputation as a sinister, shady female remained in currency—even as the living species had become increasingly rare in the wild. But whether supernatural or not, belladonna retained the Renaissance image of a dangerous woman—the femme fatale—throughout the nineteenth century; and a curious cheap late eighteenth- or early nineteenth-century print from the Bodleian Library's ephemera collection at Oxford seems to corroborate the plant's identity as a woman who traffics in the sinful Victorian underworld of prostitution and illicit sex. Presumably the man and woman in the illustration are planning to retreat into a privy-like building for sex, with another man, perhaps the woman's pimp, leading the way. Given this interpretation, the title "Antropa—Nightshade" could be a less-than-literate reference to a whore—if not exactly a "gentle" one—whose name derives from belladonna's genus.[13] With its clearly sexual subject, this print must have been rather raunchy in its day, thereby underscoring the plant's character as a shady, louche female and an outcast from decent society.

It is no wonder, then, that in the fashionably polite world of the Language of Flowers, belladonna's fate mirrored that of the Victorian fallen woman: like the living woman, it was treated as a pariah; or like the living plant, uprooted from its place altogether. In fact, belladonna's most dramatic presence in the genre occurs in France and very early, in Alexis Lucot's *Emblèmes de flore et des végétaux*, a work published in Paris in 1819, just eleven months before Charlotte de Latour's more influential *Les language des fleurs*, but evidently an important source for the latter, in which belladonna does not appear. Lucot's book is arranged as a dictionary with the flowers listed alphabetically, followed by their definitions—or as he puts it, the "emblem" for which each is representative. Belladonna, known in French as "Belle-Dame," appears, like the most dangerous of women, as the "Emblem of Despair."[14] The full entry can be translated as follows:

> The berries, which have somewhat of a resemblance to a cherry, serve in the composition of a cosmetic with which Italian women paint their faces; from this comes the name *Bella-Dona*. The same berries cause a short delirium to attack those who eat them, and they are then thrown into a furious madness which rather resembles despair: it is a Neapolitan belle who, betrayed by a lover, vows, in her despair, an implacable hatred for all men, and who, vindictive, as the saying goes, does not hesitate to sacrifice a large number of victims, if she can in her vengeance envelope the faithless one who wounded her vanity.

In other words, because the beautifying berries can also produce symptoms of madness, Lucot anthropomorphizes belladonna as the spurned courtesan so insane with "despair" that she cuts a wide swath of destruction just to take revenge on one traitorous lover. Thomas Martyn co-opted "blind Fury," a Miltonian coinage personified in *Lycidas* who "Comes . . . with th'abhorred shears, / And slits the thin-spun life" to describe the legendary wrath of the mandrake; but Milton's conflation of the Greek Furies with the third Fate, Atropos, is an even better characterization for belladonna as this vengeful woman, who, like Medea, is consumed by a rage at least as powerful as her previous passion; and she will stop at nothing in her effort to even the score.[15] Obviously, this violent woman is an aggressive, uncontrollable beast, as untamable as the most savage, predatory creatures in nature, so that Latour and her best-known English imitators like Frederic Shoberl, Henry Phillips, and Robert Tyas may have dropped Lucot's horrific entry from their flower language because it recounts an aspect of erotic love that is better left unspoken.

When "Belladonna" does appear in a handful of Victorian floral dictionaries, it usually means "silence," possibly a discreet allusion to its association with a forbidden erotic subject.[16] In some English vocabularies from the 1860s the definition is followed by an imperative "Hush!"—a rather superfluous injunction, it would seem, unless intended to have its opposite effect of arousing interest in belladonna's association with illicit sex rather than to prohibit its telling.[17] In any event "silence" is the definition—once again without further comment—that resurfaces in Kate Greenaway's now-famous 1884 version, the book that arguably marks the end of the genre in England

and thus served to stabilize a Victorian floral vocabulary for twentieth- and twenty-first-century audiences.[18] Curiously, though, in some dictionaries, the same plant is listed later in a separate entry as "Deadly Nightshade" meaning "falsehood," a reminder of the species' reputation for treachery, and perhaps intended as a counterpoise to bittersweet nightshade's usual meaning of "truth."[19] The best repository of nightshade meanings occurs in Anne Pratt and Thomas Miller's composite volume of sentimental botany entitled *The Language of Flowers; The Associations of Flowers; and Popular Tales of Flowers,* published circa 1847, whose offerings are fairly representative of the genre at mid-century. The dictionary section not only defines "Belladonna" as "silence"; "Deadly Nightshade" as "falsehood"; and "Nightshade, Bitter" as "truth"; it also offers an entry under "Nightshade" followed by the meanings "Sorcery. Scepticism. Witchcraft. Dark Thoughts."[20] All these definitions for various nightshade species had appeared or would appear elsewhere, but Pratt and Miller's list seems to be the most comprehensive—if also the most confusing for anyone seriously attempting to match a particular plant with a single, fixed meaning. Thus, while these various nightshade definitions expose the arbitrariness of the entire flower-language enterprise, they nevertheless reveal the pattern of negative associations with which we are now quite familiar, a pattern perhaps inescapable given a botanical name like "nightshade." They accentuate the reputation for evil and deceit that has traditionally plagued the entire family, while leaving "Belladonna," the one Solanaceae species explicitly endowed with a metaphorical existence as a woman, to be shrouded in ominous and mysterious secrecy—simply in "silence." The message seems to be that regardless of her beauty, this lady's wicked character demands that she be treated as a taboo (if perhaps a titillating one) by decent Victorian women who cherish the polite world of the language of flowers.[21]

Of course, Victorian popular culture was far more diverse in its tastes and proclivities than might be discerned solely from reading sentimental botanies, and belladonna's appearance in other venues and literary genres was not entirely governed by the earnest, tight-lipped propriety professed by the floriography of the era. In fact, several writers in other genres were happy to embrace belladonna's sinister reputation. For example, the great journalist, novelist, and satirist William Makepeace Thackeray mocks sentimental botany's de facto flirtation with (and inexperience of) the darker side of erotic desire in the character of one Miss Bunion, a spinster who appears in his Christmas story *Mrs. Perkins's Ball* (1846) as "the poetess, author of 'Heartstrings,' 'The Deadly Nightshade,' 'Passion Flowers,' etc.," which poems "breathe a withering passion, a smouldering despair, an agony of spirit, that would melt the heart of a drayman, were he to read them."[22]

VICTORIA'S SECRETS   113

Much more significantly, in *Vanity Fair* (1847–48), his most celebrated work and one of the greatest novels of the period, Thackeray jabs at sentimental botany again while also using it as a crucial motif, one that capitalizes on floriography's association with secret messaging in "the conduct of a love affair."[23] When George Osborne teases the corpulent nabob Joseph Sedley (recently returned from his lucrative post in the East India Company's Civil Service) about courting the novel's antiheroine, Becky Sharpe, with a nosegay, Osborne jokes, "Do they talk the language of flowers at Boggley Wollah, Sedley?" (a comment that is slightly anachronistic within the context of the novel, given that the scene takes place in 1813, several years before the fad actually began). Although Osborne is making fun of this floral mode of communication, the impecunious but opportunistic Becky, who is angling to marry Jos to avoid her fate of becoming a governess, in fact hopes (in vain) that the bouquet does contain a secret message—in the form of a "*billet-doux* hidden among the flowers."[24] Later, however, Thackeray's sarcastic treatment of courtship by nosegay takes an illicit turn when Osborne, now married to Jos's sister, Amelia, but enamored of the now also-married Becky, proposes elopement via a letter concealed in her bouquet, "coiled like a snake among the flowers."[25] Thus, while reference to the language of flowers begins lightly enough as a satire on the popularity of sentimental botany, Thackeray deploys floriography's fascination with clandestine communication to full advantage, not just as a local, comic motif but more importantly as a key plot device that reveals the sinister, hidden motives of his most reprobate characters—one ultimately employed to bring the novel to a (relatively) happy conclusion. That is, in a rare act of generosity, Becky facilitates the marriage of Amelia to her long-devoted would-be suitor Major Dobbin by showing the widow the letter Osborne had hidden in Becky's bouquet, thereby freeing Amelia from her misguided, undying devotion to her deceitful late husband.[26]

Given Thackeray's penchant for suggesting illicit sexual relations via sentimental botany, it is not surprising that his further appropriation of its terms manifests itself in the brief but memorable portrayal of a deadly nightshade made flesh, the "always jealous" Madame de Belladonna, an Italian gentlewoman who is the type of the vengeful courtesan representative of "despair" in Alexis Lucot's *Emblèmes*.[27] The beautiful but treacherous Countess of Belladonna appears late in the novel in Rome during Becky's "vagabond" period, after a disastrous affair with the debauched Marquis of Steyne and many other duplicitous dealings have driven Becky out of England. When the Countess, Lord Steyne's current mistress, discovers that Becky is a past and possibly present rival for her lover, she flies into a horrific rage; as Steyne's "confidential man" Monsieur Fiche tells Becky, "Madame de Belladonna made him a scene about you, and fired off in one of her furies." We see the Countess next riding in Steyne's barouche, a petulant floral voluptuary, "lolling on the cushions, dark, sulky, and blooming, a King Charles in her lap, a white parasol over her head, and old Steyne stretched at her side with a livid face and ghastly eyes."[28] Significantly, Thackeray concludes this episode with the information that the disreputable Steyne "died after a series of fits" in Naples with Madame de Belladonna by his side, and the author's last words about the Marquis cast the Countess's involvement in his death in a most suspicious light:

> His will was a good bit disputed, and an attempt was made to force from Madame de Belladonna the celebrated jewel called the "Jew's-eye" diamond, which his Lordship always wore on his forefinger, and which it was said that she removed from it after his lamented demise. But his confidential friend and attendant, Monsieur Fiche, proved that the ring had been presented to the said Madame de Belladonna two days before the Marquis's death; as were the bank-notes, jewels, Neapolitan and French bonds, etc., found in his Lordship's secretaire, and claimed by his heirs from that injured woman.[29]

It requires very little reading between the lines to determine that Madame de Belladonna (and probably Monsieur Fiche) had more to do with Steyne's "lamented demise" than just making off with his portable property, although this would have been reprehensible enough. That the valuables were bequeathed to the Countess just two days earlier hints that the timing of Steyne's end was not a coincidence, thereby raising the distinct possibility that his fatal "series of fits" was brought on by some agency—such as the poison that bears the Countess's name. Moreover, the case against Madame de Belladonna is strengthened by the fact that she thematically mirrors Becky herself, who, despite the latter's playful disposition and customary affability, is the novel's consummate fatal woman, every bit as adulterous, treacherous, and murderous as this toxic Italian beauty.

Indeed, Thackeray forces the comparison of these two women by presenting Becky's Roman adventure as a flashback between chapters recounting how she again meets up with and manages to seduce the weak and malleable Jos Sedley, who can't resist her fatal attractions. At the end of the novel, we learn that Jos takes Becky on as his traveling companion who nurses him "through a series of unheard of illnesses" and oversees his finances until he dies suddenly, leaving nothing but an insurance policy whose only beneficiaries are his sister, Amelia, and "his friend and invaluable attendant during sickness, Rebecca." As in the Steyne-Belladonna affair, Becky's inheritance is contested, in this case by "the solicitor of the Insurance Company [who] swore it was the blackest case that had ever come before him"; but Becky's own solicitors "declared that she was the object of an infamous conspiracy, which had been pursuing her all through her life, and triumphed finally." Thackeray ultimately describes Becky—with his tongue firmly in his cheek—as a "most injured woman" like Madame de Belladonna,[30] and the close parallel between their characters and stories indicates that in the context of *Vanity Fair's* floriographic motif, Thackeray intends Becky to be also read as a deadly nightshade, a fatal woman of the first rank.

Mid-Victorian fiction's most stunning flesh-and-blood nightshade appears in George Meredith's first major novel, *The Ordeal of Richard Feverel* (1859). The fascinating Bella, a courtesan known in London's demimonde as "Mrs. Mount," appropriately rides into the novel "driving a pair of greys," captivating the hero and ultimately helping to wreck his life. Richard at first proposes nicknaming this "glorious dashing woman" after the Roman goddess of war, "Bellona," which would also suit her relations with men; but his wise former nurse Mrs. Berry offers a corrective for the etymology of Bella's name (which she later repeats): "'My opinion is—married or not married, and wheresomever he pick her up—she's nothin' more nor less than a Bella Donna!' as which poisonous plant she forthwith registered the lady in the botanical notebook of her brain." The narrative rejoinder in the next sentence—"It would have astonished Mrs. Mount to have heard her person so accurately hit off at a glance"[31]—foreshadows the next chapter, which anatomizes Bella's identity as a nightshade temptress and witch, the terms familiar to sentimental botany and received botanical lore.

Titled "An Enchantress," the chapter depicts Bella in all her fatal, spellbinding glory as she seduces Richard into an extramarital affair. The naïve Richard has no illusions about Bella's profession, but fancies he can save her from her sins. Richard's plan doesn't stand a chance, however, for Bella, who "ought to have been an actress," deploys a devilish gift for impersonation to ensnare her victim: "Various as the Serpent of Old Nile, she acted fallen beauty, humorous indifference, reckless daring, arrogance in ruin," while the smitten Richard believes her sincere through all her changing moods and variable appearances. She even masquerades in cross-dress as a dandy named "Sir Julius," an act that "doubled [her] feminine attraction," given that "the contrast in her attire to those shooting eyes and lips, aired her sex bewitchingly." This skill at shapeshifting for sensual effect recalls what John Lindley had to say about belladonna's legendary reputation: like the plant, Bella is adept at "representing phantasms of beautiful women to the disturbed imagination"—including her exploiting the titillation of transsexual behavior. Intoxicated by this powerful dose of belladonna and often by champagne, Richard is indeed disturbed and aroused by Bella's consummately seductive wiles. After one performance he wonders, "Was she a witch verily? There was sorcery in her breath; sorcery in her hair; the ends of it stung him like little snakes." Spellbound by her dark eyes like those of the Venetian ladies for whom the plant was named ("They had a haughty sparkle when she pleased and when she pleased a soft languor circled them"), Richard casts himself in the role of "knight" in this romance ("He was a youth, and she an enchantress. He a hero; she a female will-o'-the wisp")—not realizing "that he played with fire." Indeed, just as belladonna (as we shall see) was considered by folklorists to be a "favorite of the devil," so the flames that lap around Bella are none other than hellfire, which burn with the nightshades' characteristic glow: "A lurid splendor glanced about her like lights from the pit."[32]

Thanks to *Atropa belladonna*'s feminine connotations, Victorian night-shades of the metaphorical variety were almost exclusively women, with two notable exceptions. A fictional horse by that name belonged to a legendary seventeenth-century rogue, as celebrated in the title of James Malcolm Rymer's penny dreadful *Nightshade, or Claude Duval, the Dashing Highwayman*, published circa 1865. But more in keeping with the plant's poisonous and clandestine associations, the Irish politician and militant Protestant William Johnston made *Nightshade* the title of his novel, published in 1857, which achieved some popularity, according to the Scots poet George Gilfillan, "as a manly, earnest, and faithful exposition of that dark arm of the Popish Styx, called Jesuitism; an arm which the rather requires exposition as it winds on in secret, and with tortuosities of the most inscrutable kind."[33] In other words, the novel is an anti-Jesuitical rant of the most extreme variety, and upon its publication the *Saturday Review* derided the nightshade metaphor while panning its ludicrous plot and unrestrained fearmongering.[34] The Jesuits' diabolical practices are seen at their worst when Emily, one of the twin-sister heroines, is abducted and forced into a Paris convent, where the nuns "tried to tame her by forced penances, by compelled scourgings, by confinement, by starvation"—abuses that eventually lead to her death.[35]

Johnston never refers to his title directly, but he presumably alludes to nightshades as the "poison plants" that appear in one strained metaphor foreshadowing the sisters' future: "Their lives seemed likely to run on in one dark and troubled course, as a river that would have been pure as crystal had poison plants not leant over it and cast on it their baleful shade."[36] The title's reliance on preconceived notions regarding the villainous character of nightshades suggests that the river's toxins emanate from the Jesuitical plants on the bank; but it is Johnston's imagery rather than the holy order that accounts for the novel's worst "tortuosities of the most inscrutable kind."

The prolific, popular Edward Bulwer-Lytton would also capitalize on his readers' familiarity with nightshade's reputation among sentimental plant personalities in *A Strange Story* (1861–62). Serialized in Dickens's magazine *All the Year Round, A Strange Story* is an investigation of the occult narrated by the skeptical physician Allen Fenwick, who becomes a believer in supernatural forces through the machinations of the malevolent Margrave, a charming young man wielding a magic wand and formidable mesmeric powers. After Fenwick's delicate and dreamy fiancée, Lilian, becomes the town scandal for briefly falling under Margrave's spell, she and Fenwick assess the damage by speaking in the language of flowers. When Lilian worries about a stain appearing on her hand, Fenwick assures her, "The hand is white as your own innocence, or the lily from which you take your name." To which the still-possessed Lilian replies, "Hush! You do not know my name. I will whisper it. Soft!—my name is Nightshade!" Shortly afterward Fenwick tells us, "She murmured something about Circles of Fire and a Veiled Woman in black garments; became restless, agitated, and unconscious of our presence, and finally sank into a heavy sleep."[37] In a novel full of strangeness, this is one exchange that doesn't require explanation, for the identification of "Nightshade" with a fallen woman associated with silence, falsehood, sorcery, skepticism, witchcraft, and dark thoughts is a given of sentimental botany.

Lilian is no nightshade, however; but a few years later another crony and sometime writer for Dickens, Percy Fitzgerald, would like Thackeray incarnate a deadly nightshade as a central character whose adventures would ultimately span three novels. The first in the series, *Bella Donna, or The Cross before the Name* (1864), launches the intrigues of Jenny Bell, a young woman who seems patently modeled on Becky Sharpe and who physically resembles the eponymous plant in being "round and fresh as a piece of ripe fruit."[38] Like Becky, the charming but treacherous Jenny is a woman without family or fortune who survives through her powers of seduction; and the novel advances via episodes wherein Jenny gains through her deceitful wiles the trust and affections of a series of well-heeled, easily duped gentlemen—including the father as well as the suitor of another woman. The alternate title, in fact, alludes to Jenny's desire for vengeance upon Charlotte Franklyn, her rival and nemesis—marked by a cross before Charlotte's name in her diary.[39] Labeled a "domestic Medea" and a "seductive Circe,"[40] Jenny ultimately fails in her many attempts to bewitch her victims, but she avoids retribution to continue seducing, plotting, destroying, and escaping through two more novels, until she dies wracked by poverty and disease in the last. Fitzgerald claimed that he planned all three novels from the start,[41] but the fact that *Bella Donna* was published under the pseudonym "Gilbert Dyce" and acknowledged to be Fitzgerald only after it became a bestseller suggests otherwise. There was never any question about the true identity of Jenny Bell: she is the stereotypical Belladonna, the siren and scourge of all self-respecting women, as an article about the second novel, *Jenny Bell*, in the *Saturday Review* makes clear: "By every woman of the world, every mama with eligible sons, every marriageable young lady fairly embarked upon the great ventures of woman's life, she must be set down at once as a shameless 'adventuress,' a cold calculating flirt, a bold designing hussy who ought to be scouted from all decent society."[42]

Dickens himself may have contributed his own Belladonna to Victorian literature in the person of the willful Bella Wilfer, one of two heroines in *Our Mutual Friend*, which began publication the same year as Fitzgerald's first Jenny Bell offering. Bella, however, eventually transforms from being an untamed, mercenary vixen into a thoroughly domesticated wife by the novel's end. Bella's conversion is perhaps a form of masculine wish fulfillment on Dickens's part given his attraction to Ellen Ternan, the actress who became his own fatal woman. But he seemed to have rather liked the real plant—or what he identified as such—because he speaks fondly of the roadside gardens between Genoa and Spezia "all blushing in the summer-time with clusters of Belladonna" in his travelogue *Pictures from Italy* (1844), and has his Italian character Cavaletto recall them again in *Little Dorrit* (1855–57) using the same affectionate terms.[43]

The previous examples have intimated that while "Belladonna" evokes the ironic image of a beautiful but dangerous woman, the equivalent metaphorical use of the plant's alternate name "Nightshade" or "Deadly Nightshade" (Johnston's Jesuits and Duval's horse notwithstanding) exposes a fallen woman without irony or disguise. Such at least is the case in a poem by that title in Thomas Gordon Hake's *Parables and Tales*, published in 1872. Hake was a physician best known for his care of Dante Gabriel Rossetti, whose life he saved when the poet-painter attempted suicide by overdosing on laudanum that same year;[44] and were it not for Rossetti, *Parables and Tales* would have sunk into oblivion along with thousands of other volumes of eminently forgettable Victorian verse.

The Deadly Nightshade tells the story of an alcoholic prostitute responsible for the tragic plight of her son, who is left a homeless beggar in the city after she is imprisoned (presumably for harlotry and drunkenness).[45] Intended as a contrast with *The Lily of the Valley*, a long poem about a virtuous child of the countryside whose name, "Lily of the Vale,"[46] identifies her with the innocence of nature, *The Deadly Nightshade* refers to the fallen woman by implication, but explicitly denotes the dark side of the urban world writ large, where "There still the nightshade breathes its pest / On fallen spirits not at rest,"—a world of "thieves and harlots" in which "The voice of Nature [is] heard no more."[47] The contrast thus suggests that whereas the lily of the valley evokes beneficent, unspoiled nature, the nightshade—as whore and urban malaise—represents nature against itself, a malignant force that corrupts a benign nature's healing, redemptive power. Of course, this distinction works only at a very superficial metaphorical level and lays bare the essential problem with assigning moral qualities to plants. Nevertheless, as we saw with Bulwer-Lytton's use of the lily-nightshade contrast in *A Strange Story*, the convention was widely understood and employed without question, so that even Rossetti seems to accept its terms. Rossetti, who admired Hake and promoted his poetry with positive reviews, even designed the elegant binding for Hake's book, which iconographically refers to the plants named in the paired poems. The cover features a cradle with a spade lying across it flanked by a tall belladonna plant on one side and a lily of the valley on the other, as stars rising from the cradle mingle with a crown of thorns above.[48] However, in a letter to Hake's son, Rossetti reveals the confusion inherent in the poems' metaphors when he attempts to explain the genesis of his design: "As [it] seemed to afford a sort of joint symbol—the cradle and spade for birth and death, the Lily of the Valley and Deadly Night shade as symbols chosen by the poet to represent the influences affecting them, the crown of thorns for such sacredness as may attend the struggle, and the stars finding their way thence to the cradle, for such hope as these may be."[49] Judging *Parables and Tales* by its cover can hardly give an accurate sense of the two featured poems, for Rossetti's design in effect erases the floral moralizing: that is, the lily and the nightshade are rendered as equally beautiful and vigorous plants, both unspoiled creations of the natural world.

Rossetti's sister Christina, a far more accomplished poet than Hake, nevertheless shared his penchant for moralizing by way of plant imagery, and the line "Nightshade would caress and kill me" apparently draws on the same erotic and possibly illicit associations that Hake exploits. But since this line appears in an otherwise innocent verse for children that begins by rejecting the rose ("A rose has thorns as well as honey, / I'll not have her for love or money;") in favor of holly ("But give me holly, bold and jolly, / Honest, prickly shining holly"), it's hard to interpret the plants as anything other than themselves.[50] A more serious treatment of poison fruit famously occurs in one of Christina's best poems, *Goblin Market* (1859); and in an article entitled "Covent Garden Market," Clayton Carlyle Tarr argues that the fruit in question is belladonna, based on his research concerning the London market's temptations and the wiles of its sellers.[51]

The Rossetti circle produced another, far more interesting offering in the category of nightshade literature: Oliver Madox Brown's unfinished novel *The Dwale Bluth*, published after his untimely death at the age of nineteen in 1876. The adored, precocious son of Gabriel Rossetti's good friend and former tutor, the painter Ford Madox Brown, "Nolly" had been inspired by a trip to Devonshire to write a story drawing on its scenery and lore, and his title is taken from the old Devonshire name for *Atropa belladonna*, the plant that fatefully influences the life of the heroine, Helen Serpleton, and with which she comes to identify. In the *Memoir* prefacing Brown's *Literary Remains*, we are told that "Dwale Bluth" literally translates to "Craze-bloom or Frenzy Flower, indicating the deadly nightshade,"[52] either term appropriate to describe Helen, who matures into Belladonna incarnate, a femme fatale who nevertheless has the loving sympathy of her creator, and who closely resembles the magnificent, but compromised, women in Pre-Raphaelite art.

Helen's story begins after her mother's death, when her father leaves the child with her absent-minded, scholarly uncle to be brought up at the ancestral home of the Serpletons in rural Dartmoor. Like her late Spanish mother, Helen is wild, exotic, witchy (the Serpleton housekeeper had thought the mother "a creature possessing all the malice of a sorceress if not the power of one"), and feline—she even has a "weird-looking grey cat" for her familiar—and from childhood manifests a willful, "turbulent disposition."[53] Appropriately for his story, Brown foreshadows her destiny as Belladonna, the fatal woman, with an ominous botanical metaphor: "She seemed only to require time to ripen into some sinister but beautiful flower—loving and beloved by some; but perilous, ill-meaning, and misunderstood in its relations to others."[54] The plant itself soon appears in a chapter entitled "Atropa Belladonna," which recounts how little Helen and her cousin Leah fight over a shoot of this "well-known poison-plant" found on the heath after it uproots and blows in their direction, "as if insidiously luring them to follow it." Helen's uncle, who recognizes the plant, takes it from the girls and stuffs it into his pocket, an act that gives rise to the following portentous authorial comment: "Had he thrown it away at once, as he should have done, much of this narrative would have remained unwritten."[55]

Thanks to Serpleton's rescue, the plant makes its way to their home, gets carelessly thrown out the window, and ends up taking root in a cucumber frame next to the house, where Helen eventually finds it. Following the episode on the heath, Helen had been consumed by a "vague longing" for something, which upon her discovery she immediately understands to be this plant:

> *There* was what she had been looking after, for her downcast eyes fell, with an instantaneous revival of interest, on the tendrils of the deadly nightshade, loaded with poison-flowers and berries, crawling stealthily round the inside of the frame, or crouching ignominiously along the damp ground....
>
> The Dwale Bluth is a cowardly creeper, and knows no means of rising above the earth it springs from, unless by insinuating itself among the leaves of some bolder parasite. There it now lay beneath her gaze, even throwing a grim and sinister reflection on to her dark-complexioned face, and into her eyes; there it lay at her feet, prone and helpless, as though it were entreating her to lift it.[56]

It is obvious from Brown's description that he wants to emphasize the plant's "grim and sinister," snakelike appearance, as well as its uncanny ability to cast a spell on Helen, whose brief adult life proves to be controlled by its diabolical influence—suggested here by the serpentine plant and its sorcery. However, it is also clear that such a "crawling," "crouching," creeping plant bears no resemblance to *Atropa belladonna*; rather, the plant is like *Solanum dulcamara* or bittersweet, which, except for its lack of tendrils, has all the vinelike characteristics Brown enumerates. Moreover, it seems that Brown knew the difference, for when Serpleton first examines the plant on the heath he says, "Solanum dulcama ... no ... atropa belladonna," and warns, "Look here, Nelly, and you too, Leah; they call this the deadly nightshade sometimes. You must never disturb it, or it will send you to sleep so soundly that you won't be able to wake up again. It's poison, mark you!"[57] Nevertheless, like Serpleton, Brown is insistent about the plant's identity, even to the point of ignoring the truth when his sister sent him a spray of deadly nightshade, evidently in an effort to point out his mistake. Brown had even told a potential editor that he first chose "Belladonna" as the title for the novel, but changed his mind when he learned that it had already been taken (as we know, by "Gilbert Dyce," aka Percy Fitzgerald).[58] Obviously, *Atropa belladonna* and its old-fashioned name "dwale" were the perfect metaphors for his heroine, despite the fact that a different nightshade really bore the physical qualities better suited to his thematic imagery. In any case, in the last chapters devoted to Helen's childhood, after she perversely eats the plant's berries, which "she knew perfectly well would harm her," Brown proves that he could accurately describe in great detail the symptoms of belladonna poisoning—from the dilated pupils to

122 VICTORIAN NIGHTSHADES

the delirium.[59] Helen survives thanks to an antidote provided by her uncle, but this poisoning is the defining event of her life, as we learn in the novel's final chapters.

The foreshortened conclusion of *The Dwale Bluth* skips past Helen's marriage and presumed widowhood and turns abruptly to her peripatetic conversations climbing the cliffs over the sea with her lover, a blind poet named Arthur Haenton. From childhood Helen had exhibited "a wild and irrepressible craving" for "always wearing some eccentric ornament entangled in the locks of her beautiful hair" — we first see her sporting glowworms in her tresses — and we learn that this habit manifests itself now in her adulthood with "a kind of fantastic crown, woven of deadly nightshade . . . bound round her head with delicate fillets of hair."[60] When Helen tries to explain to her lover how this practice evolved, she recounts the poisoning episode, which at first left her fearing that her hair "had all been turned into wreaths of deadly nightshade that were rooted in [her] brain," and while sleeping "its loose tendrils would maliciously try to insinuate themselves into [her] mouth to poison [her] again." Helen seemingly breaks the spell by cutting her hair, but as she tells Haenton, "it created an entire revulsion of feeling to find that I was free; and somehow, strange to say, made me take to wearing the real nightshade in my hair whenever I could procure it." In short, it seems clear that the nightshade has won complete possession of Helen — that it has indeed taken over her entire being and "rooted in her brain."[61] In fact, the locals consider her to be as "dwale" — both crazy and dangerous — as the deadly flowers she wears and fear that she is a witch capable of inflicting curses by way of the dreaded evil eye.[62]

As it turns out, Helen is dangerous only to herself and the two men who love her — Haenton and her husband, Thurlstone, who had not drowned as was believed, but who returns to discover the lovers during their tryst, just in time to hear Helen say she never really cared for the man she married. To protect her lover from her husband's wrath, Helen unhappily returns home with Thurlstone, and soon dies in a fit of "brain-fever" when "during the somnolence of her nurse, she *strangled herself with her hair*" [Brown's italics].[63] True to its folk-ballad atmosphere, the story ends when Arthur Haenton finds Helen's grave, voraciously eats the dwale bluth now growing on it, and dies, to be buried at her side.[64]

Despite *The Dwale Bluth*'s artistic shortcomings, Brown's use of the Belladonna metaphor is the most ingenious and original of all the previous examples, for he is the only writer to incorporate the actual plant into the plot. It is unfortunate that Brown chose to perpetuate the all-too-common misidentification of *Solanum dulcamara*, especially given the obvious care he took to employ botanical detail; but his fascination with the plant matches his sympathy for Helen, with both manifesting all the feral magnetism of nature's poisonous and perilous species. A comparable use of the belladonna motif would not appear until the mid-twentieth century in *The Go-Between*, when L. P. Hartley uses the plant to represent his late-Victorian heroine Marian Maudsley. The epigraph to this chapter captures the belladonna's haunting presence, irresistible attractiveness, and connection to sexual secrets that Hartley's novel brilliantly conveys.[65]

It's possible that Thomas Hardy created a final noteworthy Victorian Belladonna as fatal woman in one of the greatest late-century tragic novels, *Jude the Obscure* (1894–95); but to my knowledge, the earthy, voluptuous, and unscrupulous Arabella Donn—infamous for her persistent exploitation of the hero—has never been identified as a Belladonna, even though her full name could be a thinly disguised play on the designation. True to her character as a

FIGURE 13. Belladonna, Walter Crane, *A Floral Fantasy in an Old English Garden*. (Image from New York Public Library Digital Collections)

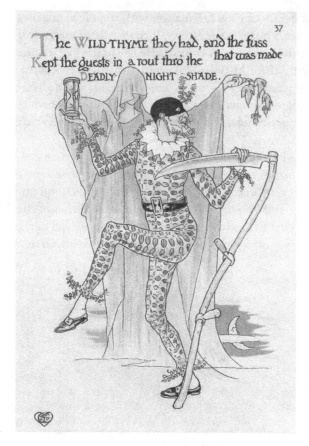

FIGURE 14. Walter Crane, *A Flower Wedding*. (Image from Smithsonian Libraries; produced by Archive.org)

coarse seductress, Arabella appropriately enters Jude's life by throwing a pig's penis at him to get his attention.[66] A miserable marriage ensues, instigated by Arabella's faking a pregnancy, and ends temporarily after Arabella emigrates to Australia; and even though they eventually divorce upon her return to England, Arabella repeatedly calls upon Jude whenever she can profit from his assistance, at last tricking him into a fatal (for him) second marriage after getting him drunk.[67] A sometime barmaid, Arabella makes liquor her toxin of choice, but there's no denying that she is Jude's poison.

Nevertheless, as the language of flowers became a quaint fashion of the past and belladonna grew in importance as a medicine, the species name's automatic identification with a fatal woman was giving way to the more historical and thus scientific (albeit nostalgic) approach to plant symbolism apparent in popular books of botanical lore like Hilderic Friend's *Flowers and Flower Lore* (1884), Richard Folkard's *Plant Lore, Legends, and Lyrics* (1884), and

T. F. Thiselton-Dyer's *The Folklore of Plants* (1889). In all three works, the language of flowers appears as a topic, but receives patronizing treatment for the often arbitrary and fanciful meanings it spawned. Even so, the vestiges of floriography remain in the charming flower books of the talented and prolific artist Walter Crane, painter of the Arts and Crafts movement and most famous as an illustrator of children books. A typical example is *A Floral Fantasy in an Old English Garden* (1899), which features whimsical verses and illustrations of anthropomorphized flowers in guises playing on their names or characteristics. In keeping with Ruskin's characterization as the "queen" of the family, Belladonna appears as a beautiful, dangerous, and evidently despairing royal personage toasted by a retinue of kingcups who "drink BELLA-DONNA" while she sits on a throne, "Clad in purple and gold so fair, / Though the DEADLY NIGHTSHADE's upon her" (see fig. 13).[68] Her last appearance, in *A Flower Wedding* (1905), is a more menacing figure of a veiled woman as Grim Reaper, seen at the end of the wedding celebration accompanying the following verse: "The Wild Thyme they had and the fuss that was made / Kept the guests in a rout thro' the deadly Night Shade." However, the representative flower held by this wraithlike image looks (alas!) like bittersweet (see fig. 14).[69]

COSMETICS AND COMPLAINTS

Belladonna's identification with seductive women in Victorian popular culture did not originate simply as a result of botanical hearsay about its use as a cosmetic in Renaissance Italy; by the mid-nineteenth century, some Victorian women were testing its powers of attraction by dilating their pupils with belladonna eyedrops for its beautifying effect. As the popular botanical writer Phebe Lankester put it in 1861, "I am told that this practice is not confined to the land of cloudless skies and southern breezes, but that in our own country the preparation is to be seen on the toilette-tables of our fashionable English ladies."[70] In the July 1, 1867, "Conversazione" section of the *Englishwoman's Domestic Magazine*, a contributor named "Dagmar" provided the specifics:

> Belladonna or atropa is extensively used in Italy to give brilliancy to the eyes. It is inserted into the eyelid at the outside corner with a bodkin; in a few moments it is diffused over the eyeball, causing the pupil to dilate so much as in some cases to cover the whole iris. This dilatation ceases directly the belladonna loses its power, and has to be renewed if the effect is still desired. Its action is most injurious; the unnatural excitement of the pupil in time produces blindness; and we know one case in which instant loss of sight was the result of the application of this poison.[71]

126   VICTORIAN NIGHTSHADES

Despite its dangers, the number of Victorian women using belladonna as eye makeup was sufficient to rouse the ire of *Punch*, which published several attacks on the practice. In 1856, an article entitled "Fine Eyes for Foolish Girls," addressed to its male readers "who take some interest" in "vain and silly individuals of the softer sex," was prompted by an ad for belladonna "put forth by certain cosmetic vendors" which it cites as follows: "A BRILLIANT EYE—Bella-Donna produces a Brilliant Eye.—The ladies of Asia hold it in high repute for its quality to give brilliancy, vivacity, and the power of fascination to the eye. Price 2s. 6d." Although *Punch* acknowledges the truth of the ad's claim ("Oh yes! *Atropa Belladonna*—deadly nightshade—produces a very brilliant eye"), the article explains that the "pupil is paralysed by the belladonna, and therefore gapes"; and though the paralysis "is but temporary" and occasional use perhaps not harmful,

> Nature generally shows her resentment of tricks practiced on the bodily frame by punishing them with disease in the part trifled with. How long a fool might go on tampering with her iris we are not prepared to say; let any young lady who is foolish enough try, for the benefit of science. But the female eye is in general sufficiently brilliant, vivacious, and fascinating, if not too much so, without recourse to belladonna: and if any stupid girl thinks her eyes are not bright enough, she had better cultivate her intellect to make them brighter.[72]

Another article, entitled "False Fine Eyes," appeared eight years later quoting the same ad—now identified as appearing in the *Morning Post*—and attacking the practice by explaining its dangers in similar terms. Once again, *Punch* appeals to its male readership to protect these vain, misguided young women: "These remarks are made not with the least expectation of persuading any young lady who is in the habit of using belladonna eyewater to leave it off, but only with the view of inducing her father, or brother, or guardian, who may happen to find a bottle of that stuff anywhere about the house, to throw it out of window."[73]

The third *Punch* assault on belladonna comes via a cartoon in 1869 concerning a segment of the female population different from young ladies presumably on the marriage market: the older woman who is still trying to broker her charms. "The Song of the Passee Belle" pictures an elaborately coifed lady, who in the euphemistic phrase of the day could be described as "of a certain age," looking somewhat pained and gazing vacantly as she plays the piano and sings about her cosmetics, with the lyrics printed in the caption: "'The Bismuthive Cream is on my brow, / I've Belladonna in my eye, / Then meet me,

THE SONG OF THE PASSÉE BELLE.

" The Bismuthive Cream is on my brow,
I 've Belladonna in my eye,
Then meet me, meet me in the evening,
When the bloom is on the wry."

FIGURE 15. "The Song of the Passée Belle," Linley Sambourne, *Punch*, vol. 57 (1869): 20. (Courtesy of HathiTrust; contributed by the University of California)

meet me, in the evening, / When the bloom is on the wry'" (see fig. 15).[74] Despite the bad pun ("wry" evidently refers to her strange expression), the message is clear: the superannuated "belle" who resorts to toxic agents in a futile attempt to beautify herself invites not only injury but insult for her folly. In any case, *Punch* makes it clear that a sizable number of Victorian women considered belladonna eyedrops one of their beauty secrets, and even Queen Victoria resorted to their use in her old age, according to Christopher Hibbert—though not for beautification, but to correct her failing eyesight rather than undergo surgery. Unsurprisingly, the result was "less than satisfactory."[75]

It's impossible to know how many Victorian women actually used belladonna as a cosmetic, but the *Punch* attacks and Phebe Lankester's comment indicate that the practice was common enough among the husband- or lover-hunting portion of the female population who moved in urban England's more fashionable circles. Even more likely, Victorian actresses, mistresses, and prostitutes exploited this beauty secret since, as the *Punch* articles reveal, belladonna cosmetics were widely advertised and readily available for sale. In contemporary literature, at least one reference to its use occurs—in *Phantom Fortune* (1883), a late novel by the author of sensationalist fiction Mary Elizabeth Braddon. Upon the urging of her chaperon, the beautiful Lesbia Haselden, recently introduced into London society, "for the first time in her life" reluctantly agrees to dilate her pupils with belladonna to revive her looks, which have suffered after a too-taxing round of social engagements and depression occasioned by mounting debt for her excessive London expenses.[76] Braddon implies that the use of belladonna is just another symptom of the jaded, modern commercial culture morally disconnected from Lesbia's home in rural, Wordsworthian Grasmere, where traditional, solid English values are represented by her sister Mary, the real heroine and moral center of the novel. In other words, belladonna use indicates that Lesbia is in effect prostituting herself: she is simply another commodity for sale on the fashionable marriage market, a Belladonna in the making.

Belladonna's intimate association with the female gender may have begun with its use as a cosmetic, but it played a more important and creditable role in another facet of Victorian women's personal lives—their health. The plant's sedative powers had never been lost on medical practitioners in search of ways to alleviate their patients' suffering; consequently, belladonna had been prescribed with some frequency as a painkiller for various ailments since the late eighteenth century. For example, an English practitioner named John Bailey published in 1818 a small treatise entitled *Observations Relative to the Use of Belladonna, in Painful Disorders of the Head and Face*, which described at length twenty cases in which moderate doses of belladonna were prescribed, with almost universal success, to treat "neuralgia facialis," better known in Victorian times by its French name, *tic douloureux*, caused by damaged facial nerves. Although this ailment, characterized by pain on one side of the face, was in most of the cases associated with the seemingly gender-neutral condition of decayed teeth, Bailey considered it a predominantly female affliction, and seventeen of the twenty cases he details concern women.[77] Moreover, these women shared a rather specific personality profile: Bailey concluded that neuralgia facialis was a form of "hemicrania" (migraine), an ailment, he said, which is "rarely seen but in delicate persons, chiefly women who are hurried into premature age by some debilitating cause peculiar to their sex."[78] Although Bailey's patients frequently suffered from belladonna's unpleasant side effects of blurred vision, dry mouth, difficulty swallowing, and dizziness, the cure was preferable to the excruciating pain occasioned by their complaint.

By mid-century belladonna figured prominently in the treatment of many female disorders; and whereas Bailey's method of administering the drug was

VICTORIA'S SECRETS   129

by tincture or extract taken internally,[79] physicians later in the century used numerous means of delivery, from pills and hypodermic injections to plasters and pessaries, liniments, ointments, and suppositories. A leading writer on women's health and diseases and founding member of the Obstetrical Society of London in 1858, the gynecologist Edward Tilt, employed belladonna alone or combined with other drugs in all these forms to treat a panoply of complaints peculiar to females at every stage of life. In *A Handbook of Uterine Therapeutics* (1863), Tilt recommended belladonna, either taken orally or given as vaginal or rectal suppositories, for painful menstruation; and he considered the application of sulphate of atropia in a glycerine ointment rubbed on the breasts as "the most powerful remedy" for bloody discharge from the nipples.[80] Atropia could also be used "with advantage" for "local neuralgia, lumbago, and sciatica."[81] In fact, belladonna's excellent reputation as a remedy for those three complaints no doubt led to the popularity of belladonna plasters sold over the counter in the latter half of the century (and to their continued use in the UK today). In the United States in the 1870s, trading cards for Mitchell's Belladonna Plasters promoted their products as "especially adapted for complaints of ladies."[82] Nevertheless, this extensive use did present hazards: in an 1872 article in the *British Medical Journal*, a Manchester physician named J. Bower Harrison reported the recent case of a woman who experienced severe symptoms of belladonna poisoning—including delirium complete with flying insects—caused by a plaster applied over raw skin on her back. Harrison cites another *BMJ* article that explains what happened: "Belladonna, remember, cannot be applied to an *abraded surface* without risk of absorption and alarming symptoms of poisoning." Harrison mentions several other relevant cases and concludes by reminding his fellow practitioners "that even the external application of belladonna is occasionally productive of mischievous effects."[83]

Such liabilities notwithstanding, Edward Tilt gave his highest praise for belladonna's efficacy when administered directly to the site of the disorder as an antispasmodic: "Externally used, I believe belladonna to be the best remedy we possess against tenesmus, whether the womb, the anus, the urethra, the nipple, or the eyelid be the seat of the forcing action. I have applied cotton-wool soaked in a solution of sulphate of atropia, to the neck of the womb, to quell the forcing pains of uterine tenesmus."[84] Referring to Tilt as "one of our best authorities on the practical use of medicines," John Harley quoted this passage in *The Old Vegetable Neurotics* to corroborate his own claim about belladonna's superior antispasmodic properties. Tilt would repay the compliment in later editions of *Uterine Therapeutics* by frequently citing Harley, whose discovery that taking belladonna and opium together strengthened the potency of both drugs while diminishing some of their negative side effects, corroborated the efficacy of Tilt's practice of "combining opium and belladonna in pills, morphia and atropia in suppositories, ointments, and in liniments."[85]

Tilt was a great believer in sedation adequate to relieve the pain and distress of symptoms coincident with menopause, and in *The Change of Life in Health and Disease,* a work based on his own clinical experience as well as statistical information gathered from five hundred British women between the ages of forty-five and fifty-five,[86] he advocates the use of many powerful drugs to treat everything from diseases of the reproductive organs to melancholia. According to Tilt, the epigastric region—the pit of the stomach—is the seat of many menopausal complaints, for which he enthusiastically recommended plasters of opium or belladonna, a combination of both, or ones "incorporating morphia and atropia with glycerine ointment," a remedy, he notes, he had advocated in *Uterine Therapeutics.*[87] In *The Change of Life,* Tilt defends his practice of prescribing such potent analgesics by saying that without sufficient relief from pain, women "will instinctively fly to stimulants, the poor to porter and gin, the rich to wine and brandy," so that taking "medicine" is better than their resorting "to deplorable habits."[88] However, the right remedy depended upon the individuality of the women being treated, and those who responded best to belladonna were evidently of the high-colored, overblown sort. Specifically, Tilt said that of the three types that women at menopause fall into—"the Plethoric, the Chlorotic [i.e., anemic], and the Nervous"—the Plethoric type profited more from taking "solanaceous sedatives" (i.e., belladonna as well as henbane, which he praised for its sedative effects) rather than opium, the drug of choice for the other two types. Curiously, Tilt's description of the "unmistakable" Plethoric type, characterized by "its turgid tissues, its over-florid countenance in a state of perspiration, its anxious expression, or half-intoxicated appearance," sounds rather like Alexi Lucot's frenzied "Belle Dame," emblematic of Despair—perhaps the fatal woman coming into focus even as a scientific and medical category.[89]

Tilt was only one of scores of practitioners writing about gynecological matters throughout the Victorian period, which witnessed an unprecedented number of innovations (if not always advances), especially in the field of obstetrics. In fact, it was during the nineteenth century that gynecology developed as a separate field, and that obstetrics became a specialty sanctioned by the medical profession. Before then, as Tilt notes, female modesty often kept women with "female" complaints—disorders of the breast or reproductive organs—from seeking the care of male practitioners;[90] and childbirth was almost exclusively the province of female midwives who delivered babies at home rather than in hospitals. Thus, what constituted women's health in these matters was by and large a secret from the male gender, including the medical profession. In the gentlemanly words of the Philadelphia professor of obstetrics and gynecology Charles Meigs, "So great, indeed, is the embarrassment arising from fastidiousness on the part either of the female herself, or of the practitioner, or both, that, I am persuaded, much of the ill success of treatment may be justly charged thereto."[91] With increasing medical knowledge among men as well as women, however, a change was bound to occur; and the shift in both personnel and place of treatment was accelerated by the introduction of the anesthetics ether and chloroform, whose use in childbirth became popular in the late forties amid considerable controversy. (Famously, Queen Victoria requested chloroform for the deliveries of her last two children—Leopold in 1853 and Beatrice in 1857—using the method that became known as "*anesthésie à la reine*": that is, she inhaled chloroform vapor intermittently during contractions but was not rendered unconscious.)[92]

But before, during, and after the forties we can be sure that all the known means of sedation—from opium to ergot to alcohol—were employed to assuage the pains of particularly difficult labors; and belladonna, applied as an ointment made from its extract, gained a position of preeminence in the first half of the century as a local anesthetic in cases of extreme rigidity of the "*os uteri*"—the mouth of the womb. This practice was introduced in France during the Napoleonic period by two well-trained experts in obstetrics and highly esteemed teachers of midwifery—interestingly, a man and a woman—Francois Chaussier and Madame Marie Lachapelle. It remained popular among French practitioners throughout the nineteenth century, championed most notably by the surgeon Alfred Velpeau, whose *An Elementary Treatise on Midwifery* (first published in 1829) was translated into English by the aforementioned American obstetrician Charles Meigs, who promoted the use of belladonna in his own practice, teaching, and writings.[93] Although Meigs was against the use of anesthesia during childbirth, he had this to say about using belladonna: "The os uteri is never to be forced. But there is a wide difference between forcing and persuading it to yield. It may be prevailed against by means of the relaxing influence of belladonna applied to the surface, aided by gentle dilatation of the hand."[94]

In England, belladonna's earliest advocate seems to have been John Tricker Conquest, whose *Outlines of Midwifery*, first published in 1820, went through many editions and translations; however, in the new edition of 1854, the editor James Winn qualified Conquest's recommendation by adding the following warning: "Belladonna is a remedy that must be used with great caution, and later experience has thrown much doubt on its power of inducing relaxation of the os uteri."[95] Winn's opinion voiced that of Francis Ramsbotham, a founder of the first Obstetric Society of London (1826) and an important pioneer in the field, whose *Principles and Practice of Obstetric Medicine and Surgery* was a primary text in England and America.[96]

Needless to say, the practitioner who introduced the use of chloroform as an anesthetic, the Scottish physician and professor of midwifery at the University of Edinburgh James Young Simpson, stood behind his own discovery for treating rigidity of the os uteri; and in his lecture notes published posthumously, he seems to question the efficacy of belladonna in such cases.[97] But Simpson as well as many other gynecologists advocated its use as a palliative for many female complaints, especially for terminal diseases like uterine cancer.[98] Simpson was also known for introducing the pessary into contemporary medical practice as a local analgesic for disorders of the uterus and vagina, and belladonna was one of the most frequent medications delivered via this method.[99]

One final use of belladonna in nineteenth-century gynecological medicine is worth mentioning, this time, however, as the cause, rather than the cure, of a woman's complaint—significantly, one coming from the Victorians' foremost female health professional, Florence Nightingale. Like leeching and cupping, the practice of blistering—applying bandages treated with caustic substances to bodily parts to raise blisters on the skin—was frequently prescribed to act as a stimulant in cases of fever; and according to her biographer Gillian Gill, in 1861 both Florence and her sister, Parthenope, were simultaneously subjected to this procedure by the same doctor. In a letter from May of that year, Florence, an invalid since the late fifties (possibly from brucellosis contracted in the Crimea), commiserates with her sister, who was ill with rheumatic fever for being, like Florence herself, "a case of poisoning by belladonna," thanks to the orders of their mutual physician; but she ultimately blames one "Mr. Brown Legrand," whom she claims "originate[d] the treatment." Florence reports that she was "groaning for 12 hours aloud" during the night from the pain caused by blistering her spine, "which is now nothing but one raw place from top to bottom." Referring again to the now-familiar headline in such incidents, she jokes, "I mean to put into the 'Lancet' 'Case of poisoning by Belladonna' & then Brown Legrand will see it & say: 'Singular case,' 'but not well reported.'"[100]

Witches' Botany

Atropa belladonna, fatal women, and the privacies of the female body converge in the subject of witchcraft, with which the Solanaceae have been identified more than any other plant family. No matter what the source—whether ancient myth, folklore, historical record, literary fiction, or current practice—the Victorians were fascinated with witches. Even though England's advances in science and technology had seriously eroded the nation's faith in religion and called into question all things supernatural, such matters nevertheless aroused curiosity, wonder, and often anxious scrutiny. Witches held a particular attraction because—unlike patently imaginary creations such as fairies or vampires—real live human beings had self-identified as witches, inspiring belief, hatred, terror, and scandalous social crises in the not-too-distant past. Moreover, contemporary newspaper articles—like the occasional series "Witchcraft in the 19th Century" published in the *Times* in 1837, 1840, and 1858 and one by the same title in an 1869 issue of Dickens's *All the Year Round*—reported minor, but recurring, disturbances implicating witches and "cunning folk," who appeared in present-day stories of credulity, suspicion, and fraud—if not of mysterious powers or marvels.[101] Bewitchings of people and livestock were "unwitched" by cunning men and women in a revolving circle of deceit and accusation, with one or both parties professing secret knowledge and magical skills. In all the stories, the reporters chastised both the alleged practitioners as well as their gullible victims or accusers for exploiting or falling prey to superstition, respectively, usually in an attempt to resolve what began as a rural squabble or domestic misunderstanding. The periodicals assumed the position taken in the last of the witchcraft acts, passed in 1736, which marked a major change in official attitude and policy by legally proclaiming witchcraft to be a delusion, one punishable for attempting to deceive others rather than for wielding supernatural powers or trafficking with the devil.[102] Nonetheless, public curiosity could not be dampened by enlightened legislation.

Of more interest than these current cases, however, were accounts of the sixteenth- and seventeenth-century witch trials; the legends of classical and mythic witches, which flourished in books, in pictures, and onstage; and new fictional tales, which surfaced everywhere.[103] For example, John Brand's ubiquitous *Observations on the Popular Antiquities of Great Britain*, first published in 1777, updated by the British Museum's head librarian Henry Ellis in 1813, and assimilated into several other publications throughout the century, provided a treasure trove of information about English witch practices, both imaginary and presumably real. The Grimm Brothers' *Fairy Tales* (first published in 1812) and Jacob Grimm's massive *Teutonic Mythology* (first published in 1835) acquainted English readers with the origins and details of Germanic witch beliefs and helped to promote even further the huge popularity of folklore among the Victorians. The antiquary Thomas Wright's *Narratives of Sorcery and Magic, from the Most Authentic Sources* (1852) offered a wide-ranging compilation of history and legend from the tenth to the eighteenth century, while the journalist Eliza Lynn Linton's *Witch Stories* (1861) — which drew heavily from Wright — recycled historical and legendary accounts of Scottish and English witches. But whether "grossest superstition," as the *Times* called it, or mysterious phenomenon, the Victorians pored over texts promising access to arcane lore and revealing secret knowledge.[104] In short, the prevailing public opinion was that witchcraft was something to investigate or titillate, not to believe in; but this general skepticism about its actual existence in no way lessened its popular appeal.

Reading about witches' herbs was a central feature of the Victorians' fascination, and not surprisingly, popular books of plant lore, the three in particular that appeared in the eighties as spinoffs of the folklore craze, play to this curiosity. Richard Folkard's *Plant Lore, Legends, and Lyrics,* Hilderic Friend's *Flowers and Flower Lore,* and to a lesser extent T. F. Thiselton-Dyer's *The Folklore of Plants* draw from many of the same sources and proffer much of the same information — Thiselton-Dyer's being largely derivative of the first two — and all devote chapters to the subjects of plants used in witchcraft and devil-worship, with belladonna and other Solanaceae, primarily mandrake and henbane, occupying an essential place in witches' pharmacopeias. The two latter works speak of belladonna's being a "favourite plant of the devil," both citing a Bohemian myth that the plant is watched over by the devil, who, in Thiselton-Dyer's words, "may be drawn from it on Walpurgis night by letting loose a black hen, after which he will run" (a curious tale that may make sense if "black hen" is a corruption or code for *Hyoscyamus niger* — black henbane).[105] Nevertheless, like the floriographies that preceded them, all three works still reflect the polite ethos of sentimental botany, offering very general comments about plants' association with sorcery and refraining from too-explicit detail regarding the more salacious aspects of the nightshades' witch story.

For example, in the chapter "Plants of the Witches" from *Plant Lore, Legends, and Lyrics,* Richard Folkard claims that "the chief strength of poor witches lies in the gathering and boiling of herbs," which were used for all sorts of potions and poisons; and the list that follows of "the most esteemed herbs for their purposes" begins with "Betony-root [a multiuse medicinal], Henbane, Mandrake, and Deadly Nightshade."[106] The list includes some thirty plants — and many more elsewhere in the chapter — but these three Solanaceae species no doubt leapt to mind not only because of their incredible efficacy as poisons and medicines, but also for their alleged magical potency. Folkard opens his chapter with the information that in Greek mythology, Hecate, who "may be fairly styled as the goddess, queen, and patroness

VICTORIA'S SECRETS   135

of Witches and sorcerers," and who (according to him) mothered the great witches Medea and Circe, "was acquainted with the properties of every herb, and imparted this knowledge to her daughters." In the list of plants "specially consecrated" to this fearsome trinity, three Solanaceae take the lead: "the Mandrake, the Deadly Nightshade, the Common Nightshade."[107] The mandrake, classified as *Atropa mandragora* by Folkard (now designated as *Mandragora officinarum*), rightfully takes first honors. A native of southern Europe, it has always been considered the quintessential witches' herb; and in the second part of Folkard's book, an alphabetical listing of six hundred plants, mandrake occupies three pages devoted to its ancient reputation as an aphrodisiac, fertility drug, lucky charm, and even "the embodiment of some unquiet or evil spirit," valued both in what might be considered amateur and professional sorcery. Like belladonna and henbane, it packs a tropane alkaloidal punch, and works as both an anesthetic and a hallucinogenic. Its identification with sorcery is so close that "in Germany, since the time of the Goths, the word *alruna* has borne the double meaning of witch and Mandrake," Folkard tells us.[108] But mandrake is a Continental plant, so its English cousin *Atropa belladonna*, with which it previously shared a genus, assumes priority on the other side of the Channel. The last Solanaceae in the list, "common nightshade," which usually refers to *Solanum nigrum* or black nightshade, was a known sedative and analgesic; but why it replaces the far more reliably potent henbane is a mystery.

Although Folkard describes deadly nightshade in the alphabetical listing as "a plant of ill omen, and one of which witches are reported to be fond," he gives little information about its association with sorcery beyond what is mentioned above except for a couple of references to Ben Jonson, in particular to his *Masque of Queens* (1609), in which a coven of hags details their nightly exploits that include herb gathering. One woman tells about pulling up a mandrake whose "groan" she detected while lying on the ground, whereas another furtively collected several witches' favorites, the two other most relevant Solanaceae among them: "'And I ha' been plucking plants among / Hemlock, Henbane, Adder's Tongue; / Nightshade, Moonwort, Libbard's Bane, / And twice by the dogs was like to be ta'en.'"[109] These *Masque of Queens* lyrics are a favorite of all three folklorists because the learned Jonson backed up his claims by heavily documenting his original text, thus providing a handy bibliography of sorcery works written up to the early seventeenth century. Jonson notes that the plants named are the "common venefical ingredients"—that is, those used in witchcraft—given by the leading Continental writers on the occult during the Renaissance, "*Paracelsus, Porta, Agrippa* and others"; but he offers no information about their specific use; nor does Folkard at this point, although in the chapter entitled "Plants of the Devil," he gives a few hints. After mentioning that "the berry of the Deadly Nightshade [is called] the devil's berry," he notes that "the plant itself is called Death's herb, and in olden times its fruit bore the name of Dwale-berry the word *dvale*, which is Danish, meaning a deadly trance."[110]

In *Flowers and Flower Lore*, the Reverend Hilderic Friend offers much of the same information as Folkard, but lists about twice as many sources and gives a more extensive, revealing, and ominous—if still rather discreet—account of belladonna's sorcerous and diabolical associations, including the "ancient belief," stated earlier in this chapter, that the plant itself "is the form of a fatal enchantress or witch, called Atropa" (a myth that seems apocryphal, given that Linnaeus is credited with assigning the third of the Fates to belladonna's name).[111] Friend also has more to say about belladonna's poisonous properties, citing, for example, the folklorist William Henderson's claim that witches are "fond of Hemlock, Nightshade, St. John's Wort, and Vervain, and infuse their juices into the baleful draughts prepared for their enemies."[112] Most significantly, however, Friend broaches the subject of belladonna's and other nightshades' greatest boon to witchcraft—the plants' capacity for helping witches to fly. Seemingly reluctant to tackle the subject head on, Friend resorts instead to "notes and extracts" taken from the aforementioned Brand's *Popular Antiquities*, which goes into some detail concerning "certain magical ointments"—allegedly provided to witches by the devil—with which they anoint themselves in order to fly (often on a broom or distaff) to their Sabbath conclaves. Although *Popular Antiquities* reveals a close acquaintance with the Renaissance literature on witchcraft, Brand nevertheless treats the subject with some disdain, citing two well-known early skeptics—"Reginald Scot speaks of the vulgar opinion of witches flying," and "Wierus [Johann Weyer] exposes the folly of this opinion, proving it to be a diabolical illusion, and to be acted only in a dream"—while supporting these comments with relevant verses from the seventeenth-century satirists Samuel Butler and John Oldham. But it is the Renaissance philosopher Francis Bacon ("Lord Verulam") who, in *Sylva Silvarum* (1627), is Brand's source for two flying-ointment recipes, the second of which includes four Solanaceae. Friend quotes Bacon as follows: "The ointment that witches use is reported to be made (among other things) of the juice of Smallage [wild celery], Wolf-bane, and Cinquefoil, mingled with the meal of fine wheat; but I suppose the soporiferous medicines are likest to do it, which are Henbane, Hemlock, Mandrake, Moonshade or rather Nightshade, Tobacco, Opium, Saffron, Poplar-leaves, and the like."[113] The good Reverend Friend discreetly leaves the matter at that, no doubt in deference to his possibly squeamish Victorian readers, whereas Brand allows Bacon to say more, filling in Friend's evasive parentheti-cal "(among other things)" with Bacon's "the fat of children digged out of their graves," a stock ingredient for flying ointments given in Inquisition reports. Bacon also elucidates later how this ointment works.[114] Bacon, after all, was writing around the turn of the seventeenth century, when the English witch craze was at its height; and by offering a rational explanation for sorcery in terms of what is known as "natural magic," he defended the (mostly) miserable hags who were being punished.

First, Bacon counsels his readers not to place too much trust in what witches say because, as he puts it, "the witches themselves are imaginative, and believe oftentimes they do that which they do not." That is, the marvels of sorcery are nothing more than the effusions of excitable minds under the influence of powerful drugs: "The great wonders that they tell, of carrying in the air, transforming themselves into other bodies, &c. are still reported to be wrought, not by incantations or ceremonies, but by ointments and anointing all over." In other words, these flights are simply ones of fancy, as Bacon ex-plains further: "For it is certain that the ointments do all (if they be laid on anything thick), by stopping of the pores, shut in the vapours, and send them to the head extremely. And for the particular ingredients of those magical ointments it is like they are opiate and soporiferous: for anointing of the fore-head, neck, feet, and backbone we know is used for procuring dead sleeps." Finally, Bacon clarifies why this "outwards" method of getting high with pow-erful hallucinogens is preferable to ingestion, for "if they were used inwards they would kill those that use them."[115]

Bacon's explanation is worth comparing to that of the chemist John Mann, who describes in his authoritative *Murder, Magic, and Medicine* (1992) how "the trio *Atropa belladonna, Hyoscyamus niger*, and *Mandragora officinarum*" operated in witches' salves: "At some stage it was discovered that if the con-stituents of the plants were combined with fats or oils they would penetrate through the skin, or could be easily absorbed via the sweat ducts (for example, the armpits), or body orifices (for example, the vagina or rectum). This al-lowed the psycho-active tropane alkaloids, especially hyoscine, to gain access to the bloodstream and brain, without passage through the gut, with the at-tendant risks of poisoning."[116] Mann reports that the effect of the ointment was hallucinogenic and sedative, producing a sensation of flight—as some twentieth-century scientist-experimenters corroborated. (And needless to say, this knowledge certainly casts new light on the involvement of the witch's broomstick for flying purposes.)

Bacon, however, is more circumspect in his account; but like his con-temporary Ben Jonson, he undoubtedly got his information from a curious

138 VICTORIAN NIGHTSHADES

Renaissance genre known as books of secrets, which graphically described in naturalistic terms the workings of magic and witchcraft. The classic of the genre (and one of Ben Jonson's sources) was the Italian Giambattista della Porta's *Magia Naturalis* (1558); but the most famous English purveyor of his information was the Reginald Scot mentioned by Brand, whose *The Discoverie of Witchcraft*, published in 1584, offers two flying-ointment recipes, the second of which includes *Solanum somniferum*, considered by Porta's contemporary the botanist Mattioli to be belladonna. (In this particular recipe, the other ingredients are "*Sium* [water parsnip], *acarum vulgare* [sweet flag or calamus], *pentaphyllon* [cinquefoil], the bloud of a flitter-mouse [bat's blood], and *oleum* [oil].") Scot quotes in full Porta's account of the witches' professed use of this magical ointment:

> They stampe all these togither, and then they rubbe all parts of their bodies exceedinglie, till they looke red, and be verie hot, so as the pores may be opened, and their flesh soluble and loose. They joine herewithall either fat, or oil in steed thereof, that the force of the ointment maie the rather pearse inwardly, and so be more effectuall. By this means (saith he) in a moone light night they seem to be carried in the aire, to feasting, singing, dansing, kissing, culling [embracing], and other acts of venerie, with such youthes as they love and desire most; for the force (saith he) of their imaginations is so vehement, that almost all that part of the braine, wherein memorie consisteth, is full of conceits.

The event described is of course the unholy Sabbath mentioned by Friend, in which according to legend, rumor, the report of the Inquisition, and evidently the alleged participants themselves, witches consorted with their leader and lover Satan in lewd and disgusting, orgiastic rituals. Neither Scot nor Porta believed in the reported rituals; but both—like Francis Bacon—believed in the power of the witches' ointments, for Scot includes Porta's titillating story that occurred when, as the latter says, "there fell into my hands a witch, who of hir owne accord did promise to fetch me an errand out of hand from farre countries." Porta brings his friends to watch what happens next:

> When she had undressed herselfe, and froted hir bodie with certeine ointments (which action we beheld through a chinke or little hole of the doore) she fell downe thorough the force of those soporiferous or sleepie ointments into a most sound and heavie sleepe: so as we did breake open the doore, and did beate hir exceedinglie; but the force of hir sleepe was such, as it tooke away from her the sense of feeling; and we departed for a time.

VICTORIA'S SECRETS

139

When Porta et al returned to quiz her, the woman reported magnificent travels, which she "impudently" insisted had taken place even though challenged by the men who witnessed her activities. Scot quotes Porta's conclusion: "This (saith he) will not come to pass with everie one, but onlie with old women that are melancholick, whose nature is extreme cold, and their evaporation small; and they both perceive and remember what they see in that case and taking of theirs."[117] Porta's account and Scot's reprise demonstrate that while feeling some pity for the women accused of witchcraft, they hold them in contempt. They accept the old woman's pharmacological skill, but condemn her belief that she has experienced magic. Despite Porta's skepticism about witchcraft, he and his friends are nevertheless willing to treat its proponents cruelly, as if such women are in truth the witches they proclaim themselves to be.

It is one of history's ironies that religion in the form of the Christian Church shared the witches' belief in witchcraft and Satan-worship, whereas the defenders of the witches' innocence were skeptics who denied the witch her claims. And Scot's work, like Porta's before him, caused a furor because of its skepticism; but both served as sourcebooks for a flurry of seventeenth-century witch dramas, several of which appropriated the ointment recipes to authenticate their representations. These include Thomas Middleton's *The Witch* (ca. 1613), which conflates Porta's two recipes in listing the ingredients for Hecate's cauldron; Thomas Shadwell's *The Lancashire Witches* (1681), loosely based on probably the most infamous of English witch trials, of which more, shortly; and as we have seen, Ben Jonson's *Masque of Queens*.[118] In all these dramas, belladonna appears as a crucial item in the witches' pharmacopoeia, although what was a recipe for a flying ointment in Scot becomes a love potion in Middleton.

Hilderic Friend's *Flowers and Flower Lore* makes clear that Scot's *Discoverie of Witchcraft*, like Bacon's *Sylvia Silvarum*, was known, if not read, by interested Victorians thanks to Brand's *Popular Antiquities*; and thus belladonna, as "Moonshade or rather Nightshade," along with henbane and mandrake, were notorious throughout the nineteenth century as essential ingredients in witches' salves—although the dirty secret of precisely how they worked required some digging to uncover. And even if the Victorians were not familiar with any of the above sources (though many were), they could learn something about witches' herbs from Harrison Ainsworth's immensely popular 1848 novel *The Lancashire Witches*, which retells—with enormous liberties—the story of the sensational episode in 1612 involving two rival families of self-proclaimed witches that led to the hanging of ten people (although the witches are burned at the stake in Ainsworth's novel). Ainsworth drew from a variety of sources including a contemporary account published in 1613 by one Thomas Potts, who appears as one of the novel's more disagreeable characters; but Ainsworth also embellishes the historical details with witch lore probably gleaned from *Popular Antiquities* and certainly from the seventeenth-century witch dramas. For example, he describes, in rather graphic if bowdlerized detail, an ointment-applying scene (discreetly involving only the woman's head and neck), several flight scenes, and a lengthy cauldron scene, in which the participants take turns adding horrific items to the pot, including many poisonous herbs, mandrake and "moonshade's deadly fruit" among them.[119] And there's a "diabolical unguent" described by Thomas Potts, citing Francis Bacon as his source, conflating Bacon's recipes, and consisting of "fat from unbaptized babes compounded with henbane, hemlock, mandrake, moonshade, and other terrible ingredients." In the novel, however, Potts posits that this salve is used (presumably) as a witch's magical cosmetic: as the possible secret for the "otherwise unaccountable beauty" of Alizon Device, the novel's tragic heroine, whose only crime is having been born to one witch and raised by another.[120] Clearly, Ainsworth plays fast and loose with all his sources—his witches really fly, consort with evil spirits, and perform all sorts of supernatural feats—but the popularity of his novel further underscores the Victorians' fascination with witch lore and suggests what sort of information about occult practices—and the nightshades' complicity in these rites—were in common currency.

The most sensational account of witchcraft written in the nineteenth century confirms the Solanaceae's primacy as the quintessential witches' herbs—what all the previous sources make apparent without ever stating explicitly. Published in 1862, *La sorcière* (literally, "The Sorceress") is a fascinating, melodramatic, baffling, and often lurid potboiler by the great French historian Jules Michelet, who describes part 1 of his work as a "long analysis, historical and psychological, of the evolution of the Sorceress down to 1300."[121] This first section is also a revisionist celebration of the Witch as the heroic Wise Woman-Midwife-Healer who dared to practice a "bold homeopathy" by mastering the skill of turning plant poisons into herbal medicines,[122] thereby becoming a rival and scapegoat for the church and state when her status among the folk appeared to threaten these institutions' political and social power. Thus, *La sorcière* is in actuality less true history and more romance (in the medieval sense of heroic adventure), a quality that extends as well to its botanical information. Nevertheless, the work offers a veritable witches' botany, one that celebrates the much-maligned nightshade family as the quintessential witches' herbs, dramatically presenting them as frightening and powerful, but also as efficacious and ultimately humane as the sorceress herself.

In fact, Michelet's approach is to equate the sorceress with the plant belladonna, thereby articulating a plot against the Solanaceae family by claiming that the fate of the witch and the fate of the nightshades are inextricably tied together. However romantic this argument is, it has at least some botanical merit, for, as we have seen, in the century when sentimental botany came into full bloom—when anthropomorphizing, feminizing, and moralizing about flowers were widespread and almost automatic tendencies—the Solanaceae indigenous to Europe and particularly belladonna were largely anathematized and eradicated because they were considered to be evil plants. In *La sorcière*, however, Michelet praises the nightshades for their now well-known therapeutic virtues, and in so doing attempts to rewrite their history as part of the vexed history of witchcraft.

Michelet argues that the Sorceress is the original healer who carries forward a female folk tradition of medical knowledge at odds both with official religion and official medical practice (a figure comparable to the so-called "cunning women" of England), who was accused of being a "witch" only when her "cure failed"; usually, however, "through a combination of respect and terror, she was spoken of as the Good Lady, or Beautiful Lady (Bella Donna)."[123] Because the power of the special knowledge she wielded threatened church and state authority, the historical persecution of witches was a programmatic attempt, especially on the part of official Christianity, to eradicate a potentially revolutionary force so intimately tied to nature that nature itself was threatened by the alliance. Appropriately, then, not only were the witches endangered species, but also the herbs they employed, the nightshades chief among them. In Michelet's words:

> Her fate resembled that which still often befalls her favourite herb, the belladonna, and other beneficent poisons she made use of, and which were antidotes of the great scourges of the Middle Ages. Children and passers-by cursed these sombre flowers, without understanding their virtues, scared by their suspicious colour. They shudder and fly the spot; yet these are the Comforting plants (*Solanaceae*), which, wisely administered, have worked so many cures and soothed so much human agony.
>
> They are found growing in the most sinister localities, in lonely, ill-reputed spots, amid ruins and rubbish heaps,—yet another resemblance with the Sorceress who utilises them. Where, indeed, could she have taken up her habitation, except on savage heaths, this child of calamity, so fiercely persecuted, so bitterly cursed and proscribed? She gathered poisons to heal and save; she was the Devil's bride, the mistress of the Incarnate Evil One, yet how much good she effected, if we are to credit the great physician of

the Renaissance! Paracelsus, when in 1527, at Bale, he burned the whole pharmacopoeia of his day, declared he had learned from the Sorceresses all he knew.[124]

As Michelet tells us, belladonna shares with the Sorceress not only a name but a habitat and destiny. His description of the plant as being cursed for its "sombre flowers" of a "suspicious colour" alludes to Linnaeus's pejorative name for the nightshades, the "Luridae," and to the latter's description of them as "suspect" because of their "stinking," "maddening and narcotic," or "entirely corrosive" properties.[125] Michelet is also correct regarding the family's habitat: those indigenous or naturalized in Europe—belladonna, bittersweet, black nightshade, henbane, mandrake, and datura—were by the nineteenth century inhabitants of disturbed areas, hedgerows, waste places, and ruins, locations that mark these plants as noxious weeds. In the hands of those ignorant of their powers they can be dangerous; but the fact that belladonna and henbane, especially, can be found near ruins indicates that they were once cultivated at convents and monasteries, presumably for their medicinal uses. Thus it is likely that Michelet's profile of the nightshades is slightly anachronistic, applying more to the early modern period and contemporaneous with the large-scale witch persecutions rather than to medieval times. In any case, certainly from Gerard's time on, the nightshades like the witch were forced to become wild and uncivilized, feral outcasts from the garden and the cultivated world.

These comments in the introduction and the definition of the Solanaceae as the "comforting plants" foreshadow chapter 9, "Satan the Healer," in which Michelet graphically describes the witch's use of nightshades for their therapeutic properties. He begins with a discussion of their efficacy as analgesics, soothing remedies for the diseases of the skin so prevalent in the Middle Ages because these conditions were (human) nature's way of reacting to the denial of the flesh demanded by official religion. "By a monstrous perversion of ideas," Michelet says, "the Middle Ages regarded the flesh, in its representative, woman (accursed since Eve), as radically impure"—a belief that "Woman herself" came to accept. Thus taught to be ashamed of their bodies and its functions, women secretly turned to the sorceress for treatment, many having to do with eruptions of the skin and painful female organs, especially the breasts. Nightshades, the herbs of choice, therefore became "appropriately known as the Solanaceae (herbs of consolation)." As in the introduction, Michelet asserts that the etymology refers to the family's healing properties, and then reinforces his defense of the nightshades in a footnote in which he laments "the ingratitude of mankind" for its rude treatment of these valuable plants: "A thousand other plants have usurped their place, a hundred exotic herbs have been preferred by fashion, while these poor, humble *Solanaceae* that saved so many lives have been clean forgotten with all the benefits they conferred. Who indeed have any memory of such things? Who recognizes the time-honored obligations men owe to innocent nature?"[126]

The next few pages provide an overview of the family and detail the way the witches employed these once highly prized herbs. Michelet notes that members of the family can be found everywhere and are thus ubiquitously available for use; but administering them was a risky business: "Audacity was required to determine the dose, it may have been the audacity of genius." Appropriately, then, Michelet organizes his discussion of individual nightshade species by starting "at the bottom of the ascending scale of their potency," mentioning the edibles aubergines and tomatoes before quickly moving to the medicinals—mulleins (no longer classified as Solanaceae), bittersweet, henbane, and finally belladonna.[127] Following a graphic and melodramatic description of the sorceress applying bittersweet to a married woman's swollen bosom to relieve "congestion and blocking of the veins and arteries," Michelet dramatizes the witch "gliding among the fallen stones of [an] old ruin" to gather "a villainous-looking herb" which turns out to be henbane, for use as "an excellent emollient, a soothing, sedative plaster, that relaxes and softens the tissues, relieves the pain, and often cures the patient" even though it is otherwise "a cruel and deadly poison."[128] He then describes the use of the arch-nightshade belladonna by the witch in her capacity as midwife:

> Another of these poisons, the belladonna, doubtless so named out of gratitude, was sovran for calming the convulsions that occur in childbirth, superadding peril to peril and terror to terror at this supreme crisis. But there! a motherly hand would slip in this soothing poison, lull the mother to sleep, and lay a spell on the door of life: the infant, just as at the present day chloroform is administered, worked out its own freedom by its own efforts, and forced its way to the world of living men.[129]

Here is Michelet's supreme and we might say primal example of the nightshades' power of consolation, embodied in its "sovran" belladonna: its sedative and antispasmodic power to soothe the pains of labor and delivery. Michelet correctly presents this usage as daring; but in a footnote he mentions that "Madame La Chapelle and M. Chaussier have returned to these practices of old-fashioned popular medicine with great advantage to their patients"—an accurate claim about current obstetrical practice intended to underscore the prescience and medical acumen of his medieval witch.[130] However, the revisionist etymology—that belladonna "was so named doubtless out of gratitude" on the part of suffering mothers—seems to have originated with Michelet himself.

Finally, Michelet celebrates belladonna's role in treating one of the strangest of medieval epidemics: the so-called "dancing mania" of mid-fourteenth-century Europe. Known as "*St. Guy's dance* (St. Vitus's dance, *chorea*)" according to Michelet, this bizarre affliction, characterized by large gatherings of people seized by epileptic convulsions before leaping up and dancing spasmodically until they collapsed from exhaustion, was of unknown origin, but considered by the nineteenth-century German physician Justus Hecker to be the mass hysteria of "a wretched and oppressed populace" in response to a period of natural and social disasters, including the recent visitation of the plague.[131] Michelet gives the date of these outbreaks as "about the year 1350," but he probably refers to ones that began in the German town of Aix la Chappelle in 1374 and spread from there to the Netherlands and Belgium later the same year. According to Hecker, Paracelsus was the first to argue that the affliction was not the work of demons, but originated from three possible natural sources: the imagination, sensual desire, and/or corporeal causes; yet is it not surprising that at the time of their occurrence, they were believed to be demoniacally inspired.[132] Asserting the homeopathic principle that "belladonna cures the convulsive dancing of the limbs by setting up another dance," Michelet surmises that the herb must have been used as a remedy because its antispasmodic properties were widely known among witches: "At the period when Sorcery and Witchcraft were at their point of highest activity and repute, the very extensive employment of the *Solanaceae*, and especially belladonna, was the most marked general characteristic of the remedial measures taken to combat this class of disease." Michelet continues, "At the great popular gatherings, the Witches' Sabbaths[,] . . . the *Witches' herb*, infused in hydromel, beer, as well as in cider, and perry, and strong drinks of the West, set the crowd dancing,—but in wanton, luxurious measures, showing no trace of epileptic violence."[133] As this comment suggests and later chapters make clear, Michelet considers the "Witches' Sabbaths" actually to have taken place—as peasant festivals of revolt against a corrupt church and an oppressive feudal system unresponsive to the people's spiritual and temporal needs. And considering their dire circumstances, it is no wonder that some people turned to Satan-worship and witchcraft as acts of protest, or that witches themselves became beloved Belladonnas for knowing how to administer powerful, pain-relieving, and reality-escaping drugs.

That Michelet's *La sorcière* was embraced by the radical French Feminists of the 1970s indicates how this text may be read sympathetically as a tale of liberated women who triumph in their secret and powerful knowledge, use it wisely, and thus become even more dangerously embroiled in secrecy and intrigue as others appeal to them for aid.[134] But we can imagine how the book—which later gives lurid accounts of Black Masses and persecutions—must have shocked many Victorians who might have dared to peep between its covers. Nevertheless, it seems possible that the work may have influenced the many sorceress paintings by Rossetti and his associates that began appearing in the mid-sixties and gained increasing popularity through the last decades of the century—at the same time that the image of the independent and socially rebellious "New Woman" was evolving.[135]

One of these paintings—Frederick Sandys's *Medea*—offers what is surely the most sensational visualization of belladonna in the Victorian era (see fig. 16). Moreover, this painting has as sensational a history as its subject: the beautiful, clever Colchian witch who used her magical skills as an herbalist-devotee of Hecate to help her beloved Jason win the Golden Fleece, but who became a jealous and ruthless wife and destroyer whose victims included her

FIGURE 16. Frederick Sandys, *Medea*. (Courtesy of Birmingham Museums Trust)

146   VICTORIAN NIGHTSHADES

brother, Jason's intended second wife, and most horrifyingly, her own two sons. Begun in 1866 at the epicenter of late Pre-Raphaelitism, Gabriel Rossetti's home and studio in Chelsea, this oil painting quickly earned a fine reputation as a work-in-progress among the circle of bohemian artists who considered Tudor House their hub; and Sandys worked to complete the painting for the Royal Academy's Summer Exhibition of 1868, for which it was accepted. However, when the Hanging Committee later rejected *Medea* without explanation but probably for its scandalous imagery, the furor roused in the press by Sandys's supporters—most notably the poet Algernon Charles Swinburne—not only forced the committee to overturn its decision the following year,[136] but helped to secure the painting's reputation as, arguably, the masterpiece of Pre-Raphaelite representations of witchcraft. An excerpt of Swinburne's defense of the painting, published in "Notes on Some Pictures of 1868," is worth repeating since it accurately describes the painting's symbolism, which includes Sandys's allusions to witches' botany:

> Pale as from poison, with blood drawn back from her very lips, agonised in face and limbs with the labour and fierce contention of old love with new, . . . the fatal figure of Medea pauses a little on the funereal verge of the wood of death, in act to pour a blood-like liquid into the soft opal-coloured hollow of a shell. . . . Her eyes are hungry and helpless, full of a fierce and raging sorrow. Hard by her, henbane and aconite and nightshade thrive and grow full of fruit and death; before her fair feet the bright-eyed toads engender their kind. Upon the golden ground behind is wrought the likeness of the ship Argo, with other emblems of the tragic things of her life. The picture is grand alike for wealth of symbol and solemnity of beauty.[137]

Swinburne's description suggests that he is writing from memory (Medea's "fair feet," for example, are not visible), but he vividly captures the salient features and mood of this painting, which in turn captures Medea in the act of witchcraft, presumably making the poison that will set fire to her rival's wedding dress. Like the strange objects that surround her, she is an alien creature, a woman defined by her sorcery, whose beauty is of an austere, defiant sort. In fact, her appearance recalls Michelet's description of "the all-puissant Medea," whom he occasionally considers to be the archetypal witch with "her wondrous deep-set eyes and the voluptuous snaky ringlets of coal-black hair that flood her shoulders."[138] Her accoutrements are legendary ones for sorcery, just as they identify her with feral nature. Sandys's Medea is the fatal woman incarnate, a powerful, frightening figure who communicates otherness even if also handsome, passionate, and visibly suffering—like Lucot's Belladonna, a vengeful woman who is the very emblem of despair. In keeping with its reputation, belladonna, flanked by henbane and aconite, is the central plant growing under Druidical oaks in the upper right-hand corner in the "wood of death"; and the glossy black berries in the foreground next to the toads are also belladonna. They form a visual pun with Medea's black eyes—perhaps dilated from belladonna use—once again hinting at the now-intrinsic connection between deadly nightshade and the witch.

Here ends my story of the Old World Solanaceae as primarily represented by the two often-confused species: the long-suffering *Solanum dulcamara* and the much-maligned *Atropa belladonna*. Despite the threat to its existence occasioned by persistent misidentification, bittersweet survived by its own tenacity as a highly adaptable plant, whereas belladonna might have been eradicated entirely had it not been saved from extinction thanks to a combination of forces at work during the latter half of the century: the species' promotion as an indispensable drug by the medical profession and consequent cultivation by the pharmaceutical industry; the growing appreciation for all forms of plant life fostered by Darwinian science; and the concomitant instinct toward preservation manifested by both professional botanists and amateur "flower-spotters." On the other hand, the evil reputation that led to belladonna's becoming a threatened species helped to give it a shady, but fascinating, presence in Victorian popular culture, most appropriately as those "phantasms of beautiful women" it was alleged to produce appeared in literature, art, and folklore, thus ensuring that the plant achieved a spiritual life as well, one captivating the nineteenth-century imagination as a fatal enchantress—the whore and the witch.

Yet even as the nightshade's Old World tale of darkness, hardship, and accursedness was playing out, important species introduced from the New World had been working to revolutionize the family's reputation; so that by the end of the century, there would be universal regard for the Solanaceae as consoling, necessary, and cultivated plants.

6

The Triumph of the Potato

> Solanum, in *Botany*, an ample genus, comprising various kinds of Nightshade, and other deadly plants, along with the esculent Tomato, Egg-plant, and even the valuable Potatoe, owes its name, according to some authors, to its comforting quality, such authors supposing the word to have originated from *solamen*. This indeed might apply to the potatoe, could that possibly have been in the contemplation of those who gave the name.
>
> —Abraham Rees, *The Cyclopedia, or Universal Dictionary of Arts, Science, and Literature*, vol. 33 [1819]

Darwin's Potatoes

WHEN THE YOUNG Victoria came to the throne in late June 1837, thus beginning the long, illustrious period in British history honored with her name, Charles Darwin was living in London, hard at work on the research that would make his name even more famous than the Queen's own.[1] Christened "Darwin's Century" by Loren Eiseley because the theory of evolution must be considered among its greatest ideas, the nineteenth witnessed an unprecedented reassessment of nature's kinship ties that was greatly advanced by Darwin's theory, which was in turn informed by his global exploration and intense curiosity about all the earth's creatures.[2] During his recent time in South America, one New World plant—one introduced to Europe sometime in the sixteenth century—had particularly captured his notice because of its homely familiarity: *Solanum tuberosum*, the edible, nutritious nightshade whose demonstrated wholesomeness was so completely at odds with the sinister reputation of its Old World relatives that its membership in the family almost defied belief, thus calling into serious question the aura of evil surrounding the nightshade name. Rather, as the Industrial Revolution mobilized a demographic shift in England from country to city that began around the middle of the eighteenth century, the common potato gained increasing prominence to become the favorite vegetable in the diet of all classes throughout the British Isles. Thus it was during Victoria's reign that an underground tuber emerged as the champion and savior of the nightshades' reputation, overcoming some initial suspicion and even contempt to become the prime embodiment of the consolatory connotations of Solanaceae, the recently minted family name. By the end of the century, the tuber's universal popularity and its close association with those domestic values that England held sacred would be the most important factors in rescuing the nightshades from the taint of their lurid Old World past.

The potato's impact on the Victorian era cannot be overstated. Larry Zuckerman, William McNeill, John Reader, and Andrew F. Smith have written superlative accounts for twenty-first-century audiences telling—to quote from the first two authors' titles—*How the Humble Spud Rescued the Western World* and "How the Potato Changed the World's History" by fueling a European population explosion with cheap, easily grown, highly productive, readily available, effortlessly prepared, and tasty sustenance, especially for impoverished and underpaid laborers—whether urban or rural—who often couldn't afford to buy or bake bread, which had traditionally been the lower classes' primary foodstuff.[3] By 1837 the potato had become the most valuable non-grain food crop in the world[4]—and certainly the nightshade family's most important member in historical, economic, and political terms, occupying the deliberations of the greatest minds of the Industrial Age. However, not many years later its essential role as the number one dietary staple for a huge segment of the British population would be brought home with a vengeance when the potato blight that swept through Europe wiped out Ireland's primary food source and with it, one million of its citizens, in the worst natural disaster of Victoria's reign. At mid-century the humble spud's stunning and tragic celebrity forced a recognition of its national importance upon the collective consciousness of Britain to such an extent that seen from a strictly vegetable point of view, the nineteenth could more correctly be known as the Potato's Century, with Darwin only one of the supporting cast in its unfolding drama. Nevertheless, the great naturalist's personal engagement with the potato at critical times over the course of the century serves as an effective vehicle for telling *Solanum tuberosum*'s tragic and ultimately triumphant Victorian story from an evolutionary perspective.

Darwin had returned from his five-year circumnavigation of the globe as a naturalist on the HMS *Beagle* in October 1836, preceded home by all manner of geological, zoological, and botanical specimens he had collected and shipped to his mentor, the Cambridge professor John Stevens Henslow, during the trip. Among these were cuttings and tubers from luxuriant plants he'd found blooming during his travels along the southwestern coast of South America that he immediately recognized as potatoes.[5] Although they were growing profusely near the beach on a remote, rain-soaked island in the Chonos Archipelago off the coast of Chile, much about them bespoke civilization: the plant was "sociable," Darwin said in his notes, forming "thick beds, in sandy, shelly soil" where the trees opened along the shore. Moreover, the plant's tubers looked, smelled, and tasted—if "rather insipid"—like the "Potatoes of Europe,"[6] that is, *Solanum tuberosum*, the New World nightshade that had become since the turn of the century so at home in English kitchens, gardens, and fields and so familiar at English tables that even with its many varieties it could hardly be mistaken. Yet the fact that the Chonos islands had never been inhabited convinced Darwin of the potatoes' wildness, and he was eager to learn what light Henslow might shed on the subject.

Darwin knew that their mutual hero, the great Prussian botanist, naturalist, and explorer Alexander von Humboldt,[7] whose works were an important part of the *Beagle*'s extensive library, thought the wild potato was native to Chile and locally called *maglia* according to the Chilean naturalist Ignatius Molina, who distinguished it from the cultivated variety by the *maglia*'s small, bitter tubers. Humboldt surmised it was from Chili that migrating tribes had spread potatoes northward toward the Equator and through the "whole Cordillera"—the Andean mountain range.[8] Molina's history was part of the *Beagle* library as well, and Darwin, who had studied Spanish for the voyage, undoubtedly checked this original source to verify Humboldt's information.[9] Another work in the ship's holdings, the English businessman Alexander Caldcleugh's *Travels in South America*, mentioned finding potatoes "in a state of nature" growing abundantly in ravines among the foothills above the coastal town of Valparaiso in central Chile, and recommended a "very interesting paper" on the subject written by "Mr. Sabine" for the Horticultural Society of London's *Transactions*—a paper that Darwin was clearly anxious to consult.[10] Darwin would meet Caldcleugh in Santiago a few months later in the *Beagle* voyage, and they no doubt discussed these potatoes.[11] Thus, the sources immediately available seemed to suggest to Darwin that he had found the original, the ur-potato, presumably the ancestor of the cultivated species, prompting him to label his specimens "*Solanum maglia*. Molina."[12] The specimens would ultimately prove to be the well-known cultivated species *Solanum tuberosum*; but unfortunately, Darwin's labeling set in motion an error that would reverberate through the horticultural world in the decades to come, a mistake memorialized in *Solanum maglia*'s becoming known late in the century as the "Darwin potato."[13]

Even so, the uncanny resemblance of the plant Darwin found to *Solanum tuberosum* continued to weigh on his mind, as revealed by his comments published three years after his return to England in his *Journal of Researches*—the book to become popularly known as *Voyage of the* Beagle, which provides the back story for *The Origin of Species* and the theory of evolution. Although "in general habit" the Chonos potatoes seemed to be "even more closely similar to the cultivated kind than is the *maglia* of Molina," Darwin writes, he goes to some lengths to demonstrate that they could not have been introduced, not only because of their remote location, but particularly because they were familiar to the disparate and "wildest Indian tribes" scattered over a vast area of western South America. As he puts it, "The simple fact of their being known and named by distinct races, over a space of 400 or 500 miles on a most unfrequented and scarcely known coast, almost proves their native existence." Besides, as he records in the *Journal*, he thought Henslow's response to his queries corroborated this opinion: "Professor Henslow, who has examined the dried specimens which I brought home, says that they are the same with those described by Mr Sabine from Valparaiso, but that they form a variety which by some botanists has been considered as specifically distinct." As a *Journal* footnote indicates, Henslow had consulted Sabine's aforementioned paper, entitled "On the Native Country of the Wild Potatoe," published in the Horticultural Society's *Transactions* in 1824.[14] As part of the larger debate among scientists and economists about the wisdom of its increasingly widespread use and cultivation, the potato had become a preoccupation with the Society in the first decades of the century; and Joseph Sabine, a well-known naturalist and the organization's secretary, was writing about plants he had grown in the Society's Chiswick garden from two tubers sent him by Caldcleugh, whose letter had identified them initially as "specimens of the Solanum tuberosum or Native Wild Potatoe of South America." But because the "roots" were small and bitter, Caldcleugh also surmised, "I am inclined to think that this plant grows on a large extent of the coast, for in the south of Chili it is found, and called by the natives *Maglia*, but I cannot discover that it is employed to any purpose."[15] In short, Caldcleugh's opinion seemed to support Darwin's first impression: that the *maglia* was *Solanum tuberosum*'s wild progenitor.

In any case, Sabine reports potting the Valparaiso tubers and then transferring the young plants to carefully prepared and well-manured soil, "earthed up" to form ridges in the manner customarily employed for growing potatoes. The results were spectacular: the two plants produced over six hundred tubers; moreover, "the flavor of them boiled was exactly that of a young Potatoe." Sabine thus determined, "They are unquestionably the Solanum tuberosum," and the accompanying color plate of a flowering plant seems to verify Sabine's opinion (see fig. 17).[16] Here seemed to be further confirmation that, like Darwin's potatoes, these were in fact native, wild plants found, according to Humboldt, in their original home. The only problem was that Darwin had labeled his specimens *Solanum maglia*, whereas Caldcleugh and Sabine had identified the parent tubers and the resulting offspring, respectively, as the familiar *Solanum tuberosum*, despite their being "the same" as Darwin's plants according to Henslow, whose rather evasive disclaimer, "that they form a variety which by some botanists has been considered as specifically distinct," confuses rather than resolves the problem of identification. This discrepancy raised the possibility that in its primitive, wild state, *Solanum tuberosum* was actually a different species—a circumstance that would support what Darwin would later call in the *Origin* species'"transmutation";[17] and indeed, the mysteries surrounding the potato's identity and origin would not, in fact, be

FIGURE 17. *Solanum tuberosum*, Transactions of the Horticultural Society of London, vol. 5 (1824), plate 11. (Image from the Biodiversity Heritage Library; contributed by Missouri Botanical Garden, Peter H. Raven Library)

THE TRIUMPH OF THE POTATO   153

unraveled until Darwin had posited his theory. But when after his death in 1882 and for the rest of the Victorian era, *Solanum maglia* came to be known in authoritative scientific circles as the "Darwin potato," the error would lead to a botanical brouhaha involving the Royal Botanic Gardens at Kew, where one of Darwin's *Beagle* specimens was housed, and to the embarrassment of its plant taxonomist John Gilbert Baker, who created the epithet.[18] As it turns out, there is a closely related wild species bearing small, bitter tubers known as *Solanum maglia* native to Chile, but it was never assigned that species name by Molina, nor was it positively identified until 1841, several years after Darwin's visit.[19]

It would take another century to prove that Darwin's Chonos potato was neither *Solanum maglia* nor its progeny, but indeed the common *Solanum tuberosum*, the species named in the late sixteenth century by the Swiss naturalist Gaspard Bauhin after its arrival in Europe and further designated by him as *esculentum*, referring to its most distinctive and desirable feature from a human point of view, its "esculent"[20]—that is, edible—tuber, which is not a "root," as Alexander Caldcleugh called it, but a swollen part of the plant's underground stem. Although plant identification had not yet achieved Linnaean orderliness in Bauhin's time, he could not have chosen a better name. He called the potato a *Solanum* because he intuitively recognized the similarity of its leaves, flowers, and berries to other plants so classified;[21] and *tuberosum* captured its spectacular novelty in European eyes. That is, the potato was the first plant in Europe to be known, grown, and valued for its tuber, a tumescent lump of carbohydrate-rich nutrients lying hidden beneath the soil that not only functions as a subterranean storehouse of nourishment for humans and animals but, if left to its own devices, ensures the plant's survival during periods of drought or cold weather. Perhaps most spectacularly, as the potato's underground cache of vital energy, a single tuber can produce one or many genetically identical plants.[22] *Solanum tuberosum*'s first great biographer, Redcliffe Salaman, takes for granted that the native South Americans who first cultivated the plant worshiped a potato spirit;[23] if so, the tuber is its embodiment, for like a chthonic deity it has subterranean powers of resurrection and procreation that indeed seem supernatural.

That Darwin did not consider the question of the Chonos potato's species resolved is suggested by the fact that he never mentions the plant's scientific binomial in the *Journal*, instead evidently dismissing the problem of identification by reporting Henslow's inconclusive disclaimer cited above. But his final words on the subject in the *Journal* reveal that major questions remained concerning the potato's origin and variations, especially given its extreme adaptability—what is now called its "environmental plasticity."[24] In Darwin's

154 VICTORIAN NIGHTSHADES

words, "It is remarkable that the same plant should be found on the sterile mountains of central Chile, where a drop of rain does not fall for more than six months, and within the damp forests of the southern islands. From what we know of the habits of the potato, this latter situation would appear more congenial than the former, as its birthplace."[25] This last sentence would prove to be wrong, but prescient nonetheless; for we now believe that although *Solanum tuberosum* originated in the high and dry Peruvian Andes where it had been domesticated for some 8,000 years, migrating south to Chile—presumably with its human caretakers—sometime before the Spanish conquest of South America, the potatoes predominantly grown in Europe after 1811 did in fact come from the Chilean landrace—that is, the locally adapted variety of the species.[26] The Chonos plants were therefore probably more like the English potatoes that Darwin knew than like the original Andean stock.

The Chonos potatoes had most likely grown from tubers carried to the island as food by natives who came there to fish. Thus, the plants were "wild" only in the sense that they had propagated and sustained themselves—essentially becoming a colony of immigrants making the best of it in their new environment after being left on an alien shore. Like human immigrants, they brought their culture with them, retaining the palatability achieved from millennia of domestication that had bred out the bitterness and much of the poison so characteristic of the nightshades. As Bauhin intuited, their scientific name indicates that potatoes belong to the same enormous genus as the Old World *Solanum dulcamara*—bittersweet—whose glycoalkaloid solanine they share as their weapon to fend off destructive predators.[27] But as human beings have taken over the potatoes' care, the plants have cooperated by lowering their guard and developing tubers that are larger, tastier, and more abundant.[28] Domestication, however, comes at a price. As Michael Pollan explains in his brilliant *The Botany of Desire*, this long process of coevolution, in which cultivator and plant have developed a mutual dependency, works for good and ill, as the Irish fiasco demonstrates.[29] Although *Solanum tuberosum*'s susceptibility to diseases had been exposed many times before and even mentioned by Joseph Sabine as a serious danger inherent in its widespread adoption, when the potato's worst enemy, the fungus-like organism *Phytophthora infestans*—better known as late blight—took the continent by storm in 1845, neither man nor potato had a defense, and the result was unparalleled disaster.

When the first attack of the disease hit the British Isles in late summer of 1845, Darwin—by then a gentleman farmer living at Down House, his sixteen-acre estate in Kent—was growing his own potatoes, which were damaged—"a good many having rotted to the ground"—but not totally

THE TRIUMPH OF THE POTATO   155

ruined, as he reported to Henslow in a letter that October. Henslow himself, now a clergyman at Hitcham, was deeply engaged with the problem in Suffolk; and Darwin alludes to "several printed notices" Henslow had sent him, probably about how to salvage food and make flour from an infected potato crop, subjects the good reverend had addressed among his parishioners. Like Henslow, Darwin was most immediately concerned about the plight of "poor people" for whom potatoes were an essential dietary staple; and as a first measure, Darwin mentions in the letter that he planned to follow Henslow's advice "about gentlefolk not buying potatoes" in order to make the short supply available to those who could not afford bread. As a case in point, Darwin cites the lamentable but certainly typical plight of one of his workers, whose stock of potatoes was almost depleted, but who could barely afford to buy wheat flour because it was so expensive, thanks to the "infamous corn-laws" that kept domestic grain prices artificially high (one of the few salutary consequences of the blight was to effect their repeal in 1846).[30]

Over the next several years the so-called Potato Murrain would occupy hundreds of columns in the *Gardeners' Chronicle and Agricultural Gazette,* England's leading journal and forum on horticultural matters whose editor John Lindley formally announced the visitation of the "fatal malady" as his opening editorial in its August 23, 1845, issue. The previous issue had reported the first instance of the blight in England from the Isle of Wight, but in a week's time the disease had devastated the entire country except for the north, leaving behind gardens and fields filled with a "putrid mass" of stinking, rotting vegetation. Lindley was convinced that the blight was "clearly traceable to the season," which, after a hot July, had turned unusually cold and wet. Nevertheless, believing that the problem was ultimately one of metaphysical proportions, he offered this fatalistic pronouncement: "As for the cure for this distemper:—there is none. One of our correspondents is angry at our not telling the public how to stop it; but he ought to consider that man has no power to arrest the dispensations of Providence. We are visited by a great calamity; which we must bear."[31] In *Potato, A History of the Propitious Esculent,* John Reader tells the story of the ensuing debate that raged over the blight's cause: even though the real source of the disease would be identified by several mycologists almost simultaneously with the first European outbreak, Lindley and like-minded botanists rejected the "fungal theory," believing instead that its proponents were observing the consequence and not the cause of the problem. The debate would not be settled definitively until 1861 when the French botanist Anton de Bary proved that the pathogen he named *Phytophthora infestans* was the culprit by inoculating healthy plants and watching them succumb to the disease. Even so, it was not until the 1880s

VICTORIAN NIGHTSHADES

that a preventative treatment known as Bordeaux Mixture, made from copper and used in French vineyards as a protection against mildew, was discovered to be effective for potato plants as well.[32]

In the meantime, when some growers casting about for ways to cut their losses began to consider obtaining wild stock in the hope of finding disease-resistant plants, Darwin's South American potatoes gained renewed attention. In the October 3, 1846, edition of the *Gardeners' Chronicle*, Darwin's cousin W. D. Fox reported in the "Home Correspondence" section on potatoes he had been raising for several years that were descendants of a tuber grown from seed potatoes Darwin collected in the spring of 1835 from the central Chilean Cordillera, following his visit to the Chonos Archipelago. Darwin had found the plants "many miles from any inhabited spot," and had concluded that they were "certainly in a state of nature." It had taken Fox's crop a few years to produce edible tubers, but the plants had then flourished until 1845 and 1846, when they suffered the same fate as his other varieties. Thus Fox reported, "I fear this decides the point as to the uselessness of procuring seed from even the fountain head—the wild stock itself."[33] Before the article's publication, Darwin had provided Fox with the details about the stock in a letter that included the following comment: "I have sometime thought of calling the attention of the readers of the Gardeners' Chronicle to the remarkable difference of climate of the Chonos islands & central Chile, in both of which places the Potato grows wild—if you think it worth while to allude to this, refer to the 1st Edit. of my Journal, if you have it."[34] Although Fox chose not to mention this information in the *Chronicle*, it was clear that Darwin was musing over the possibility that Chonos potato stock might fare better than that from central Chile since plants from the archipelago were so well adapted to wet weather—which, if not the cause, was at least complicit in the blight's spread.

Needless to say, Darwin's attention turned to vaster subjects through the late forties and the fifties; but potatoes appear again in his most massive work, the two-volume *The Variation of Animals and Plants under Domestication*, first published in 1868, whose central purpose is to give exhaustive support for his theory of natural selection by showing how it applies to variations that occur when species are cultivated—that is, when more or less controlled via human guidance. As evidence, Darwin not only conducted his own experiments, but also scoured the literature available and consulted breeders and growers throughout the world for information on everything from dogs to dahlias, comparing wild to domesticated races whenever possible.[35] Most of Darwin's firsthand information came from the plants he grew in his greenhouse and garden, where potatoes had been a subject of study since the for-

THE TRIUMPH OF THE POTATO 157

ties: even before the blight hit in 1845, Darwin had requested in February that his cousin Fox send a few potatoes from the Chilean stock "chiefly to get true seed from them, & see whether they will sport [i.e., mutate] or not readily."[36] Not surprisingly, then, when in *The Variation* he begins his discussion of *Solanum tuberosum,* Darwin recalls his South American adventures, saying, "There is little doubt about the parentage of this plant; for the cultivated varieties differ extremely little in general appearance from the wild species, which can be recognized in its native land at first glance." The blight had increased the already avid interest in creating potato cultivars immune to disease, and Darwin mentions that at least 175 potato varieties were growing at that time in Britain alone; he himself was raising eighteen different kinds. Yet despite the proliferation of varieties, Darwin reports that the plants "differed but little" from each other in all their parts except for their tubers, thus reinforcing what he had found to be true throughout his investigations: "the principle that the valuable and selected parts of all cultivated productions present the greatest amount of modification."[37] Since most potatoes were not grown from true seed but propagated asexually from the buds on tubers, any variations that occurred were of particular note; and Darwin cites several instances in which "a single bud or eye sometimes varies and produces a new variety; or, occasionally, and this is a much more remarkable circumstance, all the eyes in a tuber vary in the same manner and at the same time, so that the whole tuber assumes a new character."[38]

Minute, firsthand observations and the reports from breeders and botanists also serve as the evidence for the second, more speculative part of *The Variation,* which attempts to establish universal laws governing changes and inheritance in species, a task that Darwin freely admits is monumental and fraught with difficulty. It is ironic—and unfortunate—that he and Gregor Mendel were exact contemporaries yet did not share their work, for Darwin would then have been privy to the theory of genetics, the missing link in his grand argument. Instead, Darwin devised his own theory about what he calls "the wonderful nature of inheritance," which, as his astute biographer Janet Browne says, "plugged the gap left in the *Origin of Species*." *Pangenesis,* as Darwin termed it, attempts to explain heredity via sexual reproduction as a rather mysterious blending of parental elements, in Browne's terms: "the highly abstract notion that every tissue, cell, and living part of an organism produced minute, unseen gemmules (or what he sometimes called granules or germs) which carried inheritable characteristics and were transmitted to the offspring via the reproductive process."[39] However, regardless of the specific mechanism at work in sexual reproduction, it seemed obvious that in most cases outbreeding produced more vigorous individuals than inbreeding, thereby indicating that plants propagated vegetatively were at far greater risk of degeneration and disease. Thus the potato, "which until recently was seldom multiplied by seed," was "probably now as tender in England as when first introduced," Darwin says.[40] Moreover, "plants that produce a large number of tubers are apt to be sterile, as occurs, to a certain extent, with the common potato"; and since many varieties failed to bear flowers or fruit, some botanists believed that such plants "had lost the habit of sexual generation," a theory, Darwin says, about which "I will not venture, from the want of sufficient evidence, to express an opinion."[41] Nevertheless, Darwin notes that cross-fertilization did occur in potato varieties that were self-sterile, another clue that outbreeding, as a general rule, was beneficial for a species' continuation. And even when potato growers didn't introduce variation by way of sexual reproduction, Darwin reports that "the practice of exchanging sets [tubers] is almost everywhere followed"—an example of the good derived from changed conditions.[42]

Although *The Variation* never addresses the idea of crossing closely allied tuber-bearing *Solanum* species, during the forties the blight had significantly intensified interest in the possibility; and from his correspondence with Fox, it's clear that Darwin still considered his Chonos potatoes to be potentially better suited to the English climate than the Cordilleran ones that had proven not to be blight-resistant. Whether the Chonos plants were a separate species may still have been a question for Darwin; but he himself had shown that the line between varieties and species was extremely blurry, if not in many cases indistinguishable, especially in plants like the potato with a long history of cultivation.

It was not until after his death that two of his apostles took action at last to attempt a cross between the so-called wild South American potato Darwin had long ago identified as *Solanum maglia* and the homegrown, cultivated *Solanum tuberosum*, a move prompted by another attack of late blight in 1879. Alan Frederick, Earl Cathcart, president of the Royal Agricultural Society of England, led the charge in 1883 with an appeal to the House of Commons to publicize "a series of experiments with a view of producing new and disease-proof varieties," which Cathcart was personally sponsoring. In an 1884 article published in the Agricultural Society's *Journal*, Cathcart announced that he had enlisted J. G. Baker, a Kew botanist and taxonomist, to provide "A Review of the Tuber-Bearing Species of Solanums" and to recommend which ones might be suitable for hybridizing with *Solanum tuberosum* for disease resistance. Cathcart reports that Baker had accommodated with an article published in the *Botanical Journal of the Linnean Society* the same year, in which he corroborates Darwin's erroneous identification of the Chonos specimens housed at Kew, claiming they were "quite characteristic" of *Solanum maglia*, having compared them with plants in Kew's gardens identified as the same species. Baker then quotes at length from Darwin's *Journal* about the "very remarkable" climatic difference between the archipelago and the Cordillera as evidence that the specimens Darwin collected from the two localities actually represent two different species. In deferring to Darwin's identification of the Chonos plant as *Solanum maglia*, Baker notes that he contradicts not only Alexander Caldcleugh's and Joseph Sabine's assessment of the Valparaiso specimens in the 1820s, but also that of the great French botanist Michel Felix Dunal, the recognized authority on the *Solanum* genus. Baker then concludes, "As far as climate is concerned, it cannot be doubted that *Solanum Maglia* (or the Darwin potato as we might suitably christen it in English) would be better fitted to succeed in England and Ireland than *S. tuberosum*, a plant of a comparatively dry climate."[43]

Thus *Solanum maglia* became the prime candidate for hybridization with *Solanum tuberosum*, and Lord Cathcart engaged one of the foremost seed producers in England, Sutton and Sons of Reading, to carry out the experiment in its trial grounds—to considerable fanfare throughout the horticultural world.[44] In May 1884, Kew's director and the late Darwin's close friend Joseph Dalton Hooker announced the project in *Curtis's Botanical Magazine* with an article on *Solanum maglia*, recounting the story of the species' arrival in England and promoting the "Darwin Potato" moniker—even though Hooker himself had formerly identified Darwin's Chonos specimens as *Solanum tuberosum* forty years earlier, following his own South American expedition. John Nugent Fitch provided the accompanying color illustration, drawn from the plants to be used in the experiment, Kew specimens grown from tubers given by a "Dr. Sclater" in 1862, "which flower freely every autumn, and yield watery scarcely edible potatoes."[45] The horticultural press was abuzz with the project, especially the *Gardeners' Chronicle*, which produced articles on the subject throughout the year, at last proclaiming in its November 8, 1884, issue that "Messrs. Sutton," who have been "engaged for the first time in the history of agriculture in hybridising Potatos," had "succeeded in effecting a cross between S. Maglia and one of their best Potatos." The report was that *tuberosum* pollen had impregnated a *maglia* flower, which gave birth to three berries—an amazing feat, since the latter species had never been known to produce seed in captivity (or anywhere else, for that matter).[46] The November 29, 1884, edition of the *Times* published Arthur Sutton's letter to Lord Cathcart announcing the success, which the *Gardeners' Magazine*—the *Chronicle's* rival edited by the champion of amateur horticulturists, Shirley Hibberd—reprinted in its December 6, 1884, edition.[47] Both magazines continued to follow the story the next year. In September 1885, the *Gardeners' Magazine* published an illustration of *Solanum maglia* referring to the species by its now "familiar name," the "Darwin Potato" (see fig. 18).[48] And in October, the *Gardeners' Chronicle* proudly reported that seeds from the *tuberosum-maglia* cross had successfully delivered a "very respectable crop" of tubers varying in size, skin texture, and color, with a "flavor when cooked" that was "by no means bad." The *Chronicle* praised the growers' stunning success, proclaiming Sutton and Sons the "magicians of Reading."[49]

Unfortunately, in November the following year the *Chronicle* had to report the Messrs. Sutton's bad news: a group of experts (including Shirley Hibberd) visiting their trial grounds had determined that the parental "*maglia*" was no such thing, "but a form of the ordinary Solanum tuberosum grown at Kew for many years, without any special cultivation." The magazine stated that neither Sutton and Sons nor Lord Cathcart was responsible for the "unfortunate error," thereby implying that the authorities at Kew were to blame

FIGURE 18. "Darwin Potato," *Gardeners' Magazine*, vol. 28 (September 12, 1885): 517. (Courtesy of HathiTrust; contributed by the University of Illinois Urbana Champaign)

THE TRIUMPH OF THE POTATO 161

for supplying the wrong species. But regardless of whose fault, the *maglia* experiment had yet to be undertaken. The *Chronicle* defended Sutton and Sons' right to be called the "Magicians of Reading," however; for they had not only produced several promising *tuberosum* combinations, but also some authentic *Solanum* hybrids by crossing potatoes with tomatoes, bittersweet, and common nightshade.[50]

News of the failed *maglia-tuberosum* cross and the mistaken identity of the former species spread quickly through the Empire: the *Times* and the *Daily Telegraph* published articles in late November 1886, reprinted as far away as India for its agricultural monthly, the *Indian Forester*.[51] While these articles highlighted Sutton and Sons' experimental successes, they advanced the idea of Kew's failure, incensing Baker, who shot back in the December 4, 1886, *Chronicle* "to protest energetically" Arthur Sutton's blaming Kew for the "blunder," instead accusing Sutton and Sons of having substituted the "Kew type" of *maglia* for the "Reading type," as he chose to label it.[52] The controversy was reported as the gossip of the "Potato Tercentenary, 1586–1886,"[53] a horticultural celebration held in December at St. Stephen's Hall in London, where Baker read a paper on wild species, which acknowledged that the Kew *maglia* "looks very like the ordinary cultivated potato."[54] Precisely how the error occurred will probably always remain a mystery, but it seems most likely to have arisen with Darwin's original misidentification; and information from Redcliffe Salaman's classic, *The History and Social Influence of the Potato*, suggests that all the Kew *maglia* may have been later misidentified following Baker's lead. When in 1906 Salaman requested *Solanum maglia* tubers from Kew, he received instead yet a different species, *Solanum edinense*, learning later that Kew's *maglia* stock "had died out, and that another supposedly wild species had by accident acquired its label." (In an odd coincidence, Salaman also notes that his "late friend Arthur Sutton" had received the same species "twenty years earlier" from the Edinburgh Botanical Gardens, there misidentified as *Solanum etuberosum*.)[55] These errors of course underscore how notoriously difficult it is to distinguish among closely related solanums based on morphological evidence alone, so it is perhaps no wonder that Darwin's encounter with potatoes in South America led to *Solanum maglia's* brief rise and subsequent fall as a botanical celebrity bearing his name. In 1913 the German botanist Georg Bitter reaffirmed Baker's classification; but in 1956 Salaman's former student John G. Hawkes at last reclassified the species Darwin found in the Chonos Archipelago as *Solanum tuberosum*, permanently laying to rest *Solanum maglia's* late Victorian fame as the "Darwin Potato."[56]

The story of the *maglia* hybridization experiment similarly ends with a whimper. In 1887 Sutton and Sons succeeded in obtaining a *maglia-tuberosum* cross and continued their *maglia* experiments for the next several years; but the resulting tubers were disappointing—and according to a notice in the *Gardeners' Chronicle* in 1899, finally not worth the effort.[57] Nevertheless, Arthur Sutton gained considerable celebrity through his *Solanum* exploits to become one of the great Victorian Solanaceae enthusiasts and promoters of edible nightshades. In a paper presented before the Royal Horticultural Society in 1895, complete with a slideshow featuring the firm's latest potato varieties and its trial grounds, he recounted the *maglia* experiment; but reported that the previous year the *maglia* crop planted outdoors had been almost entirely destroyed by disease, and the hybrids had produced tubers that were "very far behind the ordinary Potato in appearance, crop, and qualities."[58] The long-cultivated *Solanum tuberosum* with its myriad varieties had prevailed.

The Cinderella of Nature

When word of Lord Cathcart's hybridization project was first making the horticultural rounds, Robert Fenn, one of England's most successful potato breeders,[59] felt called upon to offer his support, but also a warning, in a February 1884 issue of the *Journal of Horticulture and Cottage Gardener*, to which he frequently contributed. Fenn's views carried authority, for he was one of Sutton and Sons' primary suppliers—in fact, his *tuberosum* cultivars, a fledgling Sir Charles Douglas variety and the well-established Reading Russet, would provide the pollen for both the forthcoming *maglia* experiments.[60] Even so, Fenn admitted to skepticism about the plan's ultimate success, having himself failed to achieve a cross between an English *tuberosum* variety and a species from New Mexico, *Solanum fendleri*, several years earlier. What he feared most, however, was that the current rage for potato breeding would ultimately do more damage than good. The quest for a disease-resistant potato in the second half of the century had accelerated the popularity of horticultural societies and shows promising celebrity and prizes for new cultivars; and Fenn lamented the fact that a "legion of breeders" had arisen, some of whom might be tempted to engage in reckless practices of "indiscriminate crossing" that sacrificed "quality" for "size and appearance"—a problem that would only be exacerbated by this latest news of a project promoted by an English Earl and subject to scrutiny by the Royal Botanic Gardens at Kew. Or, in Fenn's more eloquent terms: "Doubtless an extra fillip will be given to the cultivation of the esculent now that the Cinderella of Nature is countenanced by nobility, and the savants have questioned our doings."[61] Fenn's words proved prophetic: as Salaman and Reader tell the story, the rush to cash in on the post-blight breeding mania ultimately led to the Great Potato

THE TRIUMPH OF THE POTATO   163

Boom of 1903–4, which eventually collapsed when a so-called new cultivar aptly named Eldorado, whose tubers sold for as much as £150 apiece, proved actually to be a recycling of an earlier, rather forgettable variety.[62]

Fenn's felicitous epithet the "Cinderella of Nature" effectively conveys the potato's upwardly mobile trajectory after its introduction to Britain in Elizabethan times to the end of Victoria's reign, during which time it fully achieved the appreciation it deserved for long and meritorious service to the dietary and culture of the British Isles. Couched in varietal terms, the tuber rose from the ashes of a turf fire as the much-maligned Lumper, the coarse but high-yielding potato of the Irish peasant that met its demise with the blight, to grace the royal table as the beloved Victoria, the most popular of cultivars according to Reader, one eminently qualified to bear the Queen's name.[63] Entering in the late sixteenth century as an exotic novelty, of interest primarily to botanists and the affluent owners of large gardens, the potato otherwise met with suspicion or indifference as a food until the mid-eighteenth century — except in Ireland where it began making itself at home almost a hundred years earlier.[64] Although Gaspard Bauhin mentioned and named *Solanum tuberosum* in 1596, it was John Gerard in England who a year later provided Europe with the potato's first published illustration — evidently drawn from a plant in his own garden — in his celebrated *Herbal*, where it also had pride of place in the frontispiece as the plant Gerard famously holds in his hand.[65] Gerard's chosen name "Potato of Virginia" reflects the tuber's superficial similarity in appearance and taste to the root of an unrelated New World species, the sweet potato (*Ipomoea batatas*), which Gerard considered to be the "common potato," having reached Europe sometime earlier. The "Virginia" designation indicated the new species' supposed place of origin, an assumption that has baffled botanists and historians ever since, but presumably caused by the potato's association with Sir Walter Raleigh, who sponsored the first English colony in North America and who allegedly introduced the plant to Ireland. Unfortunately, Gerard did not have Bauhin's skill in recognizing botanical family characteristics, so the information that the plant was a nightshade or the name *Solanum tuberosum* was not added until 1633 in the edition of the *Herbal* revised by Thomas Johnson — and by then the two potatoes had been utterly confounded. As an alternate name for the viny sweet potato, a member of the morning glory family, Gerard gave "Skyrret of Peru," thereby suggesting it was related to the "skirret," a member of the carrot family. Frequently fashioned into confections, the sweet potato had a reputation for "procuring bodily lust"; so Gerard's reference to our spud's "being likewise a food, as also a meat for pleasure" hints at similar aphrodisiacal qualities.[66] It's reasonably certain, though, that when the randy Falstaff exhorts, "Let the sky rain pota-

164 VICTORIAN NIGHTSHADES

toes" in the contemporary comedy *The Merry Wives of Windsor* (ca. 1602), he's soliciting aid in seduction via the sweet potato.[67] However, when the "last Lord Chancellor" in Dickens's *Bleak House* quips, "Such a thing might happen when the sky rained potatoes," the date of publication—1852—and the context of an endless Chancery lawsuit suggest a meaning something like "when hell freezes over," while in this instance Victorian readers would take the Shakespearean allusion to refer instead to *Solanum tuberosum*.[68]

At any rate, although our potato entered Britain under the legendary auspices of the nobility and the actual sponsorship of well-heeled herbalists like Gerard, its identification with the opposite end of the social spectrum became well established before the close of the seventeenth century. Indeed, *Solanum tuberosum*'s history has, arguably more than any other plant, been inextricably tied to class—that is to say, the lowest class—and the relatively swift, widespread adoption of the potato as a food crop by the Irish peasantry had everything to do with this connection. Salaman writes that as early as 1662 the newly formed Royal Society of London considered the potato as a possible food for the masses and advocated planting potatoes "as a protection against famine," an idea (later to prove ironic) inspired by the success of its cultivation in Ireland.[69] Both Salaman and Cecil Woodham-Smith, author of the monumental history of the Potato Famine, *The Great Hunger* (1962), contend that the tuber's acceptance in Ireland had as much to do with the country's social climate—to wit, the perennially impoverished circumstances of the Irish peasant class—as with its moist, mild weather; and both writers cite England's centuries-long subjugation and programmatic exploitation of Ireland as the primary cause of the people's ongoing misfortunes.[70] Moreover, in a striking display of power's penchant for blaming the victim, England rewarded Ireland with contempt, considering it at best a colonial outpost good only for what it could provide the Mother Country, but in any case a lawless hinterland defined by its poverty, its ignorance, its fecklessness and savagery—and from the seventeenth century on, its potatoes.[71] Grown on ridged, manured, and trenched strips often reclaimed from bogs or land not otherwise arable through backbreaking labor, but regrettably called "lazy-beds," and serving not only as essential human food but as animal fodder, the tuber came to symbolize for the English and the rest of the Western world the abject, soil-bound penury of the Irish peasant.[72]

For example, in the early part of the nineteenth century, the potato's identification with the indigence of the Irish fueled the reformer William Cobbett's extreme hatred of the tuber, which he took every opportunity to attack because of its recent rivalry with bread as the staple food of the English working class. Cobbett was the potato's most vociferous antagonist during what has

THE TRIUMPH OF THE POTATO

been called the great potato debate of the late eighteenth and early nineteenth centuries, when its value as a new staple food source and crop was being both championed and challenged.[73] Decrying the potato in his *Cottage Economy* as the root "of slovenliness, filth, misery, and slavery," Cobbett criticized the current practice among some landowners in western England of offering allotments for growing potatoes, arguing, "This has a tendency to bring English labourers down to the state of the Irish, whose mode of living, as to food, is but one remove from that of a pig, and of the ill-fed pig too." However, he expresses some confidence that "it must be some time before English people can be brought to eat potatoes in the Irish style; that is to say, scratch them out of the earth with their paws, toss them into a pot without washing, and when boiled, turn them out on a dirty board, and then sit around the board, peel the skin and dirt from one at a time and eat the inside."[74] Unfortunately, Cobbett's description was hardly an exaggeration, for among the most destitute of the Irish, the image of a squalid, one-room, often windowless, furniture-bare mud cabin choked by the smoke of a turf fire and occupied by a husband and wife in rags, a handful of naked children, a pig, and perhaps a few chickens all inside sharing a pot of potatoes—with a dung heap just outside the door—is familiar in reports from the first half of the century; it is a scene similar to one appearing in an 1846 illustration from the *Pictorial Times*.[75] And as Thackeray notes in his *Irish Sketch Book* describing a four-month trip taken in the summer of 1842, "by the side of the cottage, the potato-field always."[76]

The spud's close identification with Ireland accounts for another of its aliases, "murphy," which (according to the *OED*) surfaced in British and American slang in the nineteenth century—to be exploited satirically in 1838 after Patrick Murphy, an immigrant from Cork and author of *Murphy's Weather Almanac*, accurately predicted that January 20 would be the coldest day of that year. As Katherine Anderson explains in her study of Victorian meteorology, his lucky success led to a run on his publication; but it also inspired a host of insulting jabs in almost every conceivable medium—newspaper articles, cartoons, songs, and even a one-act farce—several of which drew derisively on the murphy-potato connection. The most insistent of these was a cartoon printed by G. S. Tregear entitled "The Man with a Weather Eye, Being a Correct Likeness of the Celebrated Mr. Murphy," which depicts an anthropomorphic dancing potato holding a shillelagh in one hand and a basket full of potatoes in the other, surrounded by more tubers lying about his feet, each one inscribed with a weather prediction—as if the sky had indeed rained potatoes.[77]

On the eve of the famine, no better example of the Irish-potato identification seen at its worst can be found than in Friedrich Engels's *The Condi-*

166   VICTORIAN NIGHTSHADES

tion of the Working Class in England in 1844, written and published in German after his two-year residence in Manchester, Britain's first industrial city, where the worst slum, aptly called "Little Ireland," housed thousands of Irish immigrants living in squalor like animals and feeding on potatoes. Describing the slum's filth and dilapidation in sickening detail, Engels nevertheless claims that the Irish are inured to this dissolute, debased existence: "The worst dwellings are good enough for them; their clothing causes them little trouble, so long as it holds together by a single thread; shoes they know not; their food consists of potatoes and potatoes only; whatever they earn beyond these needs they spend upon drink." These brutish Irish represent the "lowest stage of humanity," certainly lower than the English laborers with whom they compete for work: "The Englishman, who is still somewhat civilised, needs more than the Irishman, who goes in rags, eats potatoes, and sleeps in a pig-sty." Engels says that in all the "great towns" of England where the Irish flocked, they partake of a "potatoes only" diet—except when down and out in London, where competition for work is most intense, they live from hand to mouth on "potato parings, vegetable refuse, and rotten vegetables."[78] But despite the fact that the potato represented bottom-feeding on the human and wage scale, it was the Irish laborer's choice as well as necessity—a mere continuation of the diet he followed at home. As Engels explains, many Irish were seasonal migrants: between potato-planting in the spring and the harvest in the fall—the months when the new crop could be left unattended (hence, one interpretation of "lazy-beds") and no potatoes remained from the previous year's supply—the Irish poor hit the road, wife and children to beg and husband to look for work either in Ireland or England.[79] But whether in or out of Ireland, the potato wholly defined the ragged existence of the Irish peasantry—a Lumper proletariat, we could say—just as the Irish peasantry inevitably evoked the image of the lowly spud. It is probably not beside the point that when Engels's friend and collaborator Karl Marx in 1852 described France's somewhat similarly soil-bound, impoverished peasantry and their concomitant lack of a cohesive class ideology, he resorted to a pejorative potato metaphor: "The great mass of the French nation is formed by the simple addition of homologous magnitudes, much as potatoes in a sack form a sack of potatoes."[80]

Unfortunately, the Victorian wave of Irish immigration that Engels describes furthered the age-old hostility between Ireland and England, which originally arose from the clash between Celtic and Anglo-Saxon cultures, but greatly intensified when Protestant England attempted to "reform" Roman Catholic Ireland by military force in the sixteenth and seventeenth centuries. Irish resistance led Parliament to pass Penal Laws that succeeded in reducing most of the native population to third-class citizens by not only denying them the right to practice their religion, but also depriving them of the right to vote, to attend school, and to purchase or inherit property. The result was the creation of a huge underclass who relied for their very existence upon trading their labor for a tiny bit of land—rented on credit at an exorbitant cost—on which to live and grow food, a cottier tenantry whose lives were controlled by middlemen agents and farmers, some of whom were not much better off, being themselves often at the mercy of absentee landlords. Only an easily grown, supremely productive plant like the potato could make such a precarious existence even sustainable; yet—as if celebrating its alleged aphrodisiacal powers—sustain it did: Reader tells us that the Irish population grew from 2 million in 1700 to 8.5 million in 1845;[81] hence, the need for many to seek work in England. During that period agrarian Ireland transformed from a country of oats, barley, and pastureland to one primarily of potatoes, grown in small plots even on the most inhospitable ground. Larry Zuckerman says that by 1845 Ireland had "65,000 farms of no more than an acre, on which the spade was the only tool and the potato the only crop."[82] (The nickname "spud," which according to the *OED* dates from the mid-nineteenth century, derives from an alternate name for the digging implement.) Estimates vary, but on the cusp of the famine, it seems possible that up to 90 percent of the nation existed almost exclusively on a potato diet—with adult men eating as many as fourteen pounds of potatoes a day. As Salaman summarizes, "It is not too much to say that for close on 300 years the potato both stabilized and perpetuated the misery of the Irish masses. It was, as it were, the least common denominator of Irish life, for by reducing the cost of living to the lowest possible limit, it caused the value of labour to fall to a corresponding level, whilst it permitted, if not encouraged, an ever-growing population."[83] Indeed, as Woodham-Smith says, "The potato, not money, was the basic factor by which the value of labour was determined."[84]

In short, the potato was on the one hand a godsend and the other a curse: while it provided a plentiful source of nutrition for a nation used to living perennially on the brink of starvation, it supported and abetted the Irish people's seemingly preternatural fecundity, increasing the mouths that had to be fed—by more land subdivided into ever-smaller plots planted with potatoes. In reference to England in 1776, the great economist Adam Smith had theorized enthusiastically that "if potatoes were to become the favourite vegetable food of the common people, and if the same quantity of land was employed in their culture, as is now employed in the culture of corn, the country would be able to support a much greater population and would consequently in a very short time have it." Smith's hypothesis appeared in the first

168   VICTORIAN NIGHTSHADES

edition of Thomas Malthus's *Essay on the Principles of Population* (1798);[85] in all subsequent editions—the last in 1826—Malthus bemoaned the fact that in Ireland, Smith's speculation had come true with a vengeance: "The population is pushed much beyond the industry and present resources of the country; and the consequence naturally is, that the lower classes of people are in the most impoverished and miserable state."[86]

As Malthus intimated, a surplus population living well beyond its means and subsisting by the monoculture of a disease-prone, long-domesticated species was a recipe for disaster; and three weeks after John Lindley had announced the blight's arrival in England in the *Gardeners' Chronicle*, he was compelled to begin his September 13, 1845, editorial with the following ominous report: "We stop the Press, with very great regret, to announce that the Potato Murrain has unequivocally declared itself in Ireland. The crops about Dublin are suddenly perishing. The conversion of Potatoes into flour, by the processes described by Mr. Babington and others in to-day's Paper, becomes then a process of the first national importance; for where will Ireland be, in the event of a universal Potato rot?"[87] Cecil Woodham-Smith's *The Great Hunger* answers Lindley's question by recounting what happened in copious and horrific detail: three years of blight—1845, 1846, and 1848—resulted in a famine that lasted in some parts of the country into the 1850s, killing over a million people by starvation and disease, and forcing as many others to emigrate, primarily to England and North America.[88] It fulfilled Malthus's direst prophecy to become the textbook case concerning the consequences of overpopulation, and proffered its grim solution—the radical and ruthless purging of a surplus mass of humanity.

As it turned out, the blight of 1845 was just a preview of the utter destruction of Ireland's potato crop that took place the following summer, a one-two punch resulting from a mild winter, which allowed the pathogen to survive on diseased potatoes that had been tossed out into fields, thus infecting the next season's crop. A bitterly cold winter followed in which scores of starving peasants, evicted from their homes for inability to pay their rent, roamed the countryside and towns living off weeds, nettles, and any wild food they could scrounge while resorting to begging, rioting, and squatting anywhere they could take shelter, with many literally dying in their tracks.[89] Reports of the starving people were heartbreaking, especially the accounts of the dying children, who were reduced to skeletons and even too weak to speak or cry. Woodham-Smith describes the strange effect of starvation causing hair to grow on the children's faces as they lost the hair on their heads.[90] Contagious multitudes filled hospitals, workhouses, and streets infected with typhus, cholera, and dysentery—the upper classes, interestingly enough, being the most affected by disease.[91] And while the country was still reeling from the catastrophic fallout of two blight-ridden seasons, a third occurred in 1848, a death stroke from which it seemed likely that Ireland might never recover.[92]

From the beginning the British government was totally unprepared to deal with the disaster, but considering both its singularity and magnitude, the effort put forth by some officials was heroic and impressive, if often deeply misguided. In October of 1845 the prime minister, Sir Robert Peel, appointed a scientific commission of three experts—two chemists, his Scottish friend Lyon Playfair and the Irish scientist Robert Kane, and one botanist, our old acquaintance John Lindley—to inspect the crop and propose some course of action. After a three-week visit to Ireland, the commissioners recommended various immediate steps for salvaging what remained of the diseased potatoes, most of which strategies proved both impractical and inadequate to the task.[93] But their grim assessment of the situation—Lindley, particularly, was convinced that the damage would get worse—prompted Peel to two drastic actions: the repeal of the tariff on foreign grains, that is, the notorious Corn Laws; and the purchase of "Indian corn" (maize) from the United States to be shipped secretly to Ireland and at the ready to distribute or sell cheaply if other grain supplies became scarce or too expensive.[94] The first action cost Peel his Tory ministry for going against the conservative principles of his own party, while the second helped to stave off the kind of distress that would follow in the next administration when the Whig Lord John Russell, more strictly adhering to the liberal noninterventionist, free-trade theory of laissez-faire, advocated leaving all relief efforts to private enterprise and charities, and in the summer of 1847 transferred all responsibility for aid to the Irish Poor Law, that is, rates collected from the local landlords.[95] The assistant secretary of the Treasury, Sir Charles Trevelyan, charged with overseeing the crisis, who had executed Peel's more humane policies in the previous administration, dutifully and happily followed orders—to go down in history as the most hated Englishman in Ireland after Oliver Cromwell, the Puritan whose parliamentary army sacked the country in the seventeenth century.

In a long article published in the *Edinburgh Review* in January 1848, and reprinted that fall as *The Irish Crisis*, Trevelyan defended the British government's actions, which, besides cutting off funds and grain supplies, included phasing out public works and closing soup kitchens because such hands-on efforts invited dependency, whereas "the disease was strictly local"—thereby requiring "local remedies." Because the landowners in the west and south of Ireland had "permitted or encouraged" the explosion of a population almost exclusively dependent "upon the precarious potato,"[96] they must assume responsibility for providing relief—in other words, Irish property must pay for Irish poverty.[97] The cause of the disaster, however, was the tuber itself. Trevelyan asks early on, "But what hope is there for a nation that lives on potatoes?"[98] And answers not long afterward, as if the crisis were already over: "The only hope for those who lived upon potatoes was in some great intervention of Providence to bring back the potato to its original use and intention as an adjunct, and not as a principal article of national food."[99] Thankfully, however, "Supreme Wisdom has educed permanent good out of transient evil"[100] by visiting destruction on the offending plant. To demonstrate its faults, Trevelyan dwells at length on the inferiority of potatoes to grain as a dietary staple, noting the relative difficulty of storing and transporting potatoes as well as their greater susceptibility to the ravages of disease.

Moreover, following a long, derogatory account of the indolence, fertility, and low character of the Irish laborer, Trevelyan uses terms to describe the Lumper variety that invite a comparison between the peasant and the potato, whose shared traits identify both as the most inferior members of their respective species: "There is a gradation even in potatoes. Those generally used by the people of Ireland were of the coarsest and most prolific kind, called 'Lumpers,' or 'Horse Potatoes,' from their size, and they were, for the most part, cultivated, not in furrows, but in the slovenly mode popularly known as 'lazy beds'; so that the principle of seeking the cheapest description of food at the smallest expense of labour, was maintained in all its force."[101] Considering all the liabilities of Ireland's "potato system," as Trevelyan calls it, as well as the liabilities inherent in the vegetable itself, the British government could only do so much; thus he reiterates toward the conclusion that the crisis was thankfully solved by Divine dispensation: "So far as the maladies of Ireland are traceable to political causes, nearly every practicable remedy has been applied. The deep and inveterate root of social evil remained, and this has been laid bare by a direct stroke of an all-wise and all-merciful Providence, as if this part of the case were beyond the unassisted power of man."[102]

Presumably, "the deep and inveterate root of social evil" to which Trevelyan refers is the Irish "potato system," but ultimately the potato itself; thus, according to his argument, the widespread destruction of an accursed crop, along with the lives and livelihood of millions of people, must be construed as God's blessing on Ireland. With a rhetorical flourish, Trevelyan relieves the British government of any further responsibility for helping to solve Ireland's problems, dismissing the economic and political realities that led to the mass adoption of the "potato system" in the first place—conditions fundamentally unchanged by the crop's repeated failure. As Woodham-Smith maintains, "the most serious charge" that can be leveled against the British government was its refusal to help Ireland improve its agricultural practices, which would have included teaching the peasantry to grow other crops and perhaps more important, as John Stuart Mill argued, overseeing legislation to ensure tenants' rights and protect them from eviction simply at the pleasure of their landlords. No such assistance was given during the famine or "for decades afterwards," however, leaving the suffering Irish to fend for themselves and the peasantry to continue living precariously on potatoes.[103] Yet despite the British government's serious failure during the crisis, the policies that Trevelyan supported held sway, and in April 1848 he was knighted for his service.[104]

To be sure, Ireland had not endeared itself to England politically in the years just before the famine, thus putting a further strain on relations and certainly directing public opinion as articulated by the British press, which, as Leslie Williams says, "was the principal medium through which Irish problems were filtered and fitted into public discourse in Britain."[105] And just as Trevelyan became the focus of Ireland's anger toward the British government, so the charismatic Irish MP Daniel O'Connell served for the British press as its representative Irish bogeyman: that is, a rabble-rousing politician and landlord who wielded a too-dangerous influence over domestic affairs. Known as its "Liberator" for his efforts in achieving his country's release from the Penal Laws with the passage of the Roman Catholic Relief Act in 1829,[106] O'Connell aroused increasing animosity and even alarm in England during the early forties, when he mobilized Ireland's attempt to repeal the Act of Union, which had forcibly joined the country to Britain in 1801 as a way to quash an Irish rebellion against England during the French Revolutionary War. O'Connell had even been imprisoned briefly in 1844 for his role in organizing what the *Times* labeled as "monster meetings" of Repeal supporters, huge rallies the British government outlawed for fear they would lead to another Irish insurrection—even though O'Connell himself was a fierce advocate of nonviolence.[107] The *Times'* spokesperson Thomas Campbell Foster also charged O'Connell with hypocrisy for publicly espousing the cause of the Irish peasantry while soliciting their financial support (referred to as Repeal "Rent") and allowing his own tenants to live in squalor.[108] O'Connell had therefore become the press's scapegoat as the human embodiment of Ireland's problems, and thus a readymade target for blame and mockery when the famine began—as well as a figure satirists could conveniently conflate with the famine's cause, the blighted potato.

As was often the case, *Punch* followed the *Times'* political lead, making O'Connell the butt of numerous unflattering cartoons, often depicting him as a would-be king of Ireland demanding obeisance and rent from his hapless, gullible subjects. However, the arguably most offensive and graphically memorable caricature appeared in the December 13, 1845, issue from the

FIGURE 19. "The Real Potato Blight of Ireland," William Newman, *Punch*, vol. 9 (1845): 255. (Courtesy of HathiTrust; contributed by the University of Michigan)

pen of cartoonist William Newman, entitled "The Real Potato Blight of Ireland," featuring a smug O'Connell as a disgustingly corpulent tuber—no doubt a Lumper—seated as on a throne with a collection plate before him and wearing his infamous Repeal Cap, designed as a copy of an ancient Irish crown, which had been presented to him at one of the monster meetings (see fig. 19).[109] The subtitle, "From a Sketch Taken in Conciliation Hall," refers to the Dublin headquarters for the Repeal Association, so named for O'Connell's desire to conciliate all Irish classes and religions, but taken in England as a symbol of resentment for English involvement in Irish affairs. At this early stage of the crisis, it was still possible to make light of the situation by using the potato as a means to mock O'Connell, which this image does with insulting brilliance; but the cartoon's effectiveness depends implicitly upon England's contempt for Ireland's peasant potato culture—its "potato system," as Trevelyan would call it.

Despite the striking originality of Newman's caricature, it was not the first to envision O'Connell as a potato—an obvious connection given O'Connell's stout figure and the intimate identification of both the man and the tuber with

THE TRIUMPH OF THE POTATO 173

all that was offensive about Ireland. Shortly before the passage of the Roman Catholic Relief Act in 1829, the popular and prolific satirist William Heath (aka "Paul Pry") published a raunchier and more visually vulgar cartoon titled "A Sketch of the Great Agi-Tater," printed by Thomas McLean, which depicts a Gulliver-sized O'Connell poised over a Lilliputian landscape wearing the wig and collar of a barrister, but with his robe transmogrified into a gigantic tuber (see fig. 20).[110] The title is an obvious pun on "agitation," the favorite byword of those with anti-Irish—and more specifically in this case, anti-Catholic—sympathies to characterize O'Connell's role in organizing Irish Catholics to press for legislative reform, including O'Connell's right to sit in the British Parliament.[111] And lest the sketch's potato symbolism be misconstrued, the tuber's nether parts are labeled for the viewer: its exposed "roots of evil shooting forth," respectively representing "popery," "intolerance," "popery," and "bigitory" [sic], insinuate themselves into the "Protestant Ground" of the

FIGURE 20. "Sketch of the Great Agi-tater," William Heath. (Courtesy of the British Museum)

"Church of England," while the strategically placed, penile "Pope's Eye" discharges its offensive effluvia over the land in what can best be described as a "golden shower"—as it's known euphemistically in pornographic parlance. Although the immediate targets of the cartoon are O'Connell and Catholicism, the potato's "roots of evil shooting forth" implicate the tuber using terms that Trevelyan would later repeat. In fact, the cartoon recalls a political slogan mentioned by Salaman that was popular in the mid-eighteenth century among the southern England working class, who opposed the tuber because of its Irish Catholic affiliation: "No Potatoes, No Popery."[112]

O'Connell died in March 1847, before the famine ended and without accomplishing the Act of Union's repeal; but his posthumous appearance in *Punch* to mark his 100th birthday in the August 28, 1875, issue indicates that England was feeling considerably kinder to his memory in light of the much more violent agitation for Irish independence that had occurred since his death. In John Tenniel's cartoon titled "'Save Me From My Friends!,'" the "Shade of O'Connell," now dressed in classical or angelic garb but still wearing his familiar Repeal Cap, gestures toward a huge mob of brawling Irishmen

FIGURE 21. "'Save Me from My Friends!,'" John Tenniel, *Punch*, vol. 69 (1875): 81. (Courtesy HathiTrust; contributed by the University of Michigan)

FIGURE 22. "The Irish Cinderella and Her Haughty Sisters, Britannia and Caledonia," John Leech, *Punch*, vol. 10 (1846): 181. (Courtesy of HathiTrust; contributed by the University of California)

as he addresses "Hibernia," the symbol of Ireland, to bewail the fact that the current Home Rule movement had resulted in vicious civil strife rather than national autonomy (see fig. 21).[113] What's interesting about this cartoon is that it combines a nostalgic treatment of O'Connell with the two competing personifications of Ireland that prevailed in *Punch* after mid-century. The pugnacious Irishman wearing a top hat and tails in the mob's foreground represents the typical "Paddy" of *Punch* cartoons, a brutish hothead with simian features who is always spoiling for a fight; but Paddy's foil, the beautiful, innocent, delicate maiden Hibernia, emerged later to symbolize the magazine's more sympathetic view of the country. In fact, Hibernia made her first appearance in the spring of 1846, when the potato famine in Ireland began seriously to take hold, in John Leech's cartoon "The Irish Cinderella and Her Haughty Sisters, Britannia and Caledonia," which pictures the beauty in rags, abject beside her unattractive, uncaring, but obviously affluent siblings (see fig. 22).[114] In a much later issue of the magazine—September 23, 1882—a piece entitled "Justice to Punch and Ireland" serves as an apology for *Punch's* treatment of the country at times in the past. The article, presented as a trial before the "Lord Justice Public Opinion," features an ever-youthful Hibernia coming to the defense of Mr. Punch for his flattering portrayal of her since she began her *Punch* career as Leech's Cinderella—"a sad, gentle girl seated before an empty grate"—a positive image of Hibernia that "Mr. John Tenniel, who always made her look her very best, had perpetuated, she had almost said stereotyped, in the present."[115] Thus it happens that just as the pejorative potato in the guise of O'Connell disappeared as a tacit object of derision in the cartoons, Cinderella/Hibernia arose as its replacement; and although her cartoon image alludes to the potato only by indirection, it fortuitously corroborates Robert Fenn's "Cinderella of Nature" metaphor with its reminder of the applicability of the fairy-tale heroine's early unfortunate domestic situation to the potato's near-automatic identification with Ireland and its peasant class up to the mid-century disaster.

The Providential Potato

Trevelyan's damning rhetorical question "But what hope is there for a nation that lives on potatoes?" was certainly not calculated to win favor in Ireland; no wonder, then, that *The Irish Crisis* elicited a vehement rebuttal, one attacking Trevelyan's appeal to religion by turning his providential argument on its head. In an article in the *Dublin Review* (June 1850), Roman Catholic Cardinal Nicolas Wiseman repudiated Trevelyan's remarks as "'stump oratory'"—a "half political, half religious" rhetoric marshaled to support what was essentially an untenable, chauvinistic, and certainly irreligious position. Wiseman's comments appeared in his enthusiastic promotion of *Impediments to the Prosperity of Ireland* (1850), a treatise by the economist Neilson Hancock, who took Trevelyan to task for the failed logic of claiming (in Hancock's words) that "the potato is a curse, and the root of all our social evils" primarily because it provided cheap and plentiful sustenance, an argument so specious it "would require a miracle to convince any one who reflects on the subject."[116] Hancock's use of "miracle" here is telling, because it indicates how Trevelyan had so misunderstood and violated Ireland's worship of the potato, a devo-

THE TRIUMPH OF THE POTATO   177

tion the famine had not destroyed, but elevated to new heights. To counter Trevelyan's attempt to transmute the disaster into a providential act, Hancock quotes another economist, Mountifort Longfield, who had defended Ireland's reverence for the potato years earlier in his own *Lectures on Political Economy* (1834) by taking the opposite view of the tuber's spiritual worth, proclaiming instead that "Providence has bestowed upon the world a prolific, wholesome, and palatable vegetable," which "qualities must insure its general cultivation in all countries adapted to its growth. And it is a hard matter to believe that the introduction of this plant should naturally and almost inevitably introduce general distress."[117]

That "general distress" had occurred was undeniable; but according to Hancock, Trevelyan's argument understated the fundamental cause of the crisis, abject poverty, in order to condemn the miraculous vegetable that had assuaged poverty's misery, bringing consolation to the peasants of Ireland through its adoption by supplying cheap, high-yielding, nutritious, palatable—and thus, undeniably—providential food. As its Irish defenders indicate, the question of the potato's value—religious or otherwise—had really been settled conclusively long ago: like manna from heaven, it was a godsend to a starving nation, so that Ireland's tragedy only served to underscore what a precious and essential commodity the tuber had become throughout the British Isles. In fact, given its obvious and undeniable importance as a dietary staple in the Old World as well as the New, it would seem that "Supreme Wisdom" and Trevelyan emphatically disagreed—at least so far as the divine intelligence informed the secular taste of the British population. Despite a few vociferous detractors, "Lord Justice Popular Opinion" (to borrow *Punch's* handy personification) had ruled unequivocally on the side of the spud.

After two seasons of devastating blight, *Punch* was already coming to the potato's defense and celebrating (prematurely, it would turn out) the end of the crisis. In mid-September 1847, a cartoon entitled "Consolation for the Million.—The Loaf and the Potato" (see fig. 23) envisioned the formerly rival foodstuffs clasping hands as the Loaf exclaims, "Well! old Fellow I'm delighted to see you looking so well—why they said you had the Aphis Vastator" (referring to the aphid reported by the polymath Alfred Smee as the blight's cause in his recent book *The Potatoe Plant*). The Potato's response—"All humbug sir never was better in my life thank Heaven"—alludes to the fact that in 1847 the British Isles were experiencing a blight-free harvest, taken as evidence that the disease had run its course.[118] Besides expressing optimism about the potato's health and relief over the recent reduction in the price of bread, the cartoon's most noteworthy feature is its treatment of the two staples as comrades joining together to satisfy the most basic dietary needs of "the Million,"

FIGURE 23. "Consolation for the Million.—The Loaf and the Potato," John Leech, *Punch*, vol. 12 (1847): 95. (Courtesy of HathiTrust; contributed by the University of California)

which, as the background vignettes testify, refers to the lower economic half of London's population (the last census of 1841 counted its inhabitants at 1,870,127). The loaf, dressed in starched collar and suspenders, personifies the typical shopkeeper who sells bread to the urban middle classes, as represented in the vignette on the right by the ladies entering a bakeshop. Alternatively, the vignette on the left depicts a working-class boy buying a hot potato from one of London's most familiar street vendors, the baked-potato man, thus recalling the tuber's close association with the impoverished lower orders. However, the personified potato itself, whose neck ruff undoubtedly signifies Sir Walter Raleigh, has ascended to the English nobility—perhaps as tacit recognition of its heroic recovery and even its martyrdom, but in any case an acknowledgment of its by now well-established venerable qualities.

Two years later, *Punch* celebrated the blight's end once again—this time offering high praise for the potato under the guise of a tribute from the tuber's fellow vegetables. A cartoon entitled the "Grand Vegetable Banquet to the Potato on His Late Recovery," accompanying a faux news article written by "our Correspondent in Vegetaria," does a nice reversal on the usual order of things by picturing garden produce as guests rather than fare at a dinner party given in the tuber's honor. The cartoon features the Potato, in gentlemanly dress, standing at the head of the table and gesturing to the other appropriately attired attendees, while speaking about his life-threatening illness (see fig. 24). As the text explains, "A few leading members of the Vegetable Kingdom" had planned the affair, inviting "all the principal vegetables" to fête "that highly respected vegetable, the Potato," who, "though just out of his bed, was looking remarkably well, and wore his jacket, there being nothing to mark his recent illness, except a little apparent blackness round one of his eyes." The rest of the article (mostly through a series of culinary puns) describes the Onion's after-dinner toast and the Potato's response, both of which emphasize the tuber's high regard among his comrades. The Onion declares that during his honorable friend's illness, "he, the Onion, could say without flattery, that society had endeavoured to supply the place of the Potato in vain"; whereas the Potato, following enthusiastic applause, accepts the compliment with an account of his history ("Though I may be a foreigner, I may justly say, that I have taken root in the soil") and late ordeal, prefaced by an acknowledgment of his status: "I believe I have done as much good as any living vegetable; for, though almost always at the rich man's table, I am seldom absent from the poor man's humble board."[119] In short, the text confirms that by midcentury and in spite of the blight, the potato was not only a fully naturalized British citizen, but one recognized as the leading eminence of the Vegetable Kingdom.

This prominence had actually occurred decades before. As early as 1817 the Scots naturalist Patrick Neill, in a long entry on "Horticulture" written for the prestigious *Edinburgh Encyclopaedia*, could accurately record that the

FIGURE 24. "Grand Vegetable Banquet to the Potato on His Late Recovery," *Punch*, vol. 17 (1849): 204. (Courtesy of HathiTrust; contributed by the University of California)

potato topped all vegetables as "the most useful esculent that is cultivated" in the British Isles, and in Scotland "seen in almost every cottage garden."[120] Five years later, in his highly successful *Encyclopedia of Gardening*, England's horticultural patriarch John Claudius Loudon quoted Neill at length, adding that potatoes were now being grown as a field crop in every county of England. And as a way of broaching a subject that had occupied botanists since the potato's arrival in Europe, Loudon includes the former author's philosophical musing about the notorious kin of this important vegetable: "'Who,' Neill asks, 'could *a priori* have expected to have found the most useful among the natural family of the *Luridae*, L., several of which are deleterious, and all of which are forbidding in their aspect.'"[121]

This comment, with its reminder of Linnaeus's derogatory name for the nightshades, thus serves to foreshadow a major turning point in family history; for in 1829 Loudon would publish his massive and definitive *Encyclopedia of Plants* with the text written by Lindley, who according to Lindley's biographer William Stearn, became convinced of the superiority of the natural system of classification while working on Loudon's book—the system that gave birth to the nightshades' modern scientific name.[122] As detailed in chapter 3, the natural system considered all of a plant's morphological features, thereby affording more accurate insight—while introducing more subtlety and complexity—into the study of botanical family relations, a move destined to alter forever the personality profile of the nightshades. Although the major section of the *Encyclopedia of Plants* is arranged according to the Linnaean system, part 2 classifies genera by "orders," each with a paragraph outlining their general characteristics. Thus, under "*Solaneae*," Lindley begins with "the baneful nightshade" as "representative of this order, which participates very generally in its qualities"; but he notes later when describing the potato, "Notwithstanding the narcotic power of the roots of the Mandrake, the Belladonna, and others, those of the potato are found to contain an abundant faecula [i.e., starchy substance], which is among the most valuable food of man."[123] Lindley had acknowledged *Solanum tuberosum*'s importance in part 1 with a long discussion of its history and properties, although he was dismissive of the consolatory connotations associated with its genus, saying of *Solanum*, "By some ingenious commentators this word has been derived from *solari*, to comfort. The derivation may be possible, but the application is not evident."[124]

The next year, however, in *An Introduction to the Natural System of Botany*, Lindley seemed to temper his prejudicial assessment with this observation about the *Solaneae*: "At first sight this family would seem to offer a strong exception to the general uniformity of structure and property, containing as it does the deadly Nightshade and Henbane, and the wholesome Potato and Tomato; but a little inquiry will explain this apparent anomaly. The tubers of the Potato are well known to be perfectly wholesome when cooked, any narcotic property which they possess being wholly dissipated by heat."[125] Then, in the 1836 edition of his groundbreaking work, now simply titled *The Natural System of Botany*, Lindley adopted for the sake of uniformity with other families the name *Solanaceae*, thereby officially establishing the nightshades' modern scientific designation in England.[126] Although (as we have seen) in his *Ladies Botany* of 1838 Lindley would elaborate on the toxic properties of the *Solanum* genus, asserting that unless cooked, their fruit is always "deleterious" (including tomatoes) and that of the potato is "notoriously unwholesome," his noticeably ominous tone may be attributed to assumptions about his female readership, whose responsibility for and assumed ignorance of the safe handling of food warranted his authorial caution. Even so, he seems less worried about any danger posed by the potato's "roots," not only because of "their being cooked," but also because they were "composed almost entirely of a substance like flour, which in no plant is poisonous, if it can be separated either by heat or by washing, from the watery or pulpy matter it may lie among."[127] In short, despite Lindley's reservations about the safety of even the edible nightshades, he seems more and more willing to acknowledge (if somewhat grudgingly) the potato's wholesome properties.

Although the recognition of the potato's value and the change of the family's name in a sense coincided by chance, their concurrence nevertheless reflects the increasing scientific awareness of the intricacies of family ties as well as the mutability of species that grounded Darwin's evolutionary theory; and as we have seen, Darwin's own interest in the potato certainly contributed to the growing appreciation for the *Solanum*, the family's largest and most representative genus, and the one etymologically responsible for the family's consolatory associations. Lindley was obviously not thinking about *Solanum tuberosum* when he made his rather sardonic remark in Loudon's *Encyclopedia of Plants* about the lack of application of "*solari*," meaning "to comfort," to the nightshades; for there is no question that the potato embodied that quality. By offering "consolation to the million"—and millions more besides—the potato had proven to be as providential as bread, the exceptional nightshade that put the soul in "Solanaceae" to rehabilitate the family name.

Incontrovertible proof of the potato's ubiquitous presence in the diet of that huge segment of the Victorian population, its growing middle class, can be found in the many cookbooks that appeared in the nineteenth century. Primarily written for (and often by) women responsible for running households, these works reflected a new emphasis on the scientific approach to "domestic economy," providing specific instructions about all aspects of cookery, from information about utensils and equipment to precise measurements for ingredients used in the recipes. One of the most influential, Eliza Acton's best-selling *Modern Cookery, in All Its Branches*, first published in 1845 and "Dedicated to the Young Housekeepers of England," testifies to the fact that the Victorians, if they could afford it, overwhelmingly favored a meat-and-potatoes diet. Acton gives detailed instructions for the preparation of beef, mutton, pork, fish, poultry, game, and forcemeats, with their attendant sauces, gravies, and sides; and while a full eighty pages is devoted to the first three meats alone, the entire vegetable section consists of only thirty-four pages, five of which are given over to potatoes, which lead the section with fourteen recipes.[128] They figure as an ingredient in at least as many more dishes such as soups, stews, pasties, and pies; and there are even instructions on how to make potato flour.

Moreover, in the post-famine edition, retitled *Modern Cookery for Private Families*, Acton felt moved to introduce potatoes with "remarks on their properties and importance," in acknowledgment of their tragic history as well as their prominent place in "modern cookery." The opening sentences, however, use terms that could almost apply to the social advancement of a child from the "lower orders" elevated to gentility for possessing exceptional qualities: "There is no vegetable commonly cultivated in this country, we venture to assert, which is comparable in value to the potato, when it is of a good sort, has been grown in suitable soil, and is properly cooked and served. It *must* be very nutritious, or it would not sustain the strengths of thousands of people whose almost sole food it constitutes, and who when they can procure a sufficient supply of it to satisfy fully the demands of hunger, are capable of accomplishing the heaviest daily labour." Acton goes on to lament the potato's susceptibility to disease, and questions the wisdom of wholly relying on it for subsistence; nevertheless, she says, "we can easily comprehend the predilection of an entire people for a tuber which combines, like the potato, the solidity almost of bread, with the healthful properties of various other fresh vegetables, without their acidity; and which can be cooked and served in so many different forms."[129] These comments, like *Punch*'s in its Vegetable Banquet cartoon, make clear that the blight had enhanced the potato's reputation; and despite Acton's hints at its humble beginnings, the tuber's ascendancy to Victorian gentility was now an established fact.

The success of Acton's *Modern Cookery* inspired numerous other volumes; and as she put it in the preface to the 1855 edition, "part of them from the pens of celebrated professional gastronomers"—no doubt referring to Victorian England's two most famous chefs, Charles Francatelli and Alexis Soyer.[130]

THE TRIUMPH OF THE POTATO

The former, who briefly served in the early forties as the head chef for Queen Victoria, published *The Modern Cook* in 1846, a work featuring haute cuisine intended for the upper echelon of the middle class and including recipes like "Potato Soup à la Victoria" and "Quenelles of Potatoes," dishes designed for elegant dinner parties, whose bills of fare are listed for every month of the year.[131] The flamboyant Alexis Soyer, the head chef at London's tony Reform Club, followed suit the same year with *The Gastronomic Regenerator*, written for a similar audience, with potato recipes appearing in their French guise, such as "Pommes de Terre à la Maître d'Hôtel" and "Pommes de Terre à la Lyonnaise."[132] Both chefs also saw the value, pecuniary and otherwise, of tapping into a broader market: in 1849 Soyer brought out *The Modern Housewife, or Ménagère*, dedicated to "the Fair Daughters of Albion" and framed as a series of letters and discussion between Hortense B. and Eloise L., who, like Acton, enjoy potatoes boiled, steamed, baked, fried, and mashed, as well as in the French dishes "Potatoes à la Maître d'Hôtel" and "Lyonnaise."[133] There is also a strange entry titled the "Irish Way of Boiling," which gives directions for leaving the "bone" in the potato—that is, not fully cooking it in the middle—by draining the boiling water and allowing the potatoes to sit in a dry pot by the side of a turf fire. However, Soyer evidently considers this method primitive, noting, "Even in those families where such a common art of civilized life as cooking ought to have made some progress, the only improvement they have upon this plan is, that they leave the potatoes in the dry pot longer, by which they lose the *bone*."[134]

Presumably, Soyer's comment was based on personal observation, for in the spring of 1847 he had applied his culinary skill to creating low-cost soup recipes and designing a massive model kitchen for the starving people of Ireland, where he spent seven weeks executing his plan. Although roundly criticized in both England and Ireland for not providing sufficient nourishment, "Soyer's Famine Soups" formed the basis for *Soyer's Charitable Cookery, or The Poor Man's Regenerator*, a pamphlet he published in 1847, with some of the proceeds going to poor relief. Soyer's efforts were no doubt responsible for the government's eventually establishing a ration of one pound of meal per adult per day, distributed in the form of "stirabout," a carbohydrate-heavy porridge made of maize, oats, and/or rice, which could be used to thicken the soups.[135] Francatelli was also moved to consider the dietary needs of the less fortunate, publishing in 1852 his *Plain Cookery Book for the Working Classes*, with the now-familiar standard methods of potato preparation, including a drastically scaled-down soup recipe—without the consommé, cream, and garnish of asparagus, French beans, and quenelles that characterized the "à la Victoria" version from *The Modern Cook*.[136]

The most famous Victorian cookbook, Isabella Beeton's *The Book of Household Management* (1861), confirms the previous authors' view of the potato's essential place in genteel Victorian cookery. First published in monthly parts in her husband Samuel Beeton's periodical, the *Englishwoman's Domestic Magazine*, "Beeton's," as it came to be known, dominated the cookery-book market for the rest of the century; and although her recipes are quite similar to Acton's, her praise for the potato surpasses the former author's more reserved commendation of the tuber, as well as contradicting Soyer's disdain for Irish cooking. That is, in describing how "To Boil Potatoes in Their Jackets," Beeton begins effusively, "To obtain this wholesome and delicious vegetable cooked in perfection, it should be boiled and sent to table with the skin on. In Ireland, where, perhaps, the cooking of potatoes is better understood than in any country, it is always served so." (Her further instructions nevertheless call for peeling the potatoes quickly before bringing to table.)[137]

The extreme popularity of Beeton's book owed much to its familiar tone, but also to the wealth of authoritative information it provided. Unfortunately, Beeton was cavalier about giving credit where it was due, so that the interesting facts and related tidbits interspersed among the recipes throughout the book are quoted verbatim from other sources, usually without acknowledgment (a problem that evidently occurred with many of Beeton's recipes as well).[138] Thus, Beeton mentions Robert Hogg's *The Vegetable Kingdom* (1858) in her "General Observations on Vegetables," but quotes his comments on potatoes several times without naming her source; this is also the case with John Loudon's *Encyclopedia of Agriculture* (1826), whose comment concludes the potato section on a high note: "This valuable esculent, next to wheat, is of greatest importance in the eye of the political economist. From no other crop that can be cultivated does the public derive so much benefit; and it has been demonstrated that an acre of potatoes will feed double the number of people that can be fed from an acre of wheat."[139]

Clearly, the potato had risen above class prejudice to be a welcome addition to any table; and as Francatelli's royal soup suggests, Queen Victoria's table at Windsor figured most prominently among them. In *The Private Life of the Queen* written anonymously "By One of Her Majesty's Servants," we are told, "Her Majesty confesses to a great weakness for potatoes, which are cooked for her in every conceivable way." Moreover, "the Queen always had a passion for eating in the open air," and the author recounts an occasion when "she made a delightful luncheon" of warm broth and potatoes "she had helped to boil herself" on a fall picnic "on the moors above Balmoral," her favorite castle and getaway in the Highlands of Scotland. Back in England, the twelve acres of potatoes planted at Windsor were often not enough to supply the royal

THE TRIUMPH OF THE POTATO   185

household through a season.[140] When in the late fifties the grower William Paterson of Dundee, Scotland, brought out the Victoria variety, famed for its excellent yield and disease resistance, the Queen ordered seedlings for the royal gardens at Windsor Castle in a letter to Paterson written "wi' her ain hand," according to the report of a proud Scots farmer.[141]

Like the Queen herself, the Victoria potato held first place in a noble lineage, mothering Sutton's Magnum Bonum in the mid-seventies, which, along with John Nicholl's Champion, a variety of unknown parentage developed in the early sixties that proved its merit in the 1879 blight, became the highest-ranking tubers in the British Isles, firmly establishing a potato royalty.[142] In a pamphlet of advertisements and testimonials compiled by Paterson's widow in 1872, the Victoria is described as a "rough-skinned, handsome-shaped, flat potato" with "white flesh," purple flowers, abundant berries, and—again like her namesake—"a very prolific cropper," as well as "one of the best resisters of disease known."[143]

From the beginning the nineteenth-century sentimental botanies were unanimous in characterizing the potato as a providential blessing. In Alexis Lucot's seminal *Emblèmes de flore et des végétaux* (1819), "*Pomme de Terre*" serves as the emblem for *Bourrus-bienfaisans*, or "Gruff Beneficence," a term Lucot no doubt adapted from the popular eighteenth-century French comedy *Le bourru bienfaisant* (translated as *The Beneficient Bear*) by the Italian playwright Carlo Goldoni and featuring an irascible but generous-hearted rich uncle as the eponymous character who epitomizes crusty avuncularity when he becomes the benefactor of his financially embarrassed nephew and nubile niece. Thus Lucot likens the tuber's rough exterior but inner wholesomeness to a gruff but kindly person: "This humble vegetable, food for the rich and the poor, recalls those excellent hearts, whose appearance is heavy and also rude, whose manners are brusque and often bizarre, but whose beneficent spirit is the solace of misfortune" [my translation].[144]

As was frequently the case, Charlotte de Latour's better known *Le langage des fleurs* borrowed Lucot's meaning, but offered its own explanation, one translated in Frederic Shoberl's many English editions beginning in 1834, as follows: "The Potatoe, the peculiar vegetable of the poor, is also regarded as an emblem of beneficence. This root, lasting but for a year, escapes the monopoly of trade. Modest as true charity, the potato hides its treasures: it bestows them on the rich, and feeds the poor with them. America presented us with this useful vegetable, which has for ever banished from Europe one of the direst calamities—famine."[145] Although Shoberl (and others) would recycle de Latour's *Langage* well into the latter part of the Victorian period, the explanation for the potato's "beneficence" alludes to early nineteenth-century European history: that it "escapes the monopoly of trade" indicates the potato's exemption from the British Corn Laws, imposed on foreign grain after Napoleon's defeat in 1815.[146] Before and during the French Revolution and Napoleonic Wars the potato played a crucial role in relieving famine in France, thanks to its vigorous promotion by Antoine-Augustin Parmentier, whose efforts are celebrated in the names of many potato dishes.[147] Even after Ireland's famine exposed the fallacy of de Latour's final sentence, it would appear not only in later editions of Shoberl's *Language of Flowers*, but also in those of Robert Tyas for his several versions, published between 1836 and 1875.[148] However, beginning in 1869, Tyas would add, "How important it is to the inhabitants of the United Kingdom, those know who remember the failure of the crops in Ireland in 1846 or 1847." In this edition Tyas also changed the potato's meaning to "benevolence," which would become the familiar floriographic standard for the latter part of the century, appearing as such in Kate Greenaway's late 1884 version.[149]

The horticultural writer Henry Phillips's early *Floral Emblems* (1825) also follows Lucot and de Latour, giving "beneficence" as the potato's meaning; but his explanation is actually a greatly foreshortened description drawn from a long discussion of the potato in his *History of Cultivated Vegetables*, published in 1822. The result is quite poetic, appropriately ending with a few lines of verse by the eighteenth-century poet William King: "This root, which forms alike the poor man's bread and the rich man's luxury, is properly made the representative of beneficence. It is the *palladium* against famine, forming flour without a mill, and bread without an oven. It was first procured when 'Raleigh, with hopes of new discov'ries fir'd, / And all the depth of human wit inspir'd, / Mov'd o'er the western world in search of fame, / Adding fresh glory to Eliza's name.'"[150] Thus, whether defining it as "beneficence" or "benevolence," the floriographers consistently represent the potato as a providential gift to all mankind: if prized as "the rich man's luxury," it also satisfies the most basic human need as "the poor man's bread," the "solace of misfortune."

By defining the humble spud as the emblem of "beneficence" or "benevolence," the foregoing examples all insist on its identification with what was arguably the Victorians' highest religious ideal, Christian charity, which grounded the vision of the period's greatest writers. Etymologically, the two words, respectively meaning "doing good" and "willing good," signify the same godly quality, one that serves—in the era's best example—as the supreme virtue in all Dickens's novels and defines his most admirable characters, the many "benefactors" whose acts of generosity, whether great or small, earn them a place in Dickens's pantheon. Speaking of the Pickwick Club in Dickens's first major work, *Pickwick Papers* (1836–37), the narrator declares

THE TRIUMPH OF THE POTATO 187

that "general benevolence was one of the leading features of the Pickwickian theory," and we can be sure that the club's "theory" is Dickens's own, one he espoused for the rest of his writing life. "Benevolence" runs like a litany through the novel, with the quality mentioned specifically with regard to the hero a dozen times, so that "benevolent Mr. Pickwick" could well serve as an epithet for that kindly, spud-shaped gentleman, whose generosity of spirit grows more saintly as the narrative progresses.[151] Given the way Dickens takes for granted the potato's essential place among the most basic facts and necessities of life, he effectively endorses the floriographers' definition.

The meal Pickwick orders for the scoundrels Jingle and Trotter when they are incarcerated in the Fleet Prison is a case in point, as Sam Weller discovers upon his own offer of "some wittles": "'Thanks to your worthy governor, sir,' said Mr. Trotter, 'we have half a leg of mutton, baked, at a quarter before three, with the potatoes under it to save boiling.'" Sam soon learns that the hot food, with its thoughtful consideration of the prisoners' meager facilities, is just one instance of Pickwick's "purwidin'" (to use Sam's word), prompting praise for his employer in decidedly religious terms, as a "reg'lar thoroughbred angel."[152] Here as elsewhere and always in Dickens's works, providing a proper meal for those in need constitutes a divine act; consequently, these hot potatoes accompanying a joint of mutton—by all indications a favorite Dickens repast—unquestionably signify "benevolence."

Dickens clearly appreciated a well-cooked potato, and their ubiquitous presence in his novels is further testimony—if any more were needed—to their familiar place in the lives of most Victorians. Dickens himself seemed to know everything about potatoes. While references to potatoes and oatmeal or "stir-about" in the early novels *Oliver Twist* and *Nicholas Nickleby* respectively record the cheap and scanty fare offered in institutions feeding the poor,[153] hot potatoes served with mutton, a sizzling sausage, or perhaps a steak are the essential ingredients for a soul-satisfying, comfortable meal—in or out of Dickens's fiction. In fact, Dickens's wife, Kate, using the pseudonym Lady Maria Clutterbuck, is credited as author of the small 1851 publication *"What Shall We Have for Dinner?" Satisfactorily Answered by Numerous Bills of Fare for from Two to Eighteen Persons*—whose 164 menus include potatoes in all but a handful of cases. The Dickenses were known for their lavish dinner parties, so the potato dishes (sometimes two per meal) stand as worthy examples of the "rich man's luxury," with a couple of potato recipes appearing in the appendix, which gives instructions for less familiar dishes. The first, for potato balls, seems to have been a house specialty, one made with mashed potatoes shaped and coated with egg yolk, then browned in the oven or "before the fire." The other, "kalecannon" (aka "colcannon" or calecannon), is based on a traditional Irish dish of cabbage and potatoes that according to Salaman became an English upper-class favorite;[154] Kate's version sounds like an elegantly presented mélange of mashed potatoes and turnips mixed together with chopped carrots and greens and boiled in a buttered mold.[155]

If hot potatoes denote satisfying sustenance, their leftover counterpart can be the sign of privation, as evidenced in *The Old Curiosity Shop* (1840–41) by the "dreary waste of cold potatoes, looking as eatable as Stonehenge" that the harpy Sally Brass offers, with "two square inches of cold mutton," as a paltry excuse for a meal to her abused little servant, the clever girl that Dick Swiveller befriends, feeds, dubs the Marchioness, and eventually marries.[156] And Dickens seems to have been aware of the prevailing opinion regarding the dangers of eating tubers raw, for in *A Christmas Carol* (1843), when Scrooge is trying to rationalize the appearance of Marley's ghost, he speculates about what "disorder of the stomach" might be responsible for the delusion, with "a fragment of underdone potato" among the possibilities.[157]

Dickens was a master observer, one well acquainted with the streets of London and its denizens, so he was of course familiar with the best urban source for the spud in its role as the "poor man's bread": the baked-potato man, whose story and image are memorialized by the journalist (and co-founder of *Punch*) Henry Mayhew in his classic account of Victorian "street-folk," *London Labour and the London Poor*, first collected from articles in the *Morning Chronicle* into a three-volume masterpiece in 1851, with a fourth published ten years later. Mayhew details how the trade, seasonal from mid-August to April, employed some three hundred vendors who purchased raw spuds from greengrocers, had them baked at commercial bakehouses, then sold them on the street for a halfpenny from portable "cans" equipped with a charcoal water boiler for heat and side compartments for salt and butter (see fig. 25). The cans were often ornate, trimmed with highly polished brass, sometimes painted red and/or fitted with lamps sporting colored glass—no doubt a warm, welcome sight on a cold evening. Unsurprisingly, spuds were popular street fare with all classes (and even purchased as hand warmers), but the chief customers were members of the working class, including boys and girls. Potatoes—"large, hot, mealy fellows"—with pea soup or hot eels served as dinner for many lower-class boys who took all their meals as a moveable feast, as it were, going from one stand to another, but particularly congregating around the baked-potato can to eat and "talk over local matters, or discuss the affairs of the adjacent cab-stand" with other street familiars like the Thames watermen.[158]

The potato business was brisk during the daylight hours from 10:30 a.m. to 2:00 p.m., but most flourishing at night from 5:00 to 11:00 or midnight;[159]

FIGURE 25. "The Baked Potato Man," H. G. Hine and W. G. Mason, from Henry Mayhew, *London Labour and the London Poor*, vol. 1: 167. (Image from Wikipedia Commons; contributed by Tufts Digital Library)

and Dickens, in his collection of articles from his own days as a reporter for the *Morning Chronicle*, chooses the latter time to pay homage to the potato man's "little block-tin temple sacred to baked potatoes, surmounted by a splendid design in variegated lamps." Dickens loved walking the streets after dark, and in *Sketches by Boz* (1836), he describes a foggy, wet, miserable winter evening in a lower-class district of London, south of the Thames near the Victoria Theatre, when even the potato-can looked "less gay than usual" in the "dirt and discomfort" of the weather. By 11:00 p.m., when the incessant drizzle turned to pouring rain, the baked-potato man had understandably given up business for the night.[160]

The *OED* credits Dickens with the first reference to London's most famous potato fast food—chips—although the context refers to his other favorite city, Paris, before the French Revolution and is therefore most likely anachronistic, since frying potatoes as "chips" was largely a nineteenth-century phenomenon.[161] In *A Tale of Two Cities* (1859), using "Hunger" as an anaphora, Dickens gives a sweeping view of lower-class Paris streets in 1775 to convey the destitution of its citizens, concluding with the following sentence: "Hunger rattled its dry bones among the roasting chestnuts in the turned cylinder; Hunger was shred into atomies in every farthing porringer of husky chips of potato, fried with some reluctant drops of oil."[162] The sequence of details, no doubt based on Dickens's actual observations, suggests that the chips, like chestnuts, were sold by Paris street vendors; but in any case, the food historian Panikos Panayi notes that the practice of frying slices of potato in oil developed in France and England about the same time, either independently or as a "culinary transfer."[163] That France may have been the originator is suggested by the alternative term "French Fries," a recipe for which appeared in Eliza Warren's *Cookery for Maids of All Work*, published in England in 1856.[164] The quintessential British fast-food combination fish and chips seems to have developed a bit later, but possibly as early as 1860, although Mayhew records that fried-fish vendors—working either in stalls or as itinerant sellers—were as common in London as the baked-potato men.[165] (Dickens, once again, is honored with the early mention of a "fried-fish warehouse," one located in a London slum frequented by Fagin in *Oliver Twist*.)[166] According to Panayi, fried fish and baked potatoes were sold together from the late 1860s, before chips ultimately gained ascendancy, thereby creating a British comfort-food institution during the latter third of the century.[167]

Although most of Dickens's references have to do with food, in some cases a potato is just a potato, like the one tossed at Pickwick, along with a turnip and an egg, after he is thrown into a pound; or the basket of potatoes the simple-minded Maggy spills in the mud in *Little Dorrit* (1854–55). In the same novel, the very word serves as an aid to proper elocution and facial expression, as in Amy Dorrit's chaperone and "varnisher" Mrs. General's pronouncement that "Papa, potatoes, poultry, prunes, and prism are all very good words for the lips."[168] Finally, in Dickens's most autobiographical novel *David Copperfield* (1849–50), there is even an unfortunate boy working in Murdstone's warehouse nicknamed "Mealy Potatoes"—"on account of his complexion, which was pale and mealy."[169] The name and its owner are reminders of the tuber's humble status as the commonest and homeliest of vegetables; but as we have seen, the potato is "homely" in its alternate sense as well, meaning a genuinely domestic article, one "at home" in every Victorian's life, as fundamental as light and air.

The many Dickensian references to potatoes reiterate how deeply rooted they had indeed become in the soil of Victorian culture. Dickens, however, focused on their presence as an urban comestible and commodity, whereas other writers like George Eliot and Thomas Hardy, who were more familiar with rural life, included details about potatoes as essential crops. Anthony Trollope, who worked as a postal inspector in Ireland during the forties, is the only major novelist to address the Irish famine head on in his novels, especially *Castle Richmond* (1860); but there are more potatoes in Reverend Crawley's excellent and useful but not ornamental garden in Hogglestock Parish, from Trollope's Barsetshire novel *Framley Parsonage*, first serialized in *Cornhill Magazine* the same year.[170] Most poignant among references to growing potatoes is Thomas Hardy's description in *Tess of the d'Urbervilles* (1891) recounting the plight of the Durbeyfield family, who were late to plant their allotment, "having eaten all the seed potatoes,—that last lapse of the improvident."[171]

Hardy's most extensive potato reference, although more lighthearted, serves as an appropriate example of the nightshade family's reputation in Europe coming full circle from its accursed Old World beginnings to the consolatory connotations of Solanaceae, thanks to the benevolent and providential influence of the tuber. The scene occurs in his early masterpiece, *Under the Greenwood Tree* (1872), when the heroine, Fancy Day, consults Elizabeth Endorfield about her love problems. As the "Endor" in her surname suggests, the wise Elizabeth is rumored to be a witch, one who carries out her interrogation and gives counsel while paring potatoes, dropping each one into a bucket in a rhythmical, hypnotic fashion so minutely described that it invites comparison to casting a spell. Elizabeth admits to Fancy that she practices "witchery"; but since her methods rely solely on "common sense,"[172] it is significant that her craft involves potatoes, the most ordinary and familiar, yet avowedly miraculous nightshade, the redemptive member of a family traditionally known as witches' favorite plants.

When in *The Queen of the Air* (1869) John Ruskin famously calls the potato "the scarcely innocent underground stem of one of a tribe set aside for evil," he explains his predictable judgment of the Solanaceae "tribe" in the phrase that follows, as "having the deadly nightshade for its queen, and including the henbane, the witch's mandrake, and the worst natural curse of civilization—tobacco."[173] As his final example indicates, Ruskin considered tobacco to be his botanical archenemy, having railed against its polluting and demoralizing effects before; but his reference to the potato's being "scarcely innocent" seems more enigmatic, until a comment later in *Proserpina* (1875) reveals that he blamed Ireland's ongoing distress largely on nightshades, asserting that the country's "destiny" just might improve "if it can ever succeed in living without either the potato, or the pipe."[174] Yet despite Ruskin's wholesale censure of the family, the larger context of *The Queen of the Air* comment largely belies the notion that the potato is somehow complicit in the guilt of its relatives, for the statement appears in an analysis of the typical ingredients of a simple English country meal, one that also includes beans, bacon, onions, herbs, celery, radish, cheese, apples, nuts, and brown bread—fare that hardly suggests anything suspect or malign.[175] Instead, this humble but wholesome menu calls to mind the very innocence Ruskin questions. He could not deny that the potato, the notorious underground stem that became during the nineteenth century the source of so much consternation and pleasure, contempt and praise, held its place on the English table as the Victorians' most beloved vegetable, one that emerged from fatality like a martyred god from underground, to be resurrected and venerated by everyone for its providential and beneficent gift of food.

7

Sublime Tobacco

Now Let Us Praise the Deadliest Nightshade

> Some of our readers may not be prepared for the fact, that tobacco, though not food for either man or beast, is the most extensively used of all vegetable productions, and next to salt, the most generally consumed of all productions whatever—animal, vegetable, or mineral—on the face of the globe.
>
> —"The Most Popular Plant in the World," *Chambers's Journal* (1854)

Tobaccomania

In April 1843, His Royal Highness Prince Augustus Frederick, Duke of Sussex, age seventy, beloved uncle of Queen Victoria, died at Kensington Palace in considerable arrears, but leaving his wife some extremely valuable property—to be sold at auction to discharge his debts. Three months later Christie and Manson's made headlines for the record-breaking sale of what proved to be the duke's priciest possessions: his "Unrivalled Collection of Pipes," 220 in number, many with beautifully carved meerschaum or porcelain bowls mounted in silver or gold; more than sixty jars of "Turkey, Persian, Marakiebo and Kanaster Tobacco"; over fifty thousand "Havannah Cigars and Manila Cheroots of the Rarest Quality"; some thirty jars, canisters, and bottles of snuff; as well as other pertinent paraphernalia, including numerous ornate snuffboxes, sundry elegant tobacco pouches, and a silver-trimmed hookah. The son of "Mad" King George III and his consort, "Snuffy" Charlotte, the duke famously—and expensively—embraced his mother's passion for tobacco, although he preferred smoking pipes to other forms of delivery, his enormous cache of cigars coming mostly as gifts from various dignitaries.[1] The duke's godson later recalled an early memory of Sussex "puff[ing] at a long German pipe, and enveloped in a cloud of smoke which it was almost impossible for any eye to penetrate."[2]

The duke's excessive dedication to smoking and his gargantuan stash of related accessories certainly merit the diagnosis of "tobaccomania" proposed by the British writer Compton Mackenzie,[3] and the term applies as well to those members of the Victorian upper class who shared the duke's affliction. By the end of the century, however, tobaccomania would describe Victorian culture writ large as smoking became the signature habit of a modern age. From the time of its introduction to England around the mid-1500s, this notorious New World nightshade inspired zealotry among some members of every class and both genders, but this devotion intensified in the mid-1800s and would not begin to abate until the mid-twentieth century, when the lethal effects from its use were at last scientifically verified. In terms of the sheer number of lives it has taken, tobacco is considered the most dangerous plant in the world, the deadliest of deadly nightshades; yet, ironically, next to its New World relative the potato, it was the Solanaceae species primarily responsible for ameliorating the family's evil reputation, a plant so beloved for the narcotic properties maligned in its Old World kin that its use fostered a new appreciation for poisons in popular culture—and thus to the change of attitude toward the entire nightshade family that occurred by the end of the Victorian era.

The predominance of smoking supplies in the Duke of Sussex's collection reflects the British taste in haute tobacco usage current in the first decades of Victoria's reign. Although the French fashion of snuffing—which had arrived with the Restoration in 1660 when the Stuart family returned to the English throne from exile in France—took honors as the favorite vehicle for stylish tobacco consumption among members of both sexes in the previous century, smoking in England had never completely gone out of aristocratic favor. Tobacco's first and foremost British champion, Sir Walter Raleigh, established pipe smoking as the perennial patrician standard in the late sixteenth century, famously indulging in one last bowl before submitting his head to the block when he was executed for treason in 1618. Mackenzie notes that found among Raleigh's personal effects after his death were two clay pipes "with silver mounts, and a tobacco stopper of bone in the shape of a finger," which were the contents of a red leather pouch inscribed with a Latin quotation from Cicero: "*Comes meus fuit in illo miserrimo tempore*" ("It was my companion in that most wretched time"), referring to his years of imprisonment in the Tower of London.[4]

By Victoria's reign smoking was returning to its former fashionable ascendancy over snuffing as ornate pipes gained popularity—led by the intricately carved meerschaums (German for "sea foam" to describe the porous mineral from which the bowls were made), sometimes mounted in silver and often sporting amber mouthpieces, first manufactured in Austria and Germany. But it was the rage for expensive cigars introduced to England from Spain and Portugal following the Napoleonic Wars (1803–15) by returning British officers that significantly boosted the number of well-heeled, predominantly male smokers.[5] Nevertheless, whether pinching or puffing, the British upper class, like the Duke of Sussex himself, consecrated their habit with enthusiastic acquisition, priding themselves on purchases of expensive tobacco and elegant accoutrements; for these elite tobaccomaniacs treated their addiction with nothing less than religious fervor, spending enormous sums and (as we know now) sacrificing life and health in their devotion to the plant.

Yet tobacco worship was not confined to the British aristocracy. According to G. L. Apperson's 1916 history of smoking, the rural and urban middle classes—from country parsons and squires to merchants and tradesmen—kept up the tradition of pipe smoking throughout the eighteenth century;[6] and as the middle-class population increased in size and affluence during the Victorian era, so did the number of male tobaccophiles who smoked pipes as well as cigars, patterning their usage on that of their "betters" and in the process greatly expanding the market for expensive tobacco and related products in the latter half of the century. In his excellent study *Smoking in British Popular Culture, 1800–2000*, Matthew Hilton contends that "bourgeois-liberal" males justified smoking by becoming connoisseurs, not only by purchasing quality tobacco and accessories, but by acquiring knowledge about all aspects of tobacco's history, biology, cultivation, and manufacture—such expertise thereby legitimizing their habit as an appropriately manly pursuit. This quest for information about tobacco was in turn satisfied by the development of an enormous body of related periodical literature, which the genteel smoker could peruse while enjoying his pipe or cigar in his library or the smoking room at his club, priding himself on the fact that "he was not a 'consumer' but an 'ardent votary,' a worshipper, disciple and true friend of 'the divine lady nicotine.'"[7]

Last but not least, the lower classes seem to have embraced "drinking" tobacco (as smoking was first called)[8] from the plant's earliest days in Britain, establishing the habit as a favorite indulgence as well as an essential feature of their social rituals. Apperson says that "among humbler folk pipe-smoking had never 'gone out,'" evidenced by the time-honored tradition of laborers throughout England meeting after work for conversation and relaxation over

196   VICTORIAN NIGHTSHADES

a pint and a pipe. Pubs provided the long-stemmed clays, called churchwardens or known by brand names like Broseleys, which customers filled from a communal tobacco box for a small fee.[9] In London, by the report of that premier mid-century authority on the city's poor, Henry Mayhew, smoking was ubiquitous among the lowest classes: costermongers (i.e., street-sellers) gathered in beer shops to drink, smoke, and discuss the day's business, often amusing themselves with card playing while "all shrouded in tobacco-smoke."[10] The city dustmen (refuse collectors), not inclined to gamble, were content simply "to smoke as many pipes of tobacco and drink as many pots of beer as possible."[11] In the Irish slums Mayhew reports seeing women break from their chores to "chat and smoke away the morning"; he observed one old woman smoking a pipe "so short that her nose reached over the bowl."[12] And Mayhew describes the sailors who brought the oyster boats to Billingsgate market sitting on the docks "smoking their morning's pipe."[13] Like the old Irish woman, they usually smoked the short clay known as a "cutty," which was identified everywhere with sailors, who were largely responsible for spreading tobacco use around the world — a phenomenon celebrated in the historical novel *Westward Ho!* (1855), written by the Anglican clergyman (and committed pipe smoker) Charles Kingsley. In the novel, which recounts the adventures of the hero Amyas Leigh on expeditions with Francis Drake and Walter Raleigh, the seasoned shipman Salvation Yeo promotes the "Indians' tobacco" in terms that would be later adapted as an advertising slogan by the W. D. and H. O. Wills Tobacco Company, which named a smoking mixture in the novel's honor. On packets of "Westward Ho!" could be found a version of Yeo's testimonial: "When all things were made, none was made better than Tobacco; to be a lone man's Companion, a bachelor's Friend, a hungry man's Food, a sad man's Cordial, a wakeful man's Sleep, and a chilly man's Fire. There's no herb like it under the canopy of Heaven."[14]

Unlike the potato, which entered Europe at roughly the same time but was initially greeted with rather more curiosity than affection, tobacco — as the facts and legends surrounding Raleigh's sponsorship attest — came in smoking. Hailed as a panacea by the Spanish physician Nicholas Monardes in his *Joyfull Newes out of the Newe Founde Worlde* published in the mid-sixteenth century, tobacco at first gained favor on the Continent as a medicine,[15] but it quickly found its real niche in England as a recreational drug — to the wonder of some contemporary commentators. John Gerard's celebrated *Herbal* in fact serves as documentation for the historical moment when tobacco's English popularity diverged from the medicinal to the recreational — and the first seeds of the country's tobaccomania were planted. Gerard catalogues the many "vertues" of what he calls "Tobaco, or Henbane of Peru," which include everything from a cure for migraines and toothaches to an antidote for poisons; and he gives recipes for its use as an "outward medicine," or salve, "against tumours, apostumes [abscesses], olde ulcers, of hard curation, botches, scabbes, stinging with nettles, carbuncles, poisoned arrows, and woundes made with gunnes or any other weapon." However, Gerard reacts with some bemusement over the novel phenomenon of smoking and voices misgivings about its presumed therapeutic efficacy: "Some use to drinke it (as it is tearmed) for wantonness or rather custome, and cannot forbeare it, no not in the midst of their dinner, which kind of taking is unwholesome and very dangerous; although to take it seldome and that Physically is to be tolerated and may do some good; but I commend the syrup above this fume or smokie medicine."[16]

Even though the physiological and psychological mechanisms of addiction were not as yet understood, the power that tobacco exercised over its users was quickly apparent—as well as suspect. Gerard explains that tobacco, like the henbane with which it was classified, "bringeth forth drowsiness, troubleth the senses, and maketh a man as it were drunke by taking of the fume only." And further, "the benumming qualitie heereof is not hard to be perceived, for upon the taking of the fume by mouth there followeth an infirmitie like unto drunkenness, and many times sleepe, as after the taking of *Opium*."[17] Nevertheless, he dutifully reports what the tobacco historian Jordan Goodman claims to be two of smoking's most salutary and desirable effects—its ability to assuage hunger and to provide sedation.[18] As Gerard notes, "When the Moores and Indians have fainted either for want of food or rest, this hath been a present remedie unto them to supplie the one, and to helpe them to the other."[19]

Clearly, tobacco was a welcome addition to the Old World pharmacopoeia, and its many reputed virtues would be reason enough to explain two of its Latin names that Gerard mentions—*sacra herba* and *sancta herba*;[20] after all, its alleged ability to heal the sick accounts for its universal status as the most sacred herb of indigenous peoples throughout the Americas. But tobacco's divinity in the New World in fact originated from what Gerard and others found to be so disturbingly alien: the belief that the "drunkenness" it causes is the physical manifestation of supernatural power. That is, tobacco smoked in shamanic rituals functioned as a psychagogue, an agent providing access to a world of spirits from whence came its medical and magical potency. Gerard describes this phenomenon as follows: "The priests and Inchaunters of the hot countries do take the fume thereof untill they be drunken, that after they have lien for dead three or four houres, they may tell the people what woonders, visions, or illusions they have seene, and so to give them a propheticall direction or foretelling (if we may trust the Diuell [Devil]), of

198   VICTORIAN NIGHTSHADES

the successe of their businesse."[21] In other words, here was a plant widely deployed in its homeland for what must be considered witchcraft, a practice anathematized in Europe for its link to devil worship—a connection that Gerard comes close to asserting.

No wonder, then, that tobacco's European critics were as vociferous in their condemnation as its enthusiasts were in their praise. In England the most strident of its early opponents was King James I, who was also a noted enemy of witches, having written a condemnatory treatise entitled *Daemonologie* in 1597. His more famous *Counterblaste to Tobacco*, written in 1604, the year following his accession to the English throne, details the evils of smoking by challenging pro-tobacco arguments like those presented by Gerard. Thus, after rehearsing tobacco's alleged curative powers, James counters that if one is not ill, medicine is not only unnecessary but can actually be detrimental. Accusing smokers of a "vile" and "filthy custom" that mimics "the barbarous and beastly manners of the wild, godless, and slavish Indians"—a custom from which, as Gerard had said, smokers are "not able to forbear" once adopted— James remonstrates, "Are you not guilty of sinful and shameful lust? (for lust may be as well in any of the senses as in feeling) that although you be troubled with no disease, but in perfect health, yet can you neither be merry at an Ordinary, nor lascivious in the Stews, if you lack Tobacco to provoke your appetite to any of those sorts of recreation?" In short, its "bewitching quality" enslaved smokers to a vicious habit abetting other vicious habits, and thus to "sinning against God" in their abuse of their bodies—a heinous crime prompting James's famous, final judgment of smoking: "a custom loathsome to the eye, hateful to the nose, harmful to the brain, dangerous to the lungs, and in the black stinking fume thereof, nearest resembling the horrible Stygian smoke of the pit that is bottomless."[22]

James's hatred of tobacco didn't stop with verbal abuse: in 1604, whether as vengeance or opportunity (or both), he also raised the duty on tobacco imports by 4,000 percent, a move that did nothing to stop its flow into England, mostly from Spain, but served to encourage an already brisk smuggling trade—as well as to demonstrate further what a precious commodity tobacco had already become. The ultimate result of his action was to convince James of the wisdom of joining rather than fighting the tobacco enterprise, first by supporting the commercial Virginia Company's colonizing of North America's mid-Atlantic coast, where successful cultivation began in 1613 under the direction of John Rolfe; and later by creating a royal monopoly on tobacco imports by taking governmental control of the new American colonies. By the time of James's death in 1624, the Virginia tobacco plantations, now forcibly worked by slaves from Africa and felons from England, were proving to be an enormously lucrative source of revenue, one well worth the Crown's protection. As an additional aid to revenues, legislation was passed to ban domestic commercial tobacco cultivation; but these attempts proved unsuccessful until the Glorious Revolution deposed a second King James and brought William and Mary to the throne in 1688—after which the royal army was ordered to destroy any tobacco fields found in the Mother Country.[23] As a cash crop filling British coffers as well as fueling the economies of the other major European colonial powers, tobacco had indeed proven to be an invaluable and most sacred herb, one to which the English government had become just as addicted as many of its subjects. According to Goodman, by the mid-seventeenth century tobacco had achieved permanent status as the first "exotic" plant to become a mass-consumption commodity in Europe;[24] by the end of the nineteenth century, tobaccomania had become a world-wide phenomenon.

Throughout the globe no plant has ever been held more sacred, a fact registered in Lord Byron's famous paean to tobacco written in 1823 as part of his long poem *The Island*, a fictionalized recounting of Fletcher Christian's and his comrades' fates after their 1789 mutiny on the HMS *Bounty*. The following lines became the classic statement of tobacco veneration in the copious literature published about the plant in the nineteenth century:

> Sublime tobacco! Which from east to west
> Cheers the tar's labour or the Turkman's rest;
> Which on the Moslem's ottoman divides
> His hours and rivals opium and his brides;
> Magnificent in Stamboul, but less grand,
> If not less loved, in Wapping or the Strand;
> Divine in Hookas, glorious in a pipe,
> When tipp'd with amber, mellow, rich, and ripe;
> Like other charmers, wooing the caress,
> More dazzlingly when daring in full dress;
> Yet thy true lovers more admire by far
> Thy naked beauties—give me a cigar!

In this digression following the introduction of one Ben Bunting, an old salt who constantly smokes a "short frail pipe,"[25] Byron celebrates the fact that tobacco worship bridges vast social, geographical, and ethnic distances, from the sailor's much-touted love to the equally notorious passion of the Turk. And the same principle holds true, if on a less exotic scale, back home in England: for the London dockyard workers of Wapping and the elite theater crowds of the Strand both cherished their tobacco just as passionately as Eastern potentates or sailors in the South Pacific.

Concluding the passage, and in keeping with his particular mode of Romanticism, Byron shifts to the sexual metaphor that would become a cliché in Victorian popular culture (and of which, more later): tobacco as the seductive female, a "charmer" so desirable that some men are incapable of rebuffing her advances. But whatever her guise—whether fashionably outfitted in an expensive pipe or stripped down to nothing but bare leaf like the poet's favorite cigar—"My Lady Nicotine," as J. M. Barrie would later name tobacco, indeed proves as physically irresistible to her suitors as opium or an entire harem of beautiful women. If, as Matthew Hilton proposes, the feminizing of tobacco was a peculiarly Victorian phenomenon, then Byron's metaphor surely served as its prototype.[26] Iain Gately's engaging *Tobacco* (2001) is appropriately subtitled *A Cultural History of How an Exotic Plant Seduced Civilization* to underscore by implication how the physiological lust for tobacco easily equates with sexual desire.[27] Byron's final, feminine metaphor therefore harks back to the epithet that he assigns to tobacco at the beginning of the passage, for "sublime" not only applies in the classical sense of religious ecstasy or spiritual transcendence, but also suggests the Gothic idea of titillating danger, in that tobacco's allure—like that of a femme fatale offering illicit sex—tempts those who come under its spell almost in spite of themselves, leading them to forsake other allegiances and loves, while risking health and well-being in the process.

In other words, as both the Duke of Sussex and King James made abundantly clear, a tobacco habit, like all passionate indulgences, comes at a cost. Indeed, long before the perils of tobacco were fully recognized, users and nonusers alike had reason for their suspicions about its effects, given that initiation into its pleasures or just sharing its company can be attended by physical discomfort or possibly full-fledged pain. Even in the first report of tobacco's use published in English, the French friar André Thevet's account of his travels in Brazil in 1550, this "secret herb" so prized by the natives could make them "light in the head" if taken in excess, while posing a greater risk for novice users like the "Christians" who had "become very desirous of this perfume": "The first use thereof is not without danger, before that one is accustomed thereto, for this smoke causes sweats and weakness, even to fall into a syncope, the which I have tried myself."[28] Thevet's referring to those Europeans as "Christians" who are eager to sample tobacco calls attention to smoking's status as a heathen—and therefore ungodly—practice, an association that lived on through the Victorian age; yet tobacco's ties to transgressive or daring behavior have continually served to heighten not only its notoriety but also its appeal.

Thus Compton Mackenzie begins his wonderful *Sublime Tobacco* (1957)—which takes its title from Byron's verse—by recalling his own early adventures in smoking that affirm how adopting the habit was a male rite of passage in the late nineteenth century, one frequently begun on the sly and particularly relished for its clandestine attraction. Inspired by the teenage boys who smoked (and bullied him) at school, eight-year-old Compton and his younger brother began snitching their father's cigar butts and smoking them in "one of his curved Petersen pipes." When his father discovered several months later what was going on, he gave each of the boys a cigar—no doubt, as Mackenzie says, shortly expecting to see "that greenish pallor which in the *Boys' Own Paper* always appeared on the cheeks of young readers who defied the threat of early blindness and stunted growth," the fate which "the editor continually insisted was the inevitable result of smoking in early youth." The Mackenzie boys, however, had already sufficiently inured themselves to the initiatory unpleasantness of smoking, leaving their father only to regret "parting with two of his best cigars to no good purpose."[29]

Mackenzie's *Sublime Tobacco* could well be considered the culmination of a tobacco literary genre whose heyday began almost exactly one hundred years earlier, in the mid-nineteenth century: the laudatory history, two revealing examples of which are Andrew Steinmetz's idiosyncratic *Tobacco: Its History, Cultivation, Manufacture, and Adulterations* (1857) and, most famously and comprehensively, F. W. Fairholt's *Tobacco: Its History and Associations* (1859). The genre's demise was virtually guaranteed (although by no means ended) in 1957, with the publication of the British Medical Research Council's *Tobacco Smoking and Cancer of the Lung*, which, like the U.S. Surgeon General's Report the same year, announced the official endorsement of the scientific evidence that smoking caused lung cancer—confirmation that tobacco was indeed "sublime" in the most dangerous sense of the term.[30] It is thus highly ironic that Steinmetz's book was written in direct response to the "Great Tobacco Controversy" currently being played out in the pages of the *Lancet* medical journal in early 1857, as his subtitle, *Its Use Considered in Reference to Its Influence on the Human Constitution*, suggests. Steinmetz's book is a fascinating inside story of tobacco's sublimity in an impassioned attempt to present a physiology, psychology, and philosophy of smoking.

Steinmetz, a barrister, social historian, ex-Jesuit novitiate, and an unapologetic and self-proclaimed "inveterate smoker," explains in his preface that he was moved to defend tobacco after reading a notice in the *Times* entitled "Is Smoking Injurious to Health?" which recommended the most recent *Lancet* offering by the surgeon Samuel Solly, whose anti-tobacco diatribe in a December 1856 article had initiated the journal debate.[31] Commenting in a lecture on paralysis that smoking was likely one of its causes, Solly had paused to attack what he called "the curse of the present age," asserting, among other derogatory accusations, "I know of no *single* vice which does so much harm as smoking. It is a snare and a delusion."[32] What followed his incendiary remarks was a deluge of letters to the *Lancet* arguing the pros and cons of tobacco use, prompting Solly, in the journal's February 7, 1857, issue, to present the much longer assault "on the baneful effects of this noxious weed" that Steinmetz felt called upon to rebut.[33] Steinmetz was further incensed by Solly's endorsement of (in Steinmetz's words) "a very slovenly and ill-written philippic" entitled *Practical Observations on the Use and Abuse of Tobacco* (1854) by the Edinburgh surgeon John Lizars, Solly's fellow member of the Anti-Tobacco Society, which, like the contemporaneous temperance organizations' attacks on alcohol, campaigned against its targeted "evil" on moral and religious as well as medical and scientific grounds.[34]

With the *Times* seemingly aligned with Solly's position, Steinmetz feared the "tide," as he said, was "running against 'inveterate smokers,' indeed against the gentle weed" itself.[35] So to counter what he considered to be Solly's rude and excessive fearmongering informed by the medical profession's often admittedly suspect opinions, Steinmetz launched possibly one of the most original—and certainly curious—pro-tobacco arguments of the era, bolstered by his reading of two popular works of organic chemistry concerned with, among other things, the physiological effects of ingested substances on the human body. Although the use of science to support one's habit would become typical, Steinmetz's sui generis approach offers a rare insight into the private thoughts and experience of a committed Victorian smoker and unabashed tobaccomaniac.

First, drawing from James F. W. Johnston's *The Chemistry of Common Life* (1853), which set forth a hierarchy of human needs and desires represented by food, alcohol, and narcotics, Steinmetz cites the long history and present pervasive appeal of the two latter substances to justify their use. Second, by conflating information from two of Justus Liebig's *Familiar Letters on Chemistry* (1843) analyzing respiration and human nutrition, Steinmetz concludes that "alcoholic liquors are amongst the necessaries of life"—leading him to speculate that the same may well prove true for the narcotic tobacco. Its track record and current popularity not only verify that smoking is safe, Steinmetz claims, but also proffer evidence that tobacco fulfills some higher, albeit undisclosed, purpose. As he observes in his introduction, "For my own part, I cannot believe that so universal a habit—tending, as is proved, to increase with the increase of populations—has been and is a mere whim or fancy of self-indulging man; but rather is one of those mysterious means by which we are compelled, in spite of ourselves, or with free-will and pleasure, to subserve the great behests of Providence." Thus, Steinmetz proposes that smoking is an instrument of a divine plan, and to support this theory, he speculates that tobacco's reputed prophylactic effects—it had been widely employed as a defense against everything from the plague to cholera and effectively used as an insecticide and repellant—may guard "against numberless miasmata perpetually coming into existence in the universal economy of nature," thus providing even the anti-tobacco faction with a protective smokescreen: "Possibly indeed the detractors of tobacco may have been saved from many a malady by the conjoint indulgence of a world of smokers." Steinmetz closes these introductory remarks with an apologia no doubt intended as a self-evident truth and the most persuasive word on the subject: "Wherever nature gives a strong tendency, there must be a strong reason in the cause—and still more when she superadds a pleasure, like all other pleasures of which we are conscious, but can give no account, after enjoyment, which is the peculiarity of the smoker's pleasure—a point worthy of philosophic consideration."[36] This barely penetrable sentence thus serves as Steinmetz's rationale for the hedonistic argument of the book: it is right to follow a natural impulse, especially if it gives pleasure, for that is the sign of its rightness—no matter the impossibility of describing the particulars of the enjoyment it provides. In other words, if it feels good, do it. However, the ineffability of smoking's pleasure does not stop Steinmetz from attempting in the following sections to specify its inexplicably pleasing quality—as he does in the next chapter in positively lyrical terms: "a passing solace to the mind's unrest—a thrill of comfort, contentment, and submission, whilst calmly inhaling its mysterious cloud amidst the stern realities of life."[37]

As his title promises, Steinmetz's book delivers information on tobacco's "History, Cultivation, Manufacture, and Adulterations," beginning with the plant's arrival in Spain and Portugal from the New World, and recounting how its use not only obstinately persisted but increased globally, despite the attacks of James I, the edicts of popes, and the numerous prohibitions of various European and Middle Eastern rulers "who fined, ruined, mutilated, or decapitated their miserable subjects" in futile efforts to stop its spread. Steinmetz supports his claims with statistics recording per capita tobacco consumption in the UK and on the Continent, and reproduces the most recent tables showing the numerous tobacco varieties and their quantities imported to England.[38] Steinmetz also uses the "History" section to return to venomous jabs at the claims of his nemeses Solly (whom he accuses of "effete senility") and Lizars. To refute Solly's speculation that further increase in tobacco use would debilitate "the English character," causing it "to lose that combination of energy and solidity which has hitherto distinguished it" and thus to diminish the nation's stature, Steinmetz submits the long record of successes of England's navy, whose addiction to tobacco was well known, and cites the recent valor and hardihood of British soldiers in the Crimean War (1853–56), which had witnessed a significant surge of smoking in the military.[39]

Information on cultivation and manufacture is equally detailed, offering a minute, affectionate description of *Nicotiana tabacum* in leaf and flower, and giving an especially thorough account of its "manufacture" in England after it arrives by ship from the Americas to be made into cigars, smoking mixtures for pipes, and snuff. Adulteration, considered in the *Lancet* exchange to be one potential cause of smoking's harmfulness (and a familiar problem in the manufacture of other products for human consumption), is less of a concern with tobacco, according to Steinmetz, who asserts that it is "decidedly the least adulterated article of commerce." Snuff, oftentimes made with additives out of consumer preference, was the most likely tobacco product to be contaminated with poisonous ingredients, like iron or lead.[40]

The subjects Steinmetz names in his title and dutifully covers as outlined above represent the standard fare for pro-tobacco literature, but they occupy only about half the book. What makes it unique—besides the assaults on Solly, Lizars, and others of the anti-tobacco persuasion peppered throughout—is his attempt to provide a comprehensive defense and philosophy of smoking in the two final sections related to physiology, "The Influence of Tobacco on the Human System" and the "Medicinal Action of Tobacco," by combining chemical, medical, historical, and cultural information from numerous sources—but particularly Johnston's *The Chemistry of Common Life*—with sage advice derived from personal experience. Here and elsewhere Steinmetz acknowledges that smoking is not without risk; but as he infuses his narrative with increasingly personal revelations about his health and habits, a self-image emerges of an exemplary member of the tobacco confederacy, an elite group defined by its discriminating intellectual and aesthetic taste. Not surprisingly, then, initiation into this fraternity usually entails some difficulty and even sacrifice: "There can be no question that the first attempt at smoking reveals phenomena which plainly show that the herb divine requires her votaries to go through a certain ordeal or trial before admitting them to her favours, if they aspire to the rank of her highest functionaries." Nevertheless, when Steinmetz immediately reports that his own experience was otherwise, it is no doubt intended to suggest (with false humility) his superior status even among the tobacco elect: "I know not whether my case is singular, but I can state that I enjoyed my first cigar—some twenty years ago—as much as the one I am now smoking."[41]

Thus Steinmetz, in true shamanic fashion, essentially presents himself as a high priest of the "herb divine," one qualified to offer a "Modus Operandi of Smoking," which "has never yet been explained by the Faculty." And with apologies to the lay reader for resorting to the "mystical terms of anatomy," he sets out to school the medical profession about how tobacco differently affects the novice and seasoned smoker. Asserting "I do not believe that the poison takes effect in the lungs," and armed with his copy of Erasmus Wilson's *Practical and Surgical Anatomy,* Steinmetz offers a jargon-driven account of how smoke held too long in one's mouth irritates "the glosso-pharyngeal nerve and its branches," which in turn aggravate the throat, brain, stomach, and heart to cause the characteristic spitting, giddiness, vomiting, and fainting that can plague the novice. By contrast, the adept quickly releases the smoke through his nostrils, "where, together with oxygen, it stimulates the *olfactory nerve,*" which, Steinmetz theorizes, "produces the beneficial effects of tobacco ascribed to it by all its votaries." The nerve's proximity "to the anterior lobe of the brain" arouses the intellect, thereby replacing the risk of bodily pain with the realization of cerebral pleasure. In short, "It is the *art or method* of smoking which makes all the difference."[42]

If, as Steinmetz premises, the nasal passages play a crucial role in ensuring the smoker's enjoyment, it perhaps follows that the size of the nose has considerable impact on the quality of the experience; but while Steinmetz implicitly endorses and later embraces this conclusion, his argument at this point takes a more surprising—and suspect—turn: "The larger the surface of the mucous membrane of the nose the greater the activity of the intellect or the anterior lobe of the brain; and without a well-developed nasal organ there never was a well-developed intellect. The nose of genius, in every age, has been conspicuous—in every sphere of its numerous manifestations." Later he adds, "For the proof of this position I appeal to the portraits of all manner of intellectual celebrities, in every profession, in every department of art or science." Thus we are led to infer that this voluminous "nose of genius" is also, as it were, the most prominent feature in the profile of the ideal smoker, whose equally capacious intellect responds to the stimulus of tobacco according to its particular type of mental acuity: "Where the imaginative faculties predominate, their activity will be exalted; where the reasoning powers are predominant, they will attain greater concentration."[43] In other words, smoking enhances and expands the already keen intellect of the veteran smoker.

As if to prove Steimetz's nasal theory, the book's frontispiece features a drawing by the artist "Alfred Crowquill" (Alfred Henry Forrester) of a tobacco jar ornamented with the carved heads of men representing various nationalities, each with a different kind of pipe in his mouth, but all blessed with a protuberant proboscis (see fig. 26). The picture also includes a tobacco stopper in the shape of a finger, possibly alluding to the one belonging to Sir Walter Raleigh. The clear message of the drawing is smoking's universal popularity, a theme reiterated in the slightly altered line from the *Aeneid* to translate as "What region of the earth is not full of our fumes?"—as well as in the closing lines of the poem, "An Encomium on Tobacco," that follow, which offer a litany of effusive metaphors to celebrate the plant. The final couplet reinforces Steinmetz's idea that smoking is an aesthetic and intellectual experience of the highest order, one whose pleasure can only be processed through the brain of a being with an advanced consciousness and perhaps even a soul: "*The daintiest dish of a delicious Feast, / By taking which* Man differs from a Beast."[44]

In sum, through many curious sidetracks, observations, long quotations, self-revelations, and sage instructions, the idea that smokers are essentially born, not made, as an elite, intelligent breed at last emerges as the major argumentative thread of Steinmetz's book. Even though "there may be many who abuse tobacco," the fault lies with the unworthy person and not the plant, for "what may be use or abuse, is entirely a matter of idiosyncrasy of constitution." In other words, "there must be then a constitutional peculiarity—a certain nervous system—a particular brain, for which tobacco is intended."[45] It is as if the sacred weed chooses its worshippers, and as one of those favored, Steinmetz offers his expertise and guidance.

Steinmetz's book succeeded in gaining the attention of the *Lancet* editors, whose review in the April 4, 1857, issue referred to it as "an amusing account of tobacco," if "occasionally somewhat flippant and off-hand," demonstrating "much ability and acumen." However, the reviewer takes Steinmetz to task for assuming the "office of 'medical advisor'"—not only because he was "abusive" to members of the profession who opposed his views, but also for giving "evidence of his total unfitness for the office he has assumed." Moreover, his "immoderate smoking—a pound of mild cigars a week" marked his personal experience as "exceptional," and by no means a reliable indication of tobacco's safety.[46] Nevertheless, Steinmetz's recommendations regarding tobacco use generally agreed with the medical journal's final editorial word on the subject, which was accurately reflected in a poem by "Quid" published a few weeks later. Titled "The Lancet Verdict on the Tobacco

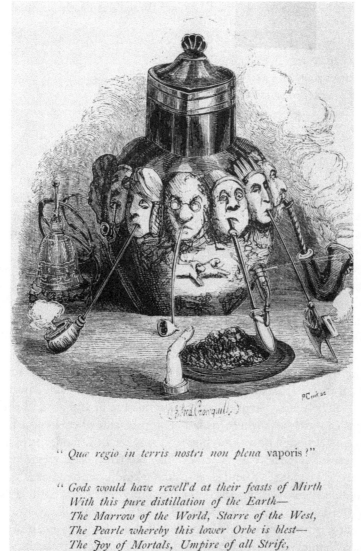

FIGURE 26. "Alfred Crowquill," frontispiece, Andrew Steinmetz, *Tobacco: Its History, Cultivation, Manufacture, and Adulterations*. (Image from Biodiversity Heritage Library; contributed by UMass Amherst Libraries)

Controversy Slightly Di-versified," its last three lines neatly summarize the journal's position:

THE LANCET does not blame tobacco's *use*,
(Except in youth, for *that* there's no excuse,)
But deprecates most strongly its *abuse*.[47]

TOBACOLOGY

The poet Quid's pun about the medical profession's "diversified" opinion on smoking captured the state of affairs at mid-century, for almost half of those responding during the *Lancet* debate considered the habit harmless, if not actually beneficial; and members of the general public who disapproved of tobacco use generally did so, like many of the doctors themselves, on moral rather than strictly medical grounds. Consequently, Steinmetz's fear of the tide running against the "gentle weed" would not be realized for another century; instead, his pro-smoking stance exemplifies the mainstream of Victorian tobacco literature and serves as a prelude to F. W. Fairholt's *Tobacco: Its History and Associations* (1859), the "classic" Victorian study of tobacco, a book written not just for tobaccophiles, but for readers wanting an objective treatment of the subject.[48]

Fairholt, an antiquarian and engraver by trade, took on the subject not only because of its intrinsic interest, but also because he had inside information, having spent the first twenty-two years of his life in and around the London tobacco warehouse where his father worked and where he was first employed. And although he was not himself a smoker, he defends the practice as "a harmless indulgence," passionately calling out the most extreme critics for their "narrowness of spirit," and even speaking kindly of Steinmetz's "agreeable little volume," while writing with an equanimity that lends authority to his words.[49] In fact, it is Fairholt's claim of being an "impartial advocate" of tobacco that sets his history apart from the vast body of Victorian pro-tobacco literature. While being "free from prejudice," his appreciative attitude, combined with the sheer volume of material that he amassed to document three hundred years of tobacco's popularity in England, helped to normalize its use as a permanent and respectable feature of Victorian popular culture.[50] And Fairholt's dispassionate treatment also extended to other members of the nightshade family.

Like Steinmetz, Fairholt draws on scientific research to support the use of tobacco, but his endorsement, which includes mention of two more nightshades at polar ends of the harmfulness spectrum, indicates the more relaxed attitude toward the whole family that the pro-tobacco lobby helped to create. After noting that botanists classified all tobacco varieties "among the *Solanaceae*, and narcotic poisons" and that "the *Atropa belladonna*, or deadly nightshade, is a member of this family," Fairholt reassures his readers that "it may be of use to the nervous to know that the common potato is in the same category." That is, just as tobacco "will produce a virulent poison—*Nicotine*," so "the Potato fruit and leaves give us *Solanine*," which he describes (quoting from Henry Prescott's *Tobacco and Its Adulterations*) as "'an acrid narcotic poison. . . . Traces of this are also found in the healthy tubers.'" From this information Fairholt concludes, "It is therefore evident that in a moderate manner, we may equally smoke our tobacco or eat our potato as regardless of the horrors that chemistry would seem at first to disclose, as when enjoying the flavor of the bitter almond, which we know to be owing to the presence of Prussic acid."[51]

Fairholt turns next to a brief account of the most common *Nicotiana* species, illustrations of which appear as the frontispiece for the book (see fig. 27). The first and most important is the tall, large-leaved, pink-flowered *Nicotiana tabacum*, named by Linnaeus and now known to be the parent of all modern commercial tobacco, which is native to South and Central America and the West Indies where it can be traced, if not to its wild state, to aboriginal cultivation. Consequently, Fairholt's designation "Virginian tobacco" could profit from clarification: the epithet derives from the fact that in 1612 the colonist John Rolfe, after disappointing attempts to cultivate the indigenous tobacco used by the natives, finally succeeded in raising a crop from *N. tabacum* seeds sent from Trinidad—thereby establishing the species, as well as the commercial and agricultural future of the Virginia colony's tobacco industry.[52]

Next in importance and the species Rolfe rejected is *Nicotiana rustica* (also named by Linnaeus), a shorter, smaller-leaved, green-flowered plant native to eastern North America—which evidently reached England first, where it could still be grown on a noncommercial scale for ornament or as an insecticide;[53] thus, as Fairholt notes, it is "a hardy annual in English gardens . . . so that by some botanists it has been termed 'common, or English tobacco'"— the latter appellation also mentioned by Gerard. However, Fairholt gives its current epithet as "Syrian Tobacco," named, so he erroneously claims, for its current habitat and use in the manufacture of Syrian, Turkish, and Latakia smoking mixtures.[54] Although Fairholt says *N. rustica* is "milder" than *N. tabacum* and "used for the more delicate cut tobaccos and cigars," *rustica* is in fact stronger and harsher; but Fairholt is presenting information that is correct to the best of his lights, and consistent with John Lindley's description of the genus in *Medical and Oeconomic Botany* (1849).[55] That the

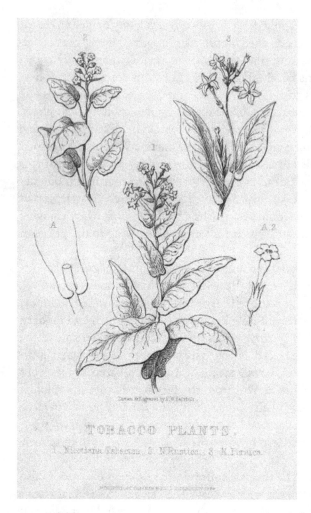

FIGURE 27. Frontispiece, F. W. Fairholt, *Tobacco: Its History and Associations*. (Image from New York Public Library; George Arents Collection)

great botanist is Fairholt's authority becomes apparent in the description of the third species featured in the illustration, *Nicotiana persica*, identified by Lindley in 1833, but which would later prove to be *N. alata*, the lovely white-flowered species cultivated as a sweet-scented ornamental rather than — as Lindley claimed — the source of the prized Shiraz tobacco of Persia.[56] Like the aforementioned Syrian smoking mixtures, Shiraz is a product of *N. tabacum*, now and in the Victorian era the source of all tobaccos manufactured for human consumption.[57]

Despite the confusion, Fairholt had clearly done his homework, and whatever his weaknesses as a botanist, his strengths as an antiquarian are legion. Fairholt's book therefore defines the genre of "tobacology," a convenient designation for a studious compendium of every aspect of the plant's history, cultivation, varieties, literature, statistical information, and methods of consumption—or in the language of Fairholt's subtitle, *Its History and Associations*. The coinage is an adaptation from Andrew Steinmetz's similar term, *tabacology*, used in his second book on smoking published in 1876, and probably adapted from the title of the Bremen physician Johann Neander's *Tabacologia* published in 1622, which employs the Latin spelling of the species name. Neander's book is one of the earliest scholarly texts on the subject, a "portly tome" according to Fairholt, wherein the plant is prescribed "for almost all the diseases of life."[58]

In keeping with these scholarly intentions, Fairholt's history begins with the Europeans' earliest encounters with the New World plant in November 1492, when two sailors in Columbus's crew sent to explore Cuba first observed the natives inhaling the smoke of a dried herb wrapped in maize; but it wasn't until 1526 that the Spaniard Oviedo, in his *Historia general de las Indias*, gave the first "clear account of smoking among the Indians of Hispaniola," explaining that the word *tabaco*, in Spanish usage as the name for the herb, in fact referred to a smoking implement rather than to the plant. Although Fairholt mentions other names for the plant, including the Brazilian word *petun*, which survives in the nightshade genus *Petunia*, variants of the Spanish *tabaco* won the day in Europe, eventually to become in English *tobacco*.[59]

Fairholt establishes tobacco's hallowed place among natives of Mexico and North America through information taken from a wide range of early and contemporary sources with illustrations, mostly of artifacts used for smoking, interspersed throughout. But his main purpose is to examine "Tobacco in Europe, and Its Literary Associations," the title and subject of the third chapter, which covers history, literary references, and illustrations of commercial artifacts, beginning with the plant's introduction "about 1560" to Spain by the physician Francisco Hernandez and to France by Jean Nicot, the ambassador to the Portuguese court who gave plants to Catherine de Medici and his name to tobacco's genus—as well as to its infamous alkaloid nicotine, first isolated in 1828. As evidence that the plant was initially prized for its medical virtues, Fairholt cites several relevant passages from early texts, including probably the most famous early literary source in English, Edmund Spenser's epic *The Faerie Queene* (1590), which names "divine Tobacco" as a healing herb collected in the woods by the huntress Belphoebe to save the wounded squire Timias.[60]

The bulk of the chapter (and indeed the book) is devoted to tobacco's popularity as a recreational drug, mostly as evidenced by celebrated indulgers in its use. Although the plant itself may have arrived in England as early as 1565, the novel practice of smoking was introduced some twenty years later by Ralph Lane, the first governor of the Virginia Colony on his return to England in 1586. Nevertheless, according to popular tradition, Sir Walter Raleigh made smoking fashionable, so much so, Fairholt says, that by the end of the century, "to take tobacco 'with a grace' was looked upon as the necessary qualification of a gentleman."[61] The numerous references to smoking that appear in contemporary plays, poetry, and prose testify to its almost immediate popularity. We learn that Ben Jonson's plays fairly reek of tobacco, offering abundant examples of the rage for smoking in taverns as well as in theaters, where "gallants" could actually sit and smoke onstage during performances.[62]

All this fanfare around the turn of the seventeenth century, which Fairholt dubs "the golden age of tobacco," was not without controversy, leading to King James's counterblast, and other diatribes as well. Fairholt goes into great detail recounting the poet Richard Braithwait's satirical invective *The Smoaking Age*, which presents a fake etymology of "Tobacco" as the offspring and namesake of the lascivious Bacchus, whom Jove punished by turning the child into a plant.[63] However, despite tobacco's frequent literary association with debauchery, Fairholt argues that its use was not "confined to the 'fast men' of the age," but maintained a steadfast hold on many sober men (and some women) of every walk of life, including Isaak Walton of *The Compleat Angler* fame, for whom smoking was solace during the mid-seventeenth-century years of the Commonwealth and the battles of the English Civil Wars. Fairholt even adopts Walton's definition for fishing, "the contemplative man's recreation," for the habit of smoking. With the Restoration and shortly afterward, during the Great Plague of London (1665), tobacco use gained increasing momentum, for "it was popularly reported, and generally believed, that no tobacconists [shopkeepers] or their households were afflicted by the pestilence." When the Glorious Revolution (1688) brought William III to the throne, "tobacco met with a patronage almost universal. Pipes grew larger then, and ruled by a Dutchman, all England smoked in peace."[64] The Dutch, of course, were legendary for their love of smoking—the only European nation at the time that could rival the English in addiction to tobacco.[65]

It was during Queen Anne's reign (1702–14) that "the custom of smoking appears to have attained its greatest height in England," even though (as we know), snuff was the fashion among members of the upper class, including the writers Alexander Pope and Jonathan Swift.[66] As evidence that pipe smoking was the preferred medium of the English divinity through the eighteenth century, Fairholt offers his rendering of a detail from William Hogarth's satirical print *A Midnight Modern Conversation* (1732), depicting a member of the clergy "in full canonicals" smoking a pipe and drinking punch. According to

Fairholt, "That it still, or till very lately, was the solace of the country parson any one acquainted with village life can tell."[67]

Fairholt's antiquarian instincts and artistic skills are on full display in documenting how the advertisement and sale of tobacco reflected its widespread impact on European culture. For example, one illustration of a "curious" contemporary tobacconist's sign features an arm attached to three hands (see fig. 28)—"the first holding snuff on a thumb, the second a pipe, the third a quid of tobacco," under which image the following couplet was inscribed: "We three are engaged in one cause; / I snuffs, I smokes, and I chaws." The same poem sometimes appeared, respectively, under "figures of a Scotchman," known for the love of so-called *sneeshin*; "a Dutchman," famed as an inveterate pipe smoker; and "a sailor," given to chewing tobacco aboard ship for convenience and fire prevention. *Chaw* also recalls the popularity of both the term and the habit among Americans, who were infamous in England for their relentless chewing and spitting.[68]

Chapter 3 ends with a brief look at famous tobaccophiles up to the present, including such (earlier) British luminaries as Isaac Newton and Thomas Hobbes; Charles Lamb, Walter Scott, and Lord Byron from the recent past; and contemporaries Thomas Carlyle, the "present Laureate Tennyson," William Thackeray, and Edward Bulwer-Lytton.[69] Among Continental writers known to partake, Fairholt mentions "George Sand, who often indulges in a cigar between the intervals of literary labour; as the ladies of Spain and Mexico delight in doing at all other intervals."[70] And Fairholt doesn't forget to mention nineteenth-century royalty, including the Duke of Sussex and

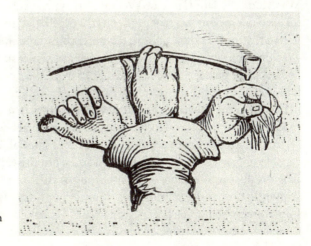

FIGURE 28. Tobacconist's shop sign, F. W. Fairholt, *Tobacco: Its History and Associations*. (Image from Biodiversity Heritage Library; contributed by Smithsonian Libraries and Archives)

George IV. These illustrious devotees and countless others of humbler circumstances lead Fairholt to praise tobacco's contribution to European culture: "Thus, from the throne to the cottage the pipe has been a solace; it has aided soldier and sailor in bearing many a hard privation."[71]

Whereas chapter 3 provides a relatively straight chronology of tobacco's European and British history, Fairholt's next chapters focus on the accoutrements of early and contemporary tobacco use throughout the world—"Tobacco-Pipes, Cigars, and the Smoker's Paraphernalia" in chapter 4 and "Snuff and Snuff-Boxes" in chapter 5. Nevertheless, like a true antiquarian, Fairholt begins with a discussion (and dismissal) of popular lore about the "'fairy pipe' of Ireland" and the similar "elfin pipes" of Scotland, relics of smoking devices sometimes found throughout the British Isles believed by the credulous to be of ancient and even of supernatural origin. Small clay pipes with tiny bowls were indeed, he thinks, the oldest tobacco pipes used in Britain; but the earliest discovered via excavation date from the Elizabethan Age, and their diminutive size had to do with the "excessive cost" of the herb at the time rather than the tiny capacity of elfin smokers. Fairholt dismisses comparable claims about smoking predating the sixteenth century in the Middle East and Asia, for "no instance can be quoted of the ancient use of any herb in the modern way tobacco is taken—that is, as a luxury, and not as a physical necessity or an intoxicating agent."[72]

Moving from such speculations to a detailed examination of old and contemporary pipes, Fairholt offers many illustrations of intricately carved bowls whose designs range from the patriotic to the satirical. One French pipe that must have either amused or offended English smokers features the hero of Waterloo, "the late Duke of Wellington," whose subsequent efforts to quash tobacco use among his troops inspired a pipe head "in which a subaltern, pipe in hand, quietly 'takes a sight' at the great commander, who is caricatured after a fashion that must have made the work a real pleasure to a Frenchman" (see fig. 29). The gesture, more colorfully known as "cocking a snook," is a time-honored show of disrespect. As Fairholt explains in a footnote, "The duke's hat receives the tobacco; the hat of the subaltern, the pipe-stem."[73]

German pipes received credit for the highest "art-workmanship," with the celebrated meerschaum considered the best material, "as it heats slowly and is capable of great absorption." Since the oil from the tobacco stained the meerschaum during smoking, pipes attaining a "rich deep brown tint" were "consequently treasured as triumphs of smoking feats," while tobacconists even sold pre-smoked items for the impatient.[74] Fairholt only briefly mentions the expensive briar-root pipe, made from the Mediterranean plant *Erica arborea* (called *bruyere* in France, from whence the first product—and the English

FIGURE 29. Duke of Wellington pipe, F. W. Fairholt. *Tobacco: Its History and Associations*. (Image from Biodiversity Heritage Library; contributed by Smithsonian Libraries and Archives)

name—derived), which according to Compton Mackenzie probably reached England around 1856, but would eventually replace the meerschaum as the most popular pipe material.[75]

Two-thirds of chapter 4 is devoted to pipe smoking, a proportion that suggests its relative importance in England compared to other modes of contemporary tobacco use: Jordan Goodman reports that around mid-century, pipe tobacco accounted for 60 percent of British consumption.[76] Even so, thanks in part to some relaxation in importation laws and duties, cigars, the perennial smoke of choice in Latin America and Spain, had been enjoying increasing popularity, as Fairholt documents. Although in 1830 they were "quite an aristocratic luxury," at the present, he says, cigars represent "fully half the quantity of tobacco smoked in our large towns." As evidence, he provides a London importer and manufacturer's list of sixty-seven varieties. Havana produced the best cigars, and "Penny Pickwicks" (named for Dickens's nonsmoking hero) were the cheapest.[77] In *London Labour and the London Poor*, Henry Mayhew attests to the mid-century urban penchant for cigar smoking while promenading the London streets when he reports that among the city's many scavengers were the "cigar-end finders or 'hard-ups.'" Most often the children of the destitute Irish, they scoured the gutters for discarded butts that they sold to entrepreneurs who dried the tobacco and sold it to the poor. The best pickings were in "the aristocratic quarters of the City and the vicinity of the theatres and casinos" where Mayhew estimated some 30,000 cigar ends were collected daily.[78]

Fairholt's chapter 4 concludes with an account of tobacco pouches, boxes, jars, and stoppers—all the peripheral accoutrements associated with the smoking habit; but in most categories, he devotes his attention to specimens dating from earlier periods. The exceptions are first, the coin-operated, pop-open tobacco boxes still used in "country-alehouses"; and second, porcelain tobacco jars, "a comparatively modern invention," for which he provides illustrations "to display the whim and fancy they occasionally exhibit," the most popular being one shaped as a girl in Regency dress whose ruffled skirt hides the lid opening.[79] Finally, most essential to smoking is of course fire, which by Fairholt's time could be produced "by many ingenious inventions" to replace the old tinder box and flint. Matches, "headed with a lump of combustible matter" and packaged in small boxes, were the most up-to-date, portable, and convenient. But also available was a device that must be the prototype of the modern lighter, consisting of a silver tube through which a cotton cord is drawn and "lit by means of a flint, elegantly fashioned from the purest stone, struck against an equally tasteful steel." However, Fairholt says, such expensive contraptions are "chiefly patronised by 'heavy swells,' who take tobacco more for the sake of ostentation than pleasure."[80]

The chapter on snuff begins with its earliest use as a "sternutatory" (sneezing agent), but focuses on its later status as a fashionable aristocratic indulgence, with its attendant rituals and accessories evidently providing as much pleasure as the snuff itself—especially since the tobacco was frequently watered down and scented with substances like musk, amber, mint, or rose leaves. Fairholt credits the Great Plague with the increase of snuffing in England, where its popularity continued through the eighteenth century, famously indulged in by George III's consort, Queen Charlotte, who made it fashionable among court ladies. Indeed, snuff more than any other mode of tobacco use became closely identified with elaborate ceremony and elegant trappings designed for show.[81] Snuffboxes were often made of gold and silver inlaid with precious stones or carved from rare materials, some touted as "relics" made of wood from "Shakespeare's mulberry tree, Nelson's ship, or Wellington's table."[82] Fairholt, however, does not mention the "indecent" or "spicy" snuffboxes picturing naked ladies or pornographic scenes that Henry Mayhew said could be bought on London streets (and hawked on the sly to "fast gents as have money to spare" or young bucks "up on a spree" from Oxbridge).[83]

Besides Pope and Swift, mentioned earlier, famous English male snuffers of the eighteenth and nineteenth centuries named by Fairholt include Samuel

Johnson, Frederick the Great, and Napoleon Bonaparte.[84] Charles Darwin, who snuffed during work hours but regulated his use by keeping his snuff jar on a mantle outside his study, could be added to their ranks.[85] Fairholt's following observation certainly applies to the great naturalist: "Among men of large intellect, snuff-taking has been rather common; it may have been felt by them as a counter-irritant to the over-worked brain."[86] Fairholt also names Samuel Johnson's biographer James Boswell and Robert Burns, two Scots who represent the British country most closely identified with snuff, so much so that from 1745—the year of the Jacobite Rebellion—the life-size wooden "figure of a Highlander helping himself to a pinch" became a familiar snuff shop sign, one that prevailed in England until at least the early 1830s.[87]

In the final chapter, entitled "The Culture, Manufacture, and Consumption of Tobacco," Fairholt returns to the plant itself, presenting in painstaking detail every step of its preparation for use by the British consumer. The intensive labor of cultivation takes place in Virginia, beginning with sowing the seeds in early spring; it ends when the cured and dried leaves have been packed in hogsheads, huge barrels each capable of holding 1,000 pounds of tobacco, which are inspected, secured for the voyage, and shipped to England in the fall. Once in the London dockyard's tobacco warehouse, the hogsheads are inspected again, and any damaged part of the shipment removed and burned in the huge cylindrical kiln "jocularly termed 'Her Majesty's tobacco-pipe.'"[88] The rest goes to the manufacturer to be made into the wide variety of products—from spun or twisted tobaccos like "pig-tail" and "negro head" to cakes of "Cavendish"—sent to shops for sale.[89]

The amount of time, effort, and expense involved in preparing tobacco for consumption underscores what an astoundingly lucrative commercial product it had become—for the grower, the shipper, the manufacturer, and of course, for the British government, whose revenue from duties in 1858 was over five million pounds. That year, Fairholt says, the British spent almost eight million pounds for tobacco and snuff; in 1853, the last year for which he gives the figures, consumption in Great Britain "amounted to 24,940,555 lbs. or 19 ounces per head to the entire population." And though British consumption did not come close to that in Asia, "where women and children smoke as well as men," the numbers sufficiently register tobacco's domestic popularity—not to mention its critical role in bolstering the national economy, a boon that even its adversaries were forced to acknowledge.[90] These impressive statistics serve as decisive evidence in Fairholt's defense of tobacco, which rests on his belief that it was a most valuable and welcome contribution to British and European culture: "As a comfort to the poor, as a luxury to the rich, tobacco unites all classes in a common pleasure";[91] and his book serves as the most comprehensive Victorian account of the plant's phenomenal importance in the British Isles and the rest of the world.

Although countless articles on tobacology would be written during the Victorian era, *Tobacco: Its History and Associations* maintained its status as the standard authority in the genre for the rest of the nineteenth century. However, Fairholt could not foresee that the incipient fashion of cigarette smoking would create a paradigm shift in English tobacco usage toward the end of Victoria's reign, permanently altering patterns of consumption for the next century and beyond while greatly increasing the nicotine-addicted population. Fairholt briefly mentions that cigarettes "are much indulged in by the ladies of South America and Spain," and he describes the little books of thin papers—the best from Valencia—that were sold for hand rolling.[92] But he makes it clear that what were essentially small paper-wrapped cigars were still an exotic novelty in England at mid-century. The radical change they effected was, however, recorded by Fairholt's self-appointed successor W. A. Penn in *The Soverane Herbe—A History of Tobacco* (1901),[93] published a few months after the Queen's death. Penn's information on cigarettes constitutes his most important contribution to the genre, providing the necessary closure for Victorian tobacology.

According to Penn, cigarettes' popularity in England originated with British officers returning from the Crimea, where the dearth of cigars led them to satisfy their desire for tobacco by adopting the smoking habit of their French and Turkish comrades. The trend at first became "fashionable among clubmen and in the higher circles," but grew over the next decade, encouraging tobacco manufacturers to begin offering ready-made varieties for sale and further accelerating the fashion's spread. Nevertheless, it wasn't until commercial rolling machines were perfected in the 1880s that cigarettes could be produced in massive quantities—"at the rate of 200 to 400 a minute"— making them widely available as well as affordable for even the most impoverished smoker.[94] Cheapness and, in comparison to pipes and cigars, the convenience of this method of tobacco consumption caused the market for cigarettes to explode by the end of the century—so much so that Penn could report, "Twenty years ago the cigarette was almost unknown outside France, Spain, Italy, and Turkey. Today it is the most popular smoke among all classes in England and America."[95]

As Jordan Goodman would later explain, other factors contributed to cigarettes' newfound popularity as well: for example, they inspired "new methods of marketing" in the form of advertising and packaging that emphasized brand recognition and appeal.[96] Mathew Hilton documents the rise in popularity of John Player and Sons' Navy Cut brand from the early 1880s via changing images of "Hero," the iconic sailor pictured on every tin or package. During the same period, the W. D. and H. O. Wills Company, armed with the newfangled Bonsack cigarette-making machine invented in America, introduced such brands as Three Castles and Gold Flake, to be followed in 1888 by Woodbines, sold in a pack of five for a penny and according to Compton Mackenzie, destined to become "the most famous cigarette in the world."[97]

But most important of all was English manufacturers' switch from the strong tobaccos traditionally used in cigars and pipes to the mild Bright tobacco, a variety—so named for its golden color—produced by flue curing, a process that changes the smoke's chemical makeup from alkaline to acidic, making it easier to inhale and thus far more addictive than any other means of ingesting tobacco because the nicotine is so quickly absorbed by the brain.[98] Penn reports that "bright Virginia, the mildest tobacco on the market," accounted for the "great majority of cigarettes now smoked," far outnumbering the Russian, Turkish, and French varieties that were first available.[99] When improvements in flue curing begun in the early 1870s conveniently coincided with newly automated production to lower prices even further,[100] all the forces were aligned to create the perfect storm of cigarette smoking in England and America that Penn records as having occurred by the close of the Victorian era.

Even though Penn acknowledges the universal triumph of the cigarette, his comments nevertheless reveal a late Victorian prejudice toward what veteran smokers considered an alien, inferior, and less-than-manly mode of consuming tobacco. The very term *cigarette*—signifying the feminine diminutive of the Spanish *cigar*—was coined in France circa 1830, to become, as Iain Gately notes, "the most commonly used French word on the planet." Its earliest recorded use comes from the French writers Honoré de Balzac and Theophile Gauthier in the thirties, followed a decade later, but most famously, by Prosper Mérimée, whose novella *Carmen* (1845) would inspire Georges Bizet's great French opera in 1875, which established its Spanish-gypsy heroine as the archetypal daring, liberated, cigarette-smoking woman. All these formative French references emphasize the cigarette's risqué and exotic Hispanic origins, thereby solidifying the practice's essential and inextricable link, in Gately's terms, to "eroticism and urban sexuality."[101]

According to Penn, for the dedicated (and doubtless male) British tobaccophile, cigarettes were usually "additional or supplementary to the pipe and cigar"—to be indulged in only when his preferred and presumably more authentic, superior, and manly methods of ingestion were inconvenient or inappropriate. By contrast, Penn's following characterization evokes the effeminate bohemian smoker of the late nineteenth-century aesthetic movement, who evidently seeks a different sort of gratification. That is, while on the one hand Penn calls the cigarette "undoubtedly the most elegant form of smoking," he almost immediately demurs before attempting to articulate the unique physiological response elicited by this type of tobacco use: "The puffing of cigarettes differs from smoking; such it can scarcely be considered. It is a form of slight excitement; it feeds rather than satisfies the appetite; it is more like, in its effects and practice, the smoking of opium than tobacco; the cigarette is a variety of the craving for absinthe and morphia."[102] Putting aside the question of whether cigarette use can be considered "smoking," Penn's description brings to mind the addictive habits of aesthetes like Charles Baudelaire, Paul Verlaine, and other members of the French avant-garde defined by their taste for novelty and sensation and their boredom with traditional mores. It also echoes the oft-quoted dictum of their British counterpart Oscar Wilde (in his guise as the character Sir Henry Wotton) from *The Picture of Dorian Gray* (1890): "A cigarette is the perfect type of a perfect pleasure. It is exquisite, and it leaves one unsatisfied. What more can one want?"[103]

In short, the cigarette's capacity for providing mild but repeated stimulation suggests it is, like the aesthetic movement itself, a peculiarly modern—and to Penn, a frivolous and distasteful—phenomenon, as his following observation makes clear: "Its popularity is a sign of the national craving for brevity, weakness and mild excitement, and of dislike for all that is solid and substantial, whether it be in food, clothes, literature, religion, or amusement. Indeed, the cigarette, denounced by the 'honest smoker' as mere flirtation with Diva Nicotina, emphasizes in one aspect the most striking phase of modern life and thought."[104] Even though Penn's remarks express some familiar Victorian prejudices, his analysis nevertheless accurately locates the cigarette's place in cultural history. However, Penn did not realize that through its insidious power to whet rather than quench desire, what may have seemed at the end of the Victorian era like a "mere flirtation with Diva Nicotina" would turn out to be tobacco's most dangerous liaison with humankind the world has known.

Penn's assessment thus anticipates the claim that the cigarette is a "crucial integer of our modernity," a statement made by Richard Klein in his fascinating study *Cigarettes Are Sublime* (1993), written as a eulogy for the habit after its demonization in the late twentieth century as one of the world's deadliest human addictions.[105] Drawing heavily from nineteenth- and twentieth-century French literature as well as his own experience, Klein eloquently explains how cigarettes are perfect exemplars of Kant's definition of the "sublime" as "negative pleasure" or "negative beauty," a concept reminiscent of Edmund Burke's and a further refinement of it, that Kant describes in *The Critique of Judgment* (in Klein's words) as "that aesthetic satisfaction which

SUBLIME TOBACCO 221

includes as one of its moments a negative experience, a shock, a blockage, an intimation of mortality."[106]

Accordingly, Klein's application of Kant's theory corrects Penn's notion of the cigarette as a "mere flirtation" with tobacco while amending Wilde's idea of the kind of pleasure smoking provides: "The 'pleasure' associated with cigarettes is negative, therefore not exactly a pleasure. One's first experience of smoking does not seem like play but like a serious act, accompanied by more dis-taste and dis-ease than the good tastes of innocent sweetness. In fact, tobacco makes one a little sick every time the poison is ingested. It announces its venomous character from the first, especially at the first puff, and subsequently as each successive puff distributes repeated jolts to the body."[107] One might think that such a visceral, negative response would seem most likely to discourage tobacco use, but Klein's point is just the opposite. As if the cigarette could somehow read the smoker's mind, each "venomous puff" delivers just what he or she desires, whether it be increased alertness or relaxation—stimulation or sedation—producing a pleasure that is all the more potent—that is in fact "sublime"—for having arrived after the requisite initial apprehension of terror, difficulty, and pain. Such "repeated jolts to the body" account for the cigarette's almost preternatural ability, if not to cure, at least to appease, whatever ails you, thereby inducing the urge for more and more. Hence Klein's profound conclusion: "All the literature of cigarette smoking, no less than the testimony of universal experience, attests that it is not in spite of their harmfulness but because of it that people profusely and hungrily smoke. The noxious character of cigarettes—their great addictiveness, and their poisonous effects—not only underlies their social benefits but constitutes the absolute precondition of their troubling, somber beauty."[108] In recognizing that smoking's poison lies at the heart of its appeal, Klein articulates the paradoxical nature of cigarettes' sublimity—"their troubling, somber beauty." But his description also recalls Byron's epithet for tobacco in any form, including as the seductress who materializes at the end of his verse in the guise of the well-dressed pipe or naked cigar to underscore the illicit quality intrinsic to smoking's pleasure. Thus, the emergence of the cigarette toward the end of the Victorian era serves as Sublime Tobacco's modern manifestation, the latest occurrence of that ancient sorceress forever identified with the nightshades, this time casting her most formidable spell.

NICOTIANA

In 1876 Andrew Steinmetz returned to his favorite subject with the anonymous publication of *The Smoker's Guide, Philosopher and Friend: What to*

Smoke—What to Smoke With—and the Whole "What's What" of Tobacco, further classifying these topics as "Historical, Botanical, Manufactural, Anecdotal, Social, Medical, etc., etc., etc." Identified on the title page simply as "A Veteran of Smokedom," Steinmetz opens his book with a statement "To the Reader," in which he acknowledges his debt to "his innumerable predecessors in Tabacology, and almost every other 'ology,'" but claims, "whilst drawing from every available source, he has taken nothing without striving to adorn it."[109] In other words, although Steinmetz presents much historical, statistical, and scientific information (including material recycled from his previous book on tobacco), *The Smoker's Guide* is not simply a "tobacology" but a wider-ranging composite on sundry smoking themes designed to entertain as much as to instruct—and thus partaking of a different, albeit related contemporary genre of tobacco-centric literature, one that aspired to the status of belles lettres. In consideration of this aesthetic goal, Steinmetz not only "adorns" his factual information with lively metaphors and other stylistic flourishes, but also includes favorite poems and anecdotes. Fairholt's *Tobacco* had included poetry and anecdote as well, but the verses and tales were incorporated explicitly to serve as documents related to the plant's history and presumably not for their intrinsic value as literary art. By contrast, Steinmetz's *Smoker's Guide* and works of its kind present both fact and fancy in numerous forms and on equal terms—but all with the tobaccophile's pleasure and amusement foremost in mind.

The first nineteenth-century work on record in this explicitly hybrid and purportedly more creative tobacco genre is an 1832 publication by one Henry James Meller,[110] a little volume (dedicated to the Duke of Sussex, "as a trifling token of veneration for his character and esteem for his taste") entitled *Nicotiana, or The Smoker's and Snuff-Taker's Companion*, containing (as the title page continues) "The History of Tobacco; Culture—Medical Qualities and the Laws Relative to Its Importation and Manufacture: with an Essay in Its Defense. The Whole Elegantly Embellished and Interspersed with Original Poetry and Anecdotes, Being Intended as an Amusing and Instructive Volume for All Genuine Lovers of the Herb." As his subtitle and description make clear, Meller, like Steinmetz, aims to delight as well as to enlighten his fellow tobaccophile with inventive, informative prose and tobacco-inspired artistic effusions. Moreover, this purpose is corroborated by his title, *Nicotiana*, which not only names the plant's genus, but also poetically suggests a miscellaneous collection of tobacco-related items, as the suffix "-ana" indicates in words like *Victoriana* and *Dickensiana*. Finally, however, the feminine ending suitably suggests a female spirit, the "phantasm" so closely associated with the nightshade family who, as we shall see, appears in tobacco fiction at the end of the century. Consequently, *Nicotiana* is a fitting designation for the heterogeneous and artful Victorian tobacco genre Meller originates and Steinmetz's *The Smoker's Guide, Philosopher and Friend* continues. And although Steinmetz does not use tobacco's genus name in this collective sense, he nevertheless enlists the term for an artistic purpose by transforming it into a literary trope. That is, throughout *The Smoker's Guide*, Steinmetz uses "Nicotiana" to personify tobacco as a wise lady bountiful who is, among other things, a "millionaire" and "a moral, material, and Governmental or financial benefactress—well deserving the gratitude of all moral, law-abiding, and loyal people."[111] Nicotiana and her largesse preside over the text like a goddess, meting out rewards to the national economy and to private citizens alike. Just as Meller's title provides the name for a new and various, designedly creative tobacco genre, so Steinmetz's Nicotiana becomes its muse.

The literary and collective connotations of the term were not lost on the antiquarian William Bragge, who compiled the most complete bibliography of tobacco books written up to its last year of publication, 1880, titled *Bibliotheca Nicotiana*. Indeed, Bragge may have noticed the applicability of the term to categorize tobacco literature writ large, so to speak, when he discovered Meller's book, which of course appears among the entries. In any case, Bragge's bibliography, which begins with the Spanish historian Oviedo's earliest account of tobacco and contains 409 entries, epitomizes the literary heterogeneity implied by the word *Nicotiana* at its broadest and most diverse. W. A. Penn would later use "Nicotiana" in a similar fashion in *The Soverane Herbe*—as the title of a chapter that covers everything from tobacco's influence on government to cigar-smoking competitions to the unique German invention of a "kapnometer," a device designed to pipe tobacco smoke ("flavoured of the finest Havana cigar") "to every room of the house."[112]

One unique and certainly influential body of publications merits particular notice as a leader in the sphere of Nicotiana, cleverly putting their own spin on the genre: the literary productions of the Liverpool tobacco firm Cope Brothers and Company, whose shrewd contributions to Victorian advertising assisted its rise in status to the second largest tobacco company in England by 1890, outpaced only by W. D. and H. O. Wills. From March 1870 until January 1881, the firm issued the monthly periodical *Cope's Tobacco Plant*, followed by the series *Smoke Room Booklets*, published from 1889 to 1893, which, in the words of the advertising historian A. V. Seaton, recycled "highlights" from the *Tobacco Plant* "with the duller bits left out."[113] While the firm's publications were only a small part of the massive body of tobacco literature appearing in the nineteenth century, they were the first to be produced by a company *selling* tobacco, thus bringing a dimension to Victorian marketing that would be exploited to the fullest in the next century and our own. In accordance with the modern principles of advertising they pioneered, the *Tobacco Plant* and *Smoke Room Booklets* were not just selling Cope's products: they were selling smoking as a fashionable practice by both targeting and creating a literate, bourgeois audience, one whose interest in a quality product could be satisfied by Cope's goods, whether of the paper or the vegetable kind.[114]

The mastermind behind these publications was the Scotsman John Fraser, the editor and man of many enthusiasms, including book collecting, phrenology, and tobacco.[115] A fellow Scotsman named John Wallace, who occasionally used the pseudonym "George Pipeshanks" in honor of Dickens's early illustrator George Cruikshank, was the artist responsible for the magnificent posters, cards, and illustrations that helped to establish the name of Cope as a leader in quality designs.[116] Launched as the self-proclaimed "organ of tobacco use and tobacco sentiment," the *Tobacco Plant* announced its intention "to interest the SMOKER, as well as the TRADER," by offering "lighter contributions—as Stories, Verse, and Anecdote which we trust will furnish the consumer with a VADE MECUM and PASTIME-BOOK for those hours of ease in which he enjoys his favourite luxury."[117] That is, Cope's publications were marketed as upscale tobacco accessories meant to enhance the leisured experience of the smoker, in this case via the additional stimulation of literate engagement. And even though Cope's numerous brands of tobacco products were targeted at a broad spectrum of smokers—in the words of one advertisement, "for all sorts and conditions of men"—the firm's literary publications were directed more narrowly toward profiling as well as grooming a specific male type: the late-Victorian gentleman whose smoking habit literally defines him. As a devotee of tobacco, he is identifiable as a thoroughly modern sophisticate whose "favourite luxury" makes him desirable company for other similarly inclined men. As the late great authority on Victorian literature and publishing history Richard Altick explained in a 1951 article written in praise of Cope's literary endeavors, its publications played to Victorian smokers' desire to cultivate an image of themselves as a "class apart," a select fraternity whose tobacco habit reflected their general good taste and gentlemanly stature. With this genteel mission in mind, the *Tobacco Plant* evolved during its eleven years of publication from an organ of trade propaganda into a serious and respectable literary journal.[118]

The booklover Fraser cast a wide net to find items of interest, publishing articles, reviews, and essays on major literary events and figures, and reprinting selections from the works of such eminences as Thomas Carlyle and John Ruskin—despite the latter's hatred of tobacco rivaling that of King James (in 1893 Ruskin successfully sued to have the unlucky *Smoke Room Booklet* No. 13, featuring "John Ruskin on Himself and Things in General," withdrawn after its publication).[119] A major force in establishing the journal's quality were the many contributions of the well-respected if saturnine poet James Thomson (author of *The City of Dreadful Night*, a poem whose despair surpasses even *The Wasteland*), who wrote articles on everything from tobacological subjects like "Tobacco Legislation in the Three Kingdoms" to reviews on any number of literary subjects and artists.[120]

A prime example is Thomson's witty act-by-act account of Bizet's *Carmen*, the opera that not only brought cigarette smoking to the Victorian stage, but also celebrated the exotic women who worked in Seville's cigar factories—famously in the person of the tragic heroine, who became an icon of liberated and dangerous female sexuality. "Tobacco at the Opera" is a serious review of the current February 1879 production of *Carmen* with an English libretto, a follow-up to the opera's successful London premiere sung in the original Italian the previous year.[121] Nevertheless, as the title of the review indicates, Thomson doesn't forget the journal's promotional mission. Thus, when describing the opening scene in which the "cigar girls" enter "smoking cigarettes" while being watched by soldiers, Thomson wonders, "Do the ladies of Messrs. Cope's factory come to work in this style, I wonder? and do the young gallants of the neighbourhood so assemble to greet their coming?" Thomson goes on to praise the opera, crediting its greatness to both Bizet's music and Mérimée's powerful characters Don Jose and Carmen, respectively: "the intense reality of the passion and despair of the sorcery-smitten hero, the dauntlessness and savage freedom of the heroine."[122] And his wonderful conclusion, besides cleverly assessing Carmen's bad-girl, treacherous charm, manages a satirical jab at the journal's favorite nemesis, the Anti-Tobacco Society—a familiar authorial move whenever the opportunity presented itself: "In compassion for the Antis, we deduce a minatory moral from this Opera of Tobacco:—A fascinating girl, who makes cigars in Seville, and who is, moreover, a gipsy, and altogether too familiar with smugglers, and who, yet further, has more than a gipsy's share of devilry in her nature, might prove troublesome in a pious tea-meeting, and is, in fact, a rather dangerous character: and no good young man should fall in love with her, if he can help it."[123]

Despite Thomson's insinuation that the "Antis" are moralistic prudes with far less appeal than the daring, animal-spirited Carmen, the Cope's firm worked to promote its own morally upright image as a progressive Liverpool business highly cognizant of its civic and social responsibilities—especially with regard to the opera's subject of women tobacco-factory workers. An advertisement appearing in several *Smoke Room Booklets* for nine different brands of Cope's Cigarettes compares the company's working conditions to

226 VICTORIAN NIGHTSHADES

those in foreign establishments via messages around the ad's border: "Made in a Model English Factory," "Not made in the Slums of Cairo or Constantinople," "Not made in Continental Prisons," and finally, "Cope's Cigarettes are made by English Girls."[124]

Promoting Cope's image as an enlightened manufacturer and employer was also the theme of novelist and former *Sunday Times* editor Joseph Hatton's effusive essay "A Day in a Tobacco Factory," first published in the January 1892 issue of the *English Illustrated Magazine* and reprinted in the fourteenth *Smoke Room Booklet* in 1893. "Romancing the Factory" would have been a more instructive title, however, for Hatton not only paints an idyllic picture of Cope's operation, but also colors his description with fanciful allusions to contemporary drama and fiction in which the heroines make love and tobacco products. Referring to "Fairholt's famous book" written before the cigarette became "a leading feature in the English tobacco factory," Hatton laments the fact that the historian had been effectively denied such a romantic subject:

> [Fairholt] missed therefore a picturesque phase of English tobacco manufacture of the present day, and with it certain attractive associations which belong to the light and airy "smoke" of the Continent and the East. He knew the *puros* [cigars] and the *papelotos* [cigarettes] of the Spaniards, and quotes the national proverb—"A paper cigarette, a glass of fresh water, and the kiss of a pretty girl will sustain a man for a day without eating"; but "Carmen," the heroine of the French librettist, and "Vjera," the heroine of the English novelist, are of these latter days, and give a touch of romance to the atmosphere of the modern tobacco factory.[125]

As Thomson had hinted in his review, the spitfire Carmen very likely bore little resemblance to any of Cope's female workers; but a young woman like Vjera, the soft-spoken, generous-hearted, self-sacrificing heroine of the tender novella *The Cigarette-Maker's Romance* (1890) by the American (not English) expatriate Marion Crawford, might well have been among their ranks, as Hatton later acknowledges ("I dare say a Vjera might be found here") in one of his several allusions to Crawford's story. Vjera, however, was one of only two women workers in a small Munich tobacconist's shop along with three men, whereas Cope's Liverpool establishment employed "about 1500 persons, a large majority being girls."[126]

According to Hatton, the great hall where some two hundred cigarette makers work is more like a "College for Girls," each one seated at a table that looks like a desk where she deftly wraps pinches of tobacco in rice paper while quietly chatting, humming, or singing.[127] This characterization of

Cope's female employees as serious, demure, and studiously focused on their tasks is corroborated by John Wallace's several accompanying illustrations, represented here by "Packing Cigarettes," "Making Cigars," and "Stripping the Leaves" (see figs. 30, 31, and 32).[128] Their sedate clothing is certainly more suggestive of what Vjera might wear and a far cry from the famously revealing dress—or lack of it—of the Seville "cigar girls" like Carmen, who were known to strip to their underwear when at work in the stifling factory.[129]

Hatton doesn't limit his day in the factory to admiring the female workers, however. Like Fairholt before him, he covers every aspect of manufacturing tobacco, from the men breaking open the hogsheads to the packaging of Cope's goods, which are at last carted away to fulfill, as Steinmetz might have said, Nicotiana's sacred and familiar mission: "to be a solace and comfort to rich and poor, to the latter sometimes meat and drink and the former 'a luxury beyond price.'"[130] Finally, in the essay's concluding paragraph Hatton reprises his romantic theme by describing the wholesome goodness of Cope's

FIGURE 30. "Packing Cigarettes," John Wallace, from Joseph Hatton, "A Day in a Tobacco Factory," *Cope's Smoke Room Booklet*, No. 14: 45. (Digitized by Henry Hughes)

FIGURE 31. "Making Cigars," John Wallace, from Joseph Hatton, "A Day in a Tobacco Factory," *Cope's Smoke Room Booklet*, No. 14: 56. (Digitized by Henry Hughes)

products in lyrical terms that recall his characterization of the firm's female workers: "In the manufacture of tobacco there is nothing that might deter the smoker. Every process through which it passes is cleanly; it goes through a course of purification which, in itself, might justify much of the eulogium of its lovers, the blackest plug the sailor chooses to chew or smoke being as sweet and pure as the finest leaf as it emerges from knife and press a golden network of imprisoned dreams."[131] It seems reasonable to infer from Hatton's words that Cope's tobacco products have somehow assumed their sweetness and purity from the virtuous qualities of the English girls who made them. But whether arising from the nature of the plant or created by some alchemical transference with the workers themselves, the resulting products are wholly imbued with romance—for their "lovers" to release at last as "a golden network of imprisoned dreams."

Hatton's allusions to contemporary fiction make clear that tobacco's appearance in the lighter literature of the Victorian period was by no means limited to the promotional publications of the Cope's brothers or tobacco-themed entertainments like those of Meller and Steinmetz. The overwhelming number of references to tobacco in nineteenth-century fiction provide such a detailed picture of contemporary trends in its reception and use that mid-twentieth- and twenty-first-century tobaccologists Compton Mackenzie, Hugh Cockerell, Richard Altick, Matthew Hilton, and Iain Gately have canonized several works as textbook cases for studying tobacco's prominent, if controversial, place in Victorian society.[132] Dickens, Thackeray, Trollope, and Conan Doyle are the four major writers of fiction who have received the most attention, with Dickens and Thackeray vying for top honors as the best sources for revealing how class, age, location, and occupation influenced tobacco's acceptance and served to determine the tobaccophile's choice of delivery system at any given moment during the period. Possibly the most famous

FIGURE 32. "Stripping the Leaves," John Wallace, from Joseph Hatton, "A Day in a Tobacco Factory," *Cope's Smoke Room Booklet*, No. 14: 58. (Digitized by Henry Hughes)

smoker in British fiction comes at the end of the century: Conan Doyle's legendary gentleman-detective Sherlock Holmes, whose "three-pipe problem" in "The Red-Headed League," first published in 1891, has since become proverbial.[133] But earlier references to all modes of tobacco use are so pervasive in novels and stories that an adequate analysis is a daunting and indeed impossible task, so that the following admittedly partial effort merely touches on all the tobacco themes addressed even by the authors I omit.

An appropriate starting place is Thackeray's *Vanity Fair*, whose setting shortly before, during, and some fifteen years after the Battle of Waterloo in 1815 memorializes the new fashion of cigar smoking popularized by British officers returning from the Peninsular War. All of Thackeray's officers smoke cigars, which occasionally serve as character-revealing props, most tellingly with regard to the novel's courtship rituals: such is the case when early in their relationship, Rawdon Crawley shares his cigar with a very game Becky Sharpe, who "loved the smell of a cigar out of doors beyond everything in the world"; and when George Osborne nonchalantly burns a love letter from his fiancée, Amelia Sedley, for a light. In an episode worthy of Freud or Groucho Marx, Thackeray makes the sexual implications obvious when, after the respective couples have married, the ever-flirtatious Becky lights Osborne's cigar, for "she knew the effect of that manoeuvre, having practiced it in former days upon Rawdon Crawley."[134] Thackeray may well have inspired the best love scene in *Jane Eyre*—the novel Charlotte Brontë published during *Vanity Fair*'s serialization in 1847–48 and dedicated to its author—in which the straitlaced heroine fondly recalls the "subtle, well-known scent" of Mr. Rochester's cigar as a "perfume" more exciting than the odors of "sweet-briar and southernwood, jasmine, pink, and rose" wafting through the garden on a romantic midsummer evening.[135]

Charles Dickens's fiction was even more popular than Thackeray's, and his first novel, *Pickwick Papers*, published serially from 1836 to 1837, conveniently picks up nineteenth-century tobacco history where *Vanity Fair* leaves off. That is, the adventures contained in the club's allegedly "posthumous" papers begin in May 1827, when thanks to its relative novelty and perhaps its association with military prowess, cigar smoking had considerable cachet among young men about town and their less affluent imitators. That it was still regarded as a slightly disreputable habit can be seen in the case of the medical student Bob Sawyer, who "had about him that sort of slovenly smartness, and swaggering gait, which is peculiar to young gentlemen who smoke in the streets by day, shout and scream in the same by night, call waiters by their Christian names, and do various other acts and deeds of an equally facetious description."[136] Sawyer, his friend Ben Allen, and the other medical students of their acquaintance are all confirmed cigar smokers, but its aficionados come in all forms, from the Bath footman who smokes his cigar through an amber tube, to the seedy inmates of the sponging house and the Fleet debtors' prison that Pickwick meets when he is incarcerated.[137] And although that kindly bon vivant acknowledges that he "is no smoker [him]self," he claims to "like it very much," and he spends most of his time in the company of tobacco users of every description, since they congregate in the smoky inns and pubs frequented by the Pickwickians.[138]

The novel's most memorable tobaccophile is the father of Pickwick's devoted servant Sam, the coachman Tony Weller, who, like the other characters employed in transporting people and goods, the cabman and bagman, smokes a pipe—in Weller's case, at every opportunity and "with great vehemence."[139] In Hablot K. Browne's (aka "Phiz") illustration entitled "The Valentine," the rotund elder Weller unforgettably stands in front of the fire in the parlor of the Blue Boar smoking a churchwarden as he advises Sam about how to compose a love letter.[140] There are a few snuffers in the novel—most notably, Pickwick's solicitor Perker—but smoking, whether of pipe or cigar, predominates.[141] In contrast to Thackeray's emphasis on upper-class manners, however, Dickens focuses on the vast middle and lower classes of English society in the late 1820s, and the haunts and habits of this social world are rife with the smoke of tobacco. No wonder, then, that cheap cigars marketed on the streets of London in the forties were known as Penny Pickwicks.[142]

With the exception of his historical novels, Dickens usually left to inference the specific dates covered in his fictions as a way to underscore the immediacy of the social problems he addresses; but for the most part the narratives span the late twenties into the forties, the period during which snuff taking had become decidedly old-fashioned and cigars were still more or less outré—the choice of "dandies," "swells," and "gents," who were, according to Richard Altick, the raffish male representatives of the upper, middle, and lower classes, respectively.[143] Exemplary representatives in the snuffing category include that notoriously slattern nurse-midwife Sairey Gamp in *Martin Chuzzlewit*, and pretentious old Mr. Turveydrop in *Bleak House*, who models his "deportment" and his tobacco habit on his royal "patron" the late Prince Regent.[144] Cigar smokers include the dissolute dandies Lord Verisopht and Sir Mulberry Hawk in *Nicholas Nickleby*; David Copperfield's friend and perfect swell James Steerforth; and *Bleak House's* street-smart gent Bart Smallweed, who "drinks and smokes in a monkeyish way," and whose "passion for a lady at a cigar-shop in the neighbourhood of Chancery Lane" no doubt began with a purchase made to satisfy the latter habit.[145] Foremost among Dickens's most memorable smokers is the evil dwarf Daniel Quilp in *The*

Old Curiosity Shop, who weaponizes his cigars by forcing his wife to sit up with him all night as he chain smokes and drinks grog.[146] In *Little Dorrit,* the eponymous heroine's proud father, long imprisoned in the Marshalsea debtors' prison, accepts and smokes cigars presented as "testimonials" from the turnkey's son John Chivery, whose father also runs a small tobacconist's shop nearby—which, we are told, was of "too modest a character to support a life-size Highlander, but it maintained a little one on a bracket on the doorpost, who looked like a fallen Cherub that had found it necessary to take to a kilt."[147] Equally unforgettable is the swell Eugene Wrayburn in *Our Mutual Friend,* whose ever-present cigar is a symbol of his arrogant indifference to everything except tobacco, an addiction he seems to renounce at last when he is redeemed by the love of Lizzie Hexam.[148]

It was left for Dickens's competitor Anthony Trollope, however, to describe one of the cigar divans that became popular in the thirties to accommodate urban gentlemen who were too polite to smoke in the streets or who lacked access to other cigar-friendly environs. Meller devotes a chapter in *Nicotiana* to these retreats, naming the six most popular in London, such as the "Oriental Divan" in Regent Street, and the "Royal Divan" in the Strand. They were comfortable lounges known for their Persian décor that provided male customers "quiet and elegant seclusion," along with coffee and newspapers to sip and peruse, respectively, while enjoying a peaceful smoke, which could be purchased in an adjoining cigar shop.[149] Thus in the first Barchester novel, *The Warden* (1855), the unworldly Septimus Harding, forced to kill time in London, happily discovers one of these havens that despite its "strong smell of tobacco, to which he was not accustomed," seemed like "a paradise" with its many books and sofas. Harding gives away the cigar required for admission, drinks his coffee, and finds the surroundings so relaxing that he inadvertently falls asleep.[150]

Even as the cigar continued to gain popularity, pipe smoking remained the default mode of tobacco use among the working classes at mid-century and beyond; and pipe smokers of every trade proliferate in Dickens's novels, from Fagin's pack of delinquent boy pickpockets in *Oliver Twist* to the "Golden Dustman" Noddy Boffin, appropriately so-called for the fortune in dust heaps he inherits from his employer in *Our Mutual Friend.*[151] Taking Dickens's most autobiographical novel *David Copperfield* for example, those pillars of Yarmouth commerce—the fisherman Mr. Peggoty and the undertaker Mr. Omer—rarely appear without their pipes, while the taciturn carrier Mr. Barkis is a model of contentment as he "philosophically smoked his pipe" after his long-desired marriage to Peggoty's sister, David's old nurse.[152] *Dombey and Son's* two most committed pipe smokers are first, the retired

SUBLIME TOBACCO

233

sailor Captain Cuttle, shopkeeper of a nautical instrument store; and second, the beggar-thief-fortune-teller "good Mrs. Brown," who constantly puffs at a "short black pipe" to become Dickens's only female character of the smoking persuasion.[153] In *Bleak House,* the former soldiers-turned-shopkeepers George Rouncewell and Matthew Bagnet enjoy a convivial pipe after dinner at the Bagnets', where on one important occasion they are joined by the shrewd but amiable detective Mr. Bucket.[154] The brickmaker in St. Albans is the novel's most poverty-stricken pipe-smoking worker.[155] *Great Expectations'* blacksmith Joe Gargery likes to take his pipe at the Three Jolly Bargemen pub, whereas Mr. Jaggers's stiff clerk Wemmick indulges only when he can relax at home in Walworth. And Pip's benefactor, the convict Magwitch, frequently takes comfort in his "short black pipe" filled with the mixture known as "negro-head," made of twisted leaves steeped in molasses.[156]

"Eastern pipes" appear in *Hard Times* and *Little Dorrit*—in the first case undoubtedly containing some opium, although euphemized as the "rare tobacco" the cad James Harthouse shares with Tom Gradgrind in order to pump him for information about Tom's sister Louisa. It is less clear in *Little Dorrit* what goes into the "Eastern pipes" Arthur Clennam, a former resident of China for twenty years, provides for himself and Mr. Pancks; however, since the session results in the two men making disastrous investments, it seems likely to have been more—or other than—tobacco.[157] In his final unfinished novel, *The Mystery of Edwin Drood,* Dickens specifies that it is "not with tobacco" that the Cloisterham choirmaster John Jasper fills his "peculiar-looking pipe," a fact corroborated later when Jasper visits the London opium den run by Princess Puffer, who makes her pipes from "old penny ink bottles."[158] These references indicate that glass pipes and hookahs were certainly exotic commodities, and evidently less likely to be used for tobacco than for opium or perhaps cannabis.

Also less prevalent in England were the "two odious practices of chewing and expectorating" tobacco, which had so grossly offended Dickens when he visited the United States in 1842 and subsequently disparaged in his travel book *American Notes for General Circulation,* published the same year.[159] *Martin Chuzzlewit* draws from this experience when the younger of the novel's two Martins makes his own trip to the States and encounters the "War Correspondent" for the *Rowdy Journal,* one Jefferson Brick, who is "unwholesomely pale" in part "no doubt, from the excessive use of tobacco, which he was at that moment chewing vigorously." Even more objectionable is Major Pawkins, who spends his time in a rocking chair, chewing and discharging nicotine-infused effluvia alternately between two brass spittoons.[160]

Finally, during Dickens's lifetime, cigarette smoking was primarily the habit of foreigners, as he records in *Bleak House* when describing the Spanish refugees living in Somers Town who smoked "little paper cigars."[161] Several years later in *Little Dorrit*, the cellmates of a Marseilles prison—the dangerous Swiss-French "citizen of the world" Rigaud and the good-hearted Italian Cavaletto—smoke cigarettes with tobacco rolled by Rigaud in "little squares of paper." From this point on in the novel, the mention of "cigarette" inevitably indicates that the villainous Rigaud is at hand, regardless of the country he is in or the alias he has assumed.[162] In the mid-forties, Dickens himself had spent several months living and traveling in Italy, Switzerland, and France— the three countries represented by the two prisoners—during which time he smoked cigarettes: "good large ones, made of pretty strong tobacco," he reported in a letter to his biographer and friend John Forster.[163] The letter, written from Switzerland in the fall of 1846, also recounts what must have been Dickens's most unusual smoking experience, dubbed by Forster a "Feminine Smoking Party" that took place in a Geneva hotel, during which two English and two American ladies shared their cigars and cigarettes with Dickens as they themselves smoked incessantly. Dickens described the younger Englishwoman, "Lady B," as he called her, leaning nonchalantly against a mantel, "her cigarette smoking away like a Manchester cotton mill." Dickens concludes, "I showed no atom of surprise; but I never *was* so surprised, so ridiculously taken aback, in my life; for in all my experience of 'ladies' of one kind and another, I never saw a woman—not a basket woman or a gypsy— smoke, before!"[164] Interestingly, it was in *Dombey and Son*, the novel Dickens was currently writing, that the pipe-smoking crone Mrs. Brown makes her appearance.

Known as something of a swell himself, Dickens smoked cigars regularly at home and in social gatherings; and in his later years, his biographer Fred Kaplan records that according to one of his amanuenses, Dickens "smoked cigarettes insatiably" at work.[165] By the year of his death, 1870, cigarettes were becoming increasingly popular among British men, and smoking was rumored to be gaining interest as well among British women—an occurrence generally viewed as a threat to conventional ideas about gender. *Punch* had certainly fostered the rumor as early as 1851, in cartoons satirizing American women's rights advocates who visited London wearing "bloomers"—the puffy pants named for Amelia Bloomer, editor of the women's temperance journal *The Lily*, who had promoted their wear. The *Punch* artists John Leech and John Tenniel both produced cartoons picturing women dressed in the new fashion and smoking cigars or cheroots—their bloomers and tobacco use sending the same message about "Woman's Emancipation," as the fictitious proto-feminist and "strong-minded American Woman" Theodosia Eudoxia Bang of Boston

FIGURE 33. "Woman's Emancipation," John Tenniel, *Punch*, vol. 21 (1851): 3. (Courtesy of HathiTrust; contributed by Princeton University)

proclaims in her letter to the magazine: "We are emancipating ourselves, among other badges of the slavery of feudalism, from the inconvenient dress of the European female. With man's functions we have asserted our right to his garb, and especially to that part of it which invests the lower extremities. With this great symbol we have adopted others—the hat, the cigar, the paletot, or round jacket." The woman in the center of Tenniel's accompanying illustration effectively conveys all these "symbols," as she lights up with the brash confidence of a swell (see fig. 33).[166] Leech produced a whole series of drawings on "Bloomerism," many of them also featuring be-trousered women smoking cigars. The two best are "Bloomerism—an American Custom" and "Something More Apropos of Bloomerism" (the latter picturing a male shopkeeper referred to as "one of the 'inferior animals'") (see figs. 34 and 35).[167]

In 1867 *Punch*'s caricatures of liberated women were brought to fictional life in the immensely popular adventure novel *Under Two Flags* by Ouida (pen name of Mary Louise Ramé). Three of the influential women in the life of the dashing hero Bertie Cecil are bold and unapologetic smokers, all of whom challenge Victorian codes of proper female behavior. The first two are Bertie's lovers: "the Zu-Zu," a very pretty but coarse ballet dancer who loves cigars and is "wild" to try "rose-scented *papelitos*"; and her married rival the aristocratic Lady Guenevere, "a coquette who would smoke a cigarette, yet a peeress who would never lose her dignity."[168] A prolific smoker himself, Bertie manages his romantic affairs with the consummate athletic skill he displays on horseback; but he cares more for tobacco than either woman, and his addiction is a frequent subject in the first part of the novel. As Ouida tells us, "Indeed, it took something as tremendous as divorce from all forms of smoking for five hours to make an impression on Bertie."[169] Nevertheless, despite his chronic "insouciance" in other matters,[170] Bertie is at heart a man of honor, who joins the French Foreign Legion in Algeria to save Lady Guenevere's reputation.

There he meets the novel's third female smoker and its most interesting character, a young vivandière, or supplier of wine and other provisions for the French army, who (to speak in the sort of similes that Ouida loves to discharge) dances like a Bacchante, rides like an Arab, curses like a sailor, drinks like a fish, and smokes like a chimney. Her name—"Cigarette"—indicates that she is intended to be the human embodiment of the cigarette's many risqué associations, including, most importantly, the gender-bending that all but the first of the above similes suggest. Yet even though Cigarette is

FIGURE 34. "Bloomerism—an American Custom," John Leech, *Punch*, vol. 21 (1851): 141. (Courtesy of HathiTrust; contributed by Princeton University)

FIGURE 35. "Something More Apropos of Bloomerism," John Leech, *Punch*, vol. 21 (1851): 196. (Courtesy of HathiTrust; contributed by Princeton University)

"more like a handsome saucy boy,"[171] Ouida's efforts to establish her character's essential femininity as well as her heroism (Cigarette dies protecting Bertie from a firing squad) reveal the author's own investment in defying Victorian strictures regarding female behavior, a trait satirized in 1881 by the *Punch* cartoonist Linley Sambourne in his famous "Fancy Portraits" series (see fig. 36).[172] The caption from *Hamlet*, "Oh, fie! 'Tis an unweeded garden," recalls Sir Francis Burnand's earlier parody in *Punch* of "Weeder's" 1865 novel *Strathmore*, whose eponymous hero is a forerunner of Bertie Cecil.[173] The cartoon features allusions to specific Ouida novels (i.e., *Two Little Wooden Shoes*, *A Dog of Flanders*, and *Strathmore*), but the hookah in the foreground and the cigarette in Ouida's hand could refer to several of her works, including *Under Two Flags*, and serve as a reminder that the novelist was an addicted "weeder" herself. Even so, Cigarette earns special status in Ouida's novels as the only example of the weed made flesh, and the smoke rings wafting upward in the cartoon could well represent the apotheosis of the heroic little vivandière.

OUIDA.

"O fie! 'tis an unweeded garden."—*Hamlet*, Act I., Scene 2.

FIGURE 36. "Punch's Fancy Portraits—No. 45. Ouida," Linley Sambourne, *Punch*, vol. 81 (1881): 83. (Courtesy of HathiTrust; contributed by Princeton University)

It seems clear that Ouida's character exposed an important truth about her namesake: the growing popularity of the cigarette among both genders from mid-century on intensified the anxiety about the relationship of tobacco to sexuality. For example, in *The Smoker's Guide, Philosopher and Friend*, Steinmetz felt called upon to assert, "There is nothing manly in the cigarette," and to repeat a familiar view certainly applicable to Bertie Cecil: "Fops and dandies prefer cigarettes, to please the ladies in general, and the pouting little pretty ones in particular." However, Steinmetz goes on to say, "But we are neither fops nor dandies."[174] Yet despite Steinmetz's endorsement of received opinions regarding smoking and masculinity, in a chapter entitled "Tobacco and the Fair Sex" he acknowledges that gender attitudes and practices were changing: "But whilst many (or most) wives, sweethearts, et cetera will now-a-days concede the gentle liberty of smoking in their pleasant presence, some of the fair ones have themselves taken to the cigarette and cigar. Nor does this appear to be in the least unbecoming, for not long ago, Sir James Hannen (of the divorce court) declared that he knew several respectable ladies who smoked." Steinmetz explains further that the prominent jurist's observation came in response to a "remarkable" case wherein a husband sued his wife for divorce in part because she smoked in bed.[175]

Nevertheless, it is not surprising that Steinmetz's insistent personification of Nicotiana as a woman throughout the *Smoker's Guide* implicitly reestablishes the love of tobacco as a heterosexual romance and a masculine privilege, a strategy employed in the most famous book in the nicotian genre to appear during Victoria's reign, *My Lady Nicotine*. First published in 1890 and written by J. M. Barrie, best known as the author of *Peter Pan*, the book is a collection of essays originally written for the *St. James Gazette*, rather loosely held together as a pseudo-autobiographical retrospective about the author's smoking days before allegedly giving up the habit upon—and as a condition of—his marriage (Barrie, in fact, did not marry until 1894). And although the first chapter, "Tobacco and Matrimony Compared," presents his rationale for preferring a wife to his beloved brier ("the sweetest ever known"), its tongue-in-cheek tone belies this decision, as does the celebration of tobacco that is the theme of the book. The chapter's final sentence, "This is the book of my dreams," thus frames Barrie's story of his bachelor smoking experiences as a nostalgic remembrance of a love affair, one filled with a sense of profound loss and irrecoverable pleasure, seriously calling into question the idea that happiness can exist in a life without tobacco.[176]

Barrie's account focuses on his relationship with his five closest friends and fellow tobaccophiles as they regularly gather in his rooms in the evening to smoke pipes filled with their favorite "Arcadia Mixture," whose fictional name underscores its association with an idyllic but imaginary paradise—a place accessible only in a vision or dream. Although enjoyed in the company of other enthusiasts, the smoking experience, by its nature necessarily individual, is similarly dreamlike; and these evenings are typically passed in silence, each participant fully immersed in his own pleasurable but private sensations. As the narrator says, "No one who smokes the Arcadia would ever attempt to describe its delights, for his pipe would be certain to go out."[177] Thus the emphasis of the book is on the bond created by the smokers' connoisseurship, their mutual if tacit appreciation for the superior quality of their tobacco and all its accoutrements: their briers, meerschaums, and clays; occasional cigars

FIGURE 37. Lady in smoke, M. B. Prendergast, from J. M. Barrie, *My Lady Nicotine*, chap. 32: 265. (Courtesy of Hathi-Trust; contributed by Harvard University)

and cigarettes; ancillary pouches, tins, jars, and smoking tables; and lights, their vestas and spills.

Given the very private nature of Barrie's love affair with tobacco, it is understandable that direct references to its persona "My Lady Nicotine" are left unspoken through most of the book; however, in the penultimate chapter, "My Last Pipe," which describes the funereal occasion marking this tragic event, the lady appears briefly, but inevitably, as a figment of Barrie's imagination. While his friends gather round to watch in silence as he takes his final drafts, Barrie recalls her fleeting apparition: "As Jimmy and the others saw only me, I tried not to see only them. I conjured up the face of a lady, and she smiled encouragingly, and then I felt safer. But at times her face was lost in smoke, or suddenly it was Marriot's face, eager, doleful, wistful."[178] Like a genie released by the lighting of a pipe, My Lady Nicotine briefly materializes into a dream come true; but when her spectral existence, transitory as smoke itself, vanishes into thin air, the smoker is wracked by desire, longing for the satisfaction tobacco provides once more.

Barrie's pipe dream of My Lady Nicotine serves as an appropriate concluding image for the story of tobacco in the nineteenth century because it verbalizes what had become a commonplace in late-Victorian nicotian literature: the passion for tobacco realized as a love affair with a beautiful woman. Moreover, this passion was frequently portrayed as smoke ascending and transmogrifying into a woman's shape; and unsurprisingly, such visualizations occur in

FIGURE 38. John Wallace, title page, *The Smoker's Garland*, part 1, Cope's Smoke Room Booklet, No. 10. (Digitized by Henry Hughes)

FIGURE 39. John Wallace, frontispiece, *The Smoker's Garland*, part 3, Cope's Smoke Room Booklet, No. 10. (Digitized by Henry Hughes)

My Lady Nicotine, with one in "My Last Pipe" to accompany Barrie's final vision (see fig. 37).[179] Copiously illustrated by M. B. Prendergast in an 1896 edition, *My Lady Nicotine* is filled with pictures of females, many of whom obviously represent the women in the failed romances of Barrie's comrades recorded in the story, one of whom even smokes a cigarette. But Prendergast distinguishes My Lady Nicotine from her rivals, first, in the illustration on the title page and elsewhere, via the box labeled "Arcadia Mixture" that she holds in her hands; and second, more generically, by her image shaped in smoke emanating from a pipe as in figure 37.

Comparable visualizations of this image can be found in John Wallace's contemporary illustrations for Cope's *Smoke Room Booklets*. On each title page of *The Smoker's Garland*, a three-part series containing poems and short

SUBLIME TOBACCO

pieces inspired by tobacco, an ethereal woman with a lyre rises in smoke from a Tyrolean-style pipe; and the frontispiece for part 3 similarly depicts a starry-crowned, spectral female ascending in smoke from the pipe of a native who cradles a flesh-and-blood woman in his lap (figs. 38 and 39).[180] The frontispiece refers to the volume's final entry, a legend recounting "Why Women Hate the Weed," whose gender politics mirror those of *My Lady Nicotine*. That is, the story, "a New Zealand myth" adapted from Andrew Lang's *Ballades of Blue China* (1880) and narrated in a tongue-in-cheek voice not unlike Barrie's, tells how tobacco comes as Nature's consolation prize to man after he has been made unhappy by the wife he had requested from the gods.[181] "Nicotia," first appearing as a woman "fair and sweet and kind," but dying after a day's dalliance, promises to return as an herb whose burning leaves will produce "an exceeding grateful vapour," capable of assuaging "all trouble and sorrow."[182] Thus, the myth recalls Barrie's dilemma of being torn between desire for a pipe and a wife, but reverses the events of his narrative: the spirit-woman of tobacco consoles *after* marriage, as a secret illicit passion satisfied by smoke.

This most celebrated image of *Nicotiana tabacum* recalls the "phantasms of beautiful women" that *Atropa belladonna*, its famously alluring but deadly relative, is fabled to produce, thereby bringing the family reputation full circle. Labeled "the most popular plant in the world" at mid-century, tobacco had truly become a global obsession by the century's end, and England's love affair with smoking figured as a central romance of the Victorian era—truly, as Altick says, it was "the favorite vice of the nineteenth century." Fifty years after the period's close, tobacco had won such devotion from its citizens that 80 percent of adult British males and 40 percent of adult females would be smokers,[183] many seriously addicted to this deeply comforting but most dangerous plant, the Solanaceae species that wholly embodies the nightshades' fatally seductive beauty.

8

Back to the Garden

Petunias, Peppers, Eggplants, and Tomatoes

> The Tomato, or Love Apple, *Lycopersicon esculentum*, has of late years passed from the rich man's garden, in which it was exclusively located, and has become a citizen of the world, the welcome guest of every household, and the subject of the poor man's special care. And no wonder, for like its cousin, the potato, it is so thoroughly useful and so accommodating in respect of the cultivation it requires, that it suits every garden and every table, and, strange to say, gratifies every taste, for it was never heard that, when fairly presented, this noble fruit displeased anyone.
>
> —Shirley Hibberd, *The Amateur's Kitchen Garden* (1877)

Bedding Out

The final chapter of the nightshades' Victorian story returns to the English garden, where four eminently friendly Solanaceae introductions would become defining features by the era's end. And once again, the Crystal Palace serves as the starting place. When the lavish reconstruction of the first Palace opened in the south London suburb of Sydenham in the summer of 1854, Joseph Paxton's post-Exhibition dream of establishing a permanent indoor "Winter Park and Garden" not only seemed to become a reality; the new site's vast outdoor pleasure ground rivaled the grandeur of the Palace itself. Among its delights were formal gardens featuring terraces and promenades whose borders and parterres were filled with tender, non-native flowers raised in the onsite greenhouses before being planted outside in solid masses each of a single vivid color.[1] In fact, by exposing the middle and especially the lower classes to this horticultural feast, Paxton's Sydenham gardens effected a revolutionary change in Victorian floriculture.[2] As "bedding out" became the rage among amateur gardeners, the petunia, an already admired ornamental nightshade, became one of the top bedders in the nation, leading to the development of countless varieties from just two species that had arrived in England only within the last thirty years.

Prompted by the wave of several brightly colored flowering plants introduced from warmer climates around the world, the practice of starting "half-hardy" species indoors until they could be transplanted outside in patterned arrangements during their blooming period had begun in the early thirties on the large country estates of the wealthy, properties possessed of greenhouses and gardeners with crews of laborers;[3] but the expense, effort, and space both indoors and out required to create and maintain huge seasonal beds made them prohibitive for small gardens. Except for a few public places like the Royal Botanic Gardens at Kew, such elaborate floral displays were the exclusive province of the rich, to be enjoyed only in their private havens, inaccessible to most of the Victorian population.

But just as the Sydenham Crystal Palace was promoted as "The Palace of the People," so its extensive grounds could lay claim to being "the people's own garden," where the public could make intimate contact with fashionable horticultural species and develop a taste for growing them as well. This democratic idea was fostered by the gardener-journalist and currently the reigning authority on bedding, Donald Beaton, floriculture expert and contributing editor to the *Cottage Gardener and Country Gentleman's Companion*, who in reporting his two-day mid-September visit to the Palace in 1854, explained that seasonal beds of colorful plants could be adapted to even the tiniest plot. Describing the Palace terraces and grounds, Beaton explained that its "whole flower-garden is laid out in the promenade system, that is, with all the flower-beds alongside the walks"—a scheme not dependent on the garden's size. In Beaton's words, "The principle can be applied, and the very shape of the beds too, *in any space whatever.*"[4] In short, while not everyone could have magnificently landscaped grounds with trees, shrubs, statues, fountains, and a lake like the Crystal Palace Park, anyone could grow the Park's delightful flowers and emulate, if on a miniature scale, the fashion of bedding out.

According to the most reliable sources, the garden historians Brent Elliott and David Stuart, there were four leading genera of bedding plants,[5] two of which, scarlet geraniums (i.e., pelargoniums) and calceolarias, filled the largest Park beds with great swaths of red and yellow flowers, choices that reflected the fashion's emphasis on primary colors. Nevertheless, the two other leading genera for bedding—petunias and verbenas—were vital for enlarging the floral spectrum by offering a wide selection of pinks, lavenders, purples, and white to the palette, while lobelias, which added blue as the third primary, were among the also-rans. Besides color, the common virtue of all these genera was their long blooming period, which ensured that once the plants were in place outside, the beds would be continuously full of flowers all summer, from late May through September.

Because the many floral options available could sometimes result in color patterns that were more glaring than attractive, Donald Beaton advocated planting center beds in parterres and other strategic areas with white flowers to neutralize the contrasts, a scheme that not only softened the overall effect but also opened the space—making even the smallest garden appear larger than it was.[6] For this purpose he found petunias to be invaluable, especially a cultivar he had bred himself, the White Shrubland petunia, named after the Suffolk country estate of Sir William Middleton where Beaton had served as head gardener. Here at the Crystal Palace, this petunia had been renamed the "Royal White," probably, as Beaton surmised, because it was procured from the Royal Botanic Gardens at Kew, where it had been successfully grown for the past six years.[7] The provenance of the name notwithstanding, by 1854 the White Shrubland and other select petunia varieties had clearly earned this recently bestowed royal status. In the thirty years since its introduction to England, this South American nightshade had become an increasingly desirable addition to ornamental gardens throughout the country; but its regal presence in the Crystal Palace Park marked the beginning of a decade-long ascendancy as one of England's most prominent bedding flowers. Like its immigrant relatives from the New World, the potato and tobacco, the petunia became a ubiquitous feature of British life during the High Victorian period; and as the only major member of the Solanaceae family to reach England during the nineteenth century, it must be considered in a horticultural sense the most exclusively *Victorian* nightshade, a South American weed welcomed and brought to bed as a universal garden sweetheart, to be found not only in the greenhouses, conservatories, and parterres of the wealthy, but also in the window boxes, hanging baskets, pots, and flower plots of rich and poor alike.

Beaton's White Shrubland was a descendant of the so-called common white petunia, as it was considered in England by mid-century. First known in Europe after the French naturalist Philibert Commerson discovered it along the Rio de la Plata in present-day Uruguay in the latter eighteenth century, the white petunia had originally been classified as a *Nicotiana* by Jean-Baptiste Lamarck. It was given its own genus *Petunia* in 1803 by Antoine de Jussieu, father of the natural system of botanical classification, who adapted the name from the South American native word *petun* for tobacco, the nightshade which, as both genus names indicate, it closely resembles.[8]

Although initially designated by a variety of confusing synonyms, *Petunia nyctaginiflora* had been greeted with universal enthusiasm by botanists for its large white flower and delightful fragrance when it arrived in England in the early 1820s; but it wasn't until the1830s, after the Scots gardener/plant-hunter James Tweedie sent seeds of a purple petunia from Brazil to the Glasgow Botanic Garden, that cross-breeding in Britain began in earnest—and with it, the Victorians' love affair with the Solanaceae family's most popular ornamental genus.

The buzz about the petunia as a promising bedder was well documented in the gardening press upon its arrival. When the horticultural eminence himself, Joseph Paxton, launched his new periodical *Paxton's Magazine of Botany* in 1834, he chose Tweedie's purple *Petunia violacea* as one of four new garden species worthy of a colored plate. This impressive introduction was accompanied by superlative praise: "There are few plants in our gardens which surpass this in brilliancy of blossoms and general beauty." Paxton reported that a whole bed planted with purple petunias had achieved a "very splendid appearance" the previous summer in the Chatsworth gardens, and he particularly recommended this method of mass bedding for showing off its flowers.[9] It didn't take long for the mating to begin; and in the magazine's next volume, the older white and the new purple species' hybrid offspring, identified as *Petunia nyctaginiflora violacea*, made its proud appearance (see fig. 40). Paxton reported that this new purple variety, which was "quite hardy, and a very desirable plant," had been raised at Chatsworth; but other plant breeders had simultaneously produced cultivars of the two petunia species, which "could be purchased for a moderate price at almost every nursery around London, and in many other places." As a case in point, the accompanying illustration came from a plant in the Botanical Garden in Manchester.[10]

In the meantime, the white *Petunia nyctaginiflora* and purple *P. phoenicea* (as it was also known) were garnering high praise from Paxton's rival horticultural authority John Loudon, who called the two species "some of the most splendid ornaments of the flower garden" in the new 1835 edition of his *Encyclopedia of Gardening,* unquestionably the era's definitive reference on the subject.[11] And when in 1840 Loudon's wife, Jane, published one of her first books on horticulture, *Instructions in Gardening for Ladies,* petunias appeared as favorites among the recently fashionable additions to the ever-increasing stock of Victorian ornamental plants. Touting petunias as having qualities reminiscent of docile domestic pets, Mrs. Loudon (as she was known) noted how easily they propagated, either by seed or cuttings; how quickly they struck root; and how they hybridized "freely with each other." These obedient habits led to her enthusiastic recommendation: "Petunias may be grown

Figure 40. *Petunia nyctaginiflora violacea*, Joseph Paxton, *Paxton's Magazine of Botany* 2 (1836): 173. (Image from Biodiversity Heritage Library; contributed by Smithsonian Libraries and Archives)

in any good garden soil; and require no particular attention as to watering, etc. In fact, they are, perhaps, the best of all plants for a lady to cultivate; as they will afford a great deal of interest and amusement, with the least possible amount of trouble."[12] With promotion like this, petunias were sure to become desirable additions to the gardens of the Victorian genteel; and in the revised and enlarged edition, *Gardening for Ladies and Companion to the Flower Garden* (1843), Mrs. Loudon notes the speed with which the petunias' popularity had spread: "Perhaps no plants have made a greater revolution in horticulture than the Petunias. Only a few years ago they were relatively unknown, and now there is not a garden, or even a window, that can boast of flowers at all, without one."[13] As if to validate Mrs. Loudon's claims, the petunia began appearing in "the Language of Flowers" with the meaning "Your presence soothes me," a floral persona reflecting its obedient, petlike qualities.[14]

The testimonials of Paxton, the Loudons, and other horticultural heavy-weights no doubt had their effect, and the petunias' regular appearance in articles, advertisements, and correspondence in Britain's leading horticultural journal, the *Gardeners' Chronicle*, from its inception in 1841 gives proof of how rapidly this nightshade genus had become a permanent fixture of Victorian floriculture. This popularity was aided of course by the cooperation of the petunias themselves, their adaptability matched only by their willingness to breed like rabbits. With their soft stems they could be fastened down to create a uniform surface of blossoms for a bed or a border, or just as easily trained to a trellis. They were also useful for hanging baskets or pots for winter displays in a greenhouse. Yet a greenhouse wasn't necessary, for though they could not tolerate frost, petunias didn't require any special equipment or other accommodation when grown as annuals. As Mrs. Loudon said, all species were "nearly hardy," so that it was even possible to sow seeds outdoors in March or April if planted in a sheltered area, protected simply by putting "a few dead leaves over the bed if the weather should be severe."[15]

By mid-century, petunia seeds and plants in scores of pink, rose, purple, white, striped, and double-flowered varieties could be purchased in-store or mail-ordered from many seedsmen or nurseries in London and throughout the country, with suppliers as far-flung as Youell and Company in Great Yarmouth to Bass and Brown in Sudbury.[16] If not purchased from a commercial source, cuttings could be had from fellow gardeners, whose numbers, like the petunia varieties themselves, were rapidly proliferating. In certain quarters, however, all this familiarity was also breeding some contempt. Although petunias unquestionably flowered and flourished under the expert care of many professional gardeners, word from other parts of the huge and various Victorian horticultural world belies Jane Loudon's assurances about the success with which petunias could always be cultivated. For example, George Glenny complained about the glut of inferior cultivars flooding the market, reporting in the 1843 volume of the short-lived *Gardener and Practical Florist* that many of the "endless varieties" developed from the original white and purple progenitors had "great faults, and there is not a flower more requires weeding of worthless sorts." Glenny's disdain for petunias extended to the plant's habit of growth, which he describes disappointingly as "at first bushy and handsome—but it grows very rapidly, soon becomes straggling and ugly, and requires to be put back or supported by a frame or sticks." However, he doesn't consider training to a vertical support to be a happy solution, for "they are by no means graceful climbing plants, even at their best; and gardeners, who have endeavoured to show their skill, by making a specimen cover a large space, have rather exhibited mechanical ingenuity, which is no part of their province, than upheld their reputation in their own profession."[17]

Despite these disparaging remarks, Glenny's critical assessment requires some context. First, he was primarily writing for the working-class audience of "practical florists" named in the magazine's title, growers who may have sold flowers, but who had developed their skill with plants as amateurs, raising specimens for competitions sponsored by floral societies organized as social clubs. As Margaret Willes and Anne Wilkinson explain in their respective recent and valuable histories focusing on British working-class gardens and gardeners, florists were traditionally artisans who enjoyed cultivating showy plants as a hobby, often specializing in a particular flower, like tulips or carnations; indeed, Glenny himself had been a watchmaker before turning to journalism, inspired by his own success as a prizewinning, master florist.[18] His evaluation of the petunia thus reflects a bias against annual bedders, plants best appreciated in the aggregate outdoors rather than enjoyed as individual specimens like the conventional "florists' flowers," which looked stunning when exhibited as potted plants or cut for decorative floral arrangements. Even so, Glenny acknowledged that the petunia was "a gay flower, capable of improvement," and in the book considered to be the florists' bible, *The Standard of Perfection for the Properties of Flowers and Plants* (1847), he listed the following essential features for the perfect, prize-worthy petunia: "strong stems and a close habit—large, thick, round, and flat flowers; abundance of bloom, while short and handsome." Glenny couldn't resist a final cantankerous quip in assessing one last quality: "The colour is a matter of taste; but such is the fancy of people these days that a new ugly colour would be thought more of than an old handsome one."[19]

A second, possibly more important, factor to keep in mind with regard to Glenny's critique of the petunia was his outspoken contempt for those pro-petunia professionals who were the target of his jab about growers whose "mechanical ingenuity" exceeded their horticultural skill. In other words, he was attacking those elite members of the gardening world responsible for popularizing the petunia as a bedding plant in the first place, his warfare earning him the not-so-affectionate nickname of "the horticultural hornet" among these recognized authorities. Glenny's animosity also extended to botanists, whose assumption of superior knowledge rankled him on both a professional and personal level, as he demonstrates in a comment concerning the petunia's scientific name: "Such is the state of glorious confusion into which modern botanists have brought things with their silly antics, that when Mr. Tweedie sent home the purple variety, Dr. Hooker called it *Salpiglossis integrifolia*; Professor Don, *Nierembergia phoenicia*; and Doctor Lindley, *Petunia violacea*. Yet these are the people who pretend to teach the uninitiated how to know plants."[20]

His animus notwithstanding, Glenny was accurately voicing a dissatisfaction with petunias experienced by some dedicated growers. In an 1861 article written for the *Floral World and Garden Guide*, a periodical devoted to amateur gardening, James Holland gives an account of the petunia's reception during the preceding thirty years that squares with Glenny's assessment: "For a long time after its first introduction the petunia was looked upon as almost worthless; indeed, it has been compared to a *'mean weed'*—a comparison not to be wondered at, looking back some few years at the flimsy appearance of the flower and the wretched foliage of the best varieties that were produced." Holland, a professional gardener and award-winning exhibitor at horticultural shows, was not writing to disparage the petunia, however, but to celebrate its apotheosis as a prizewinning florists' flower during the past several years.[21] His remarks, like Glenny's, reflect the accomplished florist's sense of superior judgment regarding true horticultural value, faulting bad taste and lack of skill among amateurs and professionals alike for the petunia's often weedy, mongrel behavior.

Holland thus offers instructions on how to grow "Hybrid Dwarf Bedders," petunias that display the "dwarf, shrubby, and compact habit" that was "so much wanted for bedding purposes as well as for pot culture, being alike desirable either for the conservatory or for beds and borders."[22] Holland knew whereof he spoke, for notices in the *Gardeners' Chronicle* and the *Gardener's Weekly Magazine* verify his success in fostering the petunia craze by developing at least six varieties of championship florists' petunias during the early sixties, starting with the rose-colored Queen, which won a First Class certificate in 1860 at the prestigious Royal Botanic Society's July garden show in Regent's Park, where it was exhibited by the prominent London nurserymen Messrs. E. G. Henderson and Son, who were marketing Holland's creation.[23] In short, Holland's success in raising prizewinning florist varieties gave proof that the petunia was indeed a first-class ornamental highly adaptable to any number of horticultural uses, thereby reaffirming what the elite members of the professional gardening world had originally concluded upon its introduction to England some thirty years earlier.

Holland's experience brings the history of the petunia's Victorian reception full circle because it registers the progress of the genus from a fashionable bedder in the parterres of the aristocracy to an all-purpose, ubiquitous standard in the gardens and florist shows of the working class—an ornamental grown and beloved by the whole of the British population. Moreover, Holland's Queen in particular among his cultivars takes the petunia's story back to the Crystal Palace Park, where its parent, Donald Beaton's Shrubland Rose petunia, shared the spotlight with his aforementioned White Shrubland, whose regal status had recently been proclaimed.

If Beaton deserves top honors for launching the petunia's rise to universal popularity by developing the first best bedding variety, Shirley Hibberd, the editor of the *Floral World and Garden Guide* and foremost champion of Victorian amateur gardening, must be credited with helping to bring that popularity to fruition in the decade following the opening of the Sydenham Crystal Palace. Beginning with *The Town Garden: A Manual for the Management of City and Suburban Gardens* (1855), Hibberd's many books and articles offered clear, detailed, and enthusiastic hands-on instructions to encourage would-be urban gardeners despite their modest means, lack of experience, and the untoward circumstances of "city smoke, dust, and scarcity of leisure," not to mention the crowded conditions and shrinking space for private gardens in metropolitan London. In *The Town Garden* petunias received only brief mention; but three years later, they were highlighted as one of sixteen flowers in *Garden Favourites* (1858), proof that many petunia varieties were thriving in the beds, pots, baskets, and windows of all social classes everywhere in the country.

Hibberd was quite catholic in his horticultural tastes, but the attention he gives to old standards and florists' flowers like tulips, roses, and carnations makes clear that he considered bedding out rather a fad, one ultimately less desirable than older garden fashions that gave priority to perennial plants of all kinds, from trees and shrubs to foliage plants as well as flowers. Even so, he freely acknowledges the attraction of the most popular bedders like petunias, though they lacked the romance of long-standing sentimental associations: "They are exotics of somewhat recent introduction, and have not yet been woven into poetic lays, or consecrated to any special service by the muses. All that they have to recommend them is their exquisite beauty, and especially as to colour, and this indeed is quite enough, for what is a modern geometric garden without a display of Verbenas and Petunias?" As this comment implies, Hibberd's praise of the petunia was somewhat measured, calling it "certainly a valuable bedder" and "not altogether unworthy of high culture" as a florists' flower; but he also demurred, "As a pot-plant the poverty of its foliage is a great drawback." Nevertheless, Hibberd particularly recommended a double white variety, the Imperialis, as a "splendid," pot-worthy cultivar along with eighteen more, including the Shrubland White and two descendants of the Shrubland Rose, the Countess of Ellesmere and Marquis de la Ferté. His eleven choices for the best bedders included the Shrubland Rose, which he considered "still good, though beaten by Favourite," another cultivar (from the Messrs. Henderson) described as an improved variety developed from Beaton's acclaimed, "model" petunia.[24]

By 1866, Hibberd was reporting in the *Floral World* on the change in horticultural fashion that he claimed had begun to take place. In an article titled "Reaction against the Bedding Mania," he noted, "The public are less mad about bedders, and are awakening to a love of flowers"—that is, to an appreciation of floral "forms and characters" rather than their appearance as masses of "mere polychromatic patterns." Although Hibberd declares, "We believe in bedding, and always did; in its place and well done, it is the grandest of all possible embellishments," he asserts that "promenade displays" are not appropriate for "small private gardens," where they are "as much out of taste as liveried servants and a military band" performing at dinner inside a small house. The fault of bedding was its insistent overuse to the exclusion of so many other useful and ornamental plants, often with labor-intensive and/or unattractive consequences: "in order that geraniums may be planted and verbenas may be pegged down, and that bedded petunias may be held up against the wind to prevent their being blown to rags." Consequently, he and other "reflective horticulturists" had at times protested "against making every little garden a bad imitation of the terraces at the Crystal Palace."[25] In spite of these reservations, Hibberd continued to champion the petunia as a worthy all-purpose ornamental, affectionately calling it in 1871 an "old favourite" and naming several cultivars that were excellent either for potting or bedding, the latter category including the Countess of Ellesmere and the Shrubland Rose.[26]

What may be considered Hibberd's last word on the petunia came in 1879, when it was featured as one of forty entries in *Familiar Garden Flowers*, and acclaimed as "one of the cheapest and grandest of annuals," whose many "showy single white, purple, and striped kinds" are capable of composing "a sumptuous bed" for the "country garden." Despite its admittedly sticky, "unhandsome" foliage and "ungainly" habit of growth, "a sheet of petunias in full flower is a glorious sight"; and Hibberd notes its superiority in hardiness to the most popular bedders, geraniums and the calceolarias. Finally, as if to compensate for his former lament about the petunia's lack of poetic associations, he bestows it with its own mythology. Playing off John Loudon's name for the first purple variety, *Petunia phoenicea*, which alluded to the purple dye that the Phoenicians made from mollusks, Hibberd waxes eloquent: "By the involutions of language, this plant takes us round by way of South America to the eastern shores of the Mediterranean, for it is a Phoenician flower, and rightly named, and we are bound to connect it with the intelligent sailor race who brought the ideas and the gold of the east to the southern and western coasts of this country, and took away in exchange the tin of Cornwall, and the wealth of timber and the suitableness of these isles for colonisation."[27] Not only had petunias found a permanent home in England; thanks to Hibberd they now had a romantic story as a pedigree.

Meanwhile, Hibberd was right that other horticulturists had been reacting against the bedding mania, and chief among them was William Robinson, who has been credited with effecting the next revolution in English garden design, one that would last through the end of the Victorian period and well into the twentieth century. Another gardener-journalist, Robinson began contributing to the *Gardeners' Chronicle* in the late sixties, going on to become a dominant force in horticultural publishing when he launched *The Garden* magazine in 1871. But it was in *The Wild Garden* (1870) that he first fully articulated what might be called his back-to-nature approach to landscaping, which replaced parterres and other formal features with foliage plants, herbaceous borders, rockwork, ivied walls, and mixed beds of old-fashioned perennials and native wild flowers. Robinson was not opposed to the use of exotics, so long as they were hardy enough to become naturalized; but he rejected the stiff formality of geometric beds filled with massed half-hardy annuals like petunias, aspiring instead to the lush vegetation of the English cottage garden, as sentimentalized in the late nineteenth-century paintings of Miles Birket Foster and Helen Allingham featuring thatched-roofed houses overgrown with roses and ivy and bordered by mixed beds of perennial flowers.[28]

Robinson gave petunias their due, however, and in the first part of *The English Flower Garden*, in which he laid out his own theory of design as superior to the "gaudiness" of "summer bedding gardening," he noticeably omits petunias when he criticizes use of the other bedding standards, "scarlet Geraniums, yellow Calceolarias, blue Lobelias, or purple Verbenas," arguing that "the constant repetition" of primary colors was "nauseating [to] even those with little taste in gardening matters." Moreover, he minces no words in attacking the formal beds at the Crystal Palace Park, declaring, "At Sydenham we have the greatest modern example of the waste of enormous means in making hideous a fine piece of ground."[29] And as if to strike at the very heart of the bedding fashion, Robinson includes a section entitled "The New Flower-Garden at Shrubland Park" as a particularly instructive example of how "the great bedding-out garden, the 'centre' of the system" could be transformed according to his return-to-nature principle. In place of the old "rigid system" of parterres "filled with plants of a few decided colours, principally yellow, white, red, and blue," the new Shrubland gardens boasted permanent beds of "fine, hardy plants" with "Tea Roses and Carnations" predominating. Nevertheless, in addition to the perennials, Robinson acknowledged that some venerable bedding standards remained, like verbenas and, most notably, "such old favourites as Petunia Countess of Ellesmere, a delightful small kind, much prettier than the big flabby things in fashion now."[30] Happily, in the Shrubland garden that Beaton had created, the petunia and his legacy lived on.

Robinson could afford to be insulting about Sydenham and disparaging of Shrubland's former gardens, for both Joseph Paxton and Donald Beaton had died by 1865 when the vogue for bedding out had begun to wane. But petunias would not be denied their enduring, widespread popularity; and in the encyclopedic second part of *The English Flower Garden*, the entry written by William Wildsmith gives what might be considered a final summation of the esteem the petunia had secured in Victorian floriculture: "In certain positions, some Petunias produce a charming effect in masses; and all are suited for large vases, for baskets of mixed plants, for low trellises, and for planting under windows and walls." Moreover, some seedlings "are now so good that they are frequently planted in mixed borders for cutting." Even so, Wildsmith's list of the five best bedders (the Countess of Ellesmere among them) offers only a tiny sample of the many colors and variegated patterns currently available. In writing about the "Evolution of the Petunia" in the 1890s, the celebrated American horticulturist Liberty Hyde Bailey would marvel at the myriad forms that had developed from the "original species in little more than half a century."[31] In short, this beautiful and benign nightshade had achieved lasting popularity as a desirable addition to any English garden, an ornamental as comfortably at-home and familiar as almost any flower you could name.

The Kitchen Garden

While the petunia was finding a permanent place in ornamental plots of every size and description throughout England, three other nightshades of even greater cultural and commercial importance had been making their way into Victorian gardens as well, even though they had in fact been known and grown in England since the sixteenth century, all three appearing in Gerard's *Herbal* as exotic novelties. But over the course of Victoria's reign, tomatoes, peppers, and, to a lesser extent, eggplants—also known at the time, respectively, as love apples, capsicums, and aubergines—were increasingly cultivated in kitchen gardens for use at table as delicious esculents, the first two especially prized as essential ingredients in popular sauces and pickles. Like the potato, tobacco, and the petunia, tomatoes and peppers originated in the Americas, whereas the eggplant had the distinction of being the only truly edible Old World nightshade, coming—at least in its last stop before Europe—

from the Middle East, brought by the Arabs to the Iberian Peninsula. In any case, Renaissance Spain served as the port of entry from whence these three nightshades spread to most of Continental Europe and the British Isles.

The pepper (*Capsicum annum*), the first edible nightshade to arrive from the New World, quickly gained the distinction of being the most important spice to originate in the Americas; and its common English name suggests a good bit about its history. Discovered by Europeans when Columbus reached the West Indies in search of a water route to Asia, peppers had long been employed as indispensable condiments by the Caribbean natives in a way similar to the Old World's use of black pepper (*Piper nigrum*), an unrelated plant indigenous to Asia, renowned as the "king of spices" and one of the desired trophies of Columbus's voyage.[32] Thus the pungency of nightshade peppers understandably appealed to the taste of Spanish and Portuguese explorers who then carried the plants around the globe, where in parts of the climate-friendly Old World like India and Africa, they soon became naturalized citizens.[33] By the sixteenth century, these nightshades, so cosmopolitan that they were now called "Indian Peppers" in reference to their supposed Southeast Asian origin, were appearing in several of the most important herbals published in Europe, those of the Italian Pietro Mattioli, the German Leonhart Fuchs, and the two great Low Country botanists Rembert Dodoens and Charles de L'Écluse (or Carolus Clusius in Latin), who provided accurate illustrations of a few varieties, the latter three authors sharing many of the same woodcuts. But the herbalists also included a pedigree that attempted to reconcile the plant with species in early materia medica, those of Actuarius, Avicenna, and Pliny, to whom it was unknown. The result was that reliable descriptions of the plant and naturalistic woodcuts—along with a good deal of misinformation—had arrived together in England by the time Gerard wrote (and partly plagiarized) his *Herbal* in 1597.

Gerard, however, knew what he calls the "Ginnie [Guinea] or Indian Pepper" firsthand from two varieties he grew in his extensive garden in Holburn, and he followed the Continental herbalists in using the Latin name *Capsicum*, which he credits to Actuarius—presumably from *capsa*, meaning "box" in reference to the seed-filled pods. Gerard also recognized the plants' affinity to other nightshades, commenting that the "long codded" variety had leaves "not unlike" garden nightshade, whereas the other capsicum had small, round, and red "cods" that looked "verie like unto the berries of *Dulcamara*." He described the fruit of the first sort poetically as "of a brave colour glittering like red corall," but bemoaned the fact that he had not been able to bring his plants to full ripeness "in the unkindely yeeres that are past," while hoping for better results "when God shall sende us a hot and temperate yeere." He nevertheless gave the requisite instructions for capsicum cultivation: sow the seeds in a bed of hot horse dung, transplant young plants to pots that can be moved at will into the sunniest locations, and bring them indoors for fruit-bearing in the fall.[34]

Gerard's comments make it clear that growing capsicums in England was a labor-intensive and unpredictable task, one unlikely to be undertaken by most of his fellow gardeners. Even so, the plant's fruit—no doubt in the form of dried pods imported from the Continent—was already familiar in England, at least to Londoners, for Gerard notes that it is "verie well known in the shops at Billingsgate by the name of Ginnie pepper, where it is usually to be bought." However, he seems to have some personal aversion to consuming capsicums, for he notes that both varieties' seeds and cods are "verie sharpe and biting," having "the taste of pepper, but not the power or vertue, notwithstanding in Spaine and sundrie parts of the Indies they do use to dresse their meate therewith." Moreover, he shares the ominous report about its "malitious qualitie, whereby it is an enimie to the liver & other of the entrails; Avicen writith it killeth dogs." On the other hand, he also included both positive and conflicting information about the plants' therapeutic qualities: peppers warmed the stomach, aided "greatly the digestion of meates," and when applied externally, rid the throat and face of various scrofulous eruptions—the former condition referred to as the "Kings Evil."[35] In sum, Gerard primarily valued capsicums as attractive novelties for his garden rather than as esculents or medicines; but their increasing importance on the Continent is reflected in Thomas Johnson's revised and enlarged edition of the *Herbal* published in 1633, twenty-one years after Gerard's death. Taking images and new information about peppers grown in Italy from Clusius's posthumously published *Curaposter*, Johnson provided three new woodcuts of whole capsicum plants as well as close-up illustrations of individual pods in a range of shapes and sizes from twelve different varieties, categorized according to whether the pods grow upright or hang down, and demonstrating the huge plasticity to be found in this single species.[36]

Although Johnson considered capsicums to be plants that would "hardly brooke our climate," their beauty, if not their taste, seems to have kept them growing in the more opulent English gardens over the next hundred years.[37] One notable exception was capsicum's cultivation for consumption in the Deptford kitchen gardens of John Evelyn, the prolific diarist, horticulturist, and advocate of a vegetarian diet, who reported on years of experience raising and preparing fresh produce in his delightful *Acetaria: A Discourse of Sallets* (1699). Evelyn prefaced his remarks with a caveat followed by a recommendation: "*Indian Capsicum*, superlatively hot and burning, is yet by the *Africans* eaten with *Salt* and *Vinegar* by it self, as an usual Condiment; but wou'd be of dangerous consequence with us; [it] being so much more of an acrimonious and terribly biting quality, which by Art and Mixture is notwithstanding render'd not only safe, but very agreeable in our *Sallet.*" Evelyn gives directions for what became popularly known as cayenne: a mixture of dried, minced pods baked with flour and ground into a powder that could serve as a "very proper Seasoning, instead of vulgar pepper." Evelyn also mentions a vinegar flavored with the bruised pod of "Guiney-pepper" as one ingredient in a zesty salad dressing made of oil, vinegar, citrus, mustard, and boiled egg yolks.[38]

It was not until the eighteenth century that capsicums secured a modest place in the kitchen gardens of England, their increased cultivation coinciding with the arrival of the large, sweet bell pepper, which was more to British taste than the hot varieties that were first introduced. By the 1730s Philip Miller, the horticultural mastermind at the Chelsea Physic Garden, was reporting in his classic *Gardeners Dictionary* that although the plant's fruit was "at present of no great use in England," it made "one of the wholesomest pickles in the world"—no doubt referring to the large round capsicum he listed among the eighteen "sorts" he knew. Like Evelyn, Miller warned about eating the "very acrid" fruit of other varieties raw, for it would "cause an extraordinary great Pain in the Mouth and Throat of such Persons as are not accustom'd to eat of it"; but he also described the powder made from capsicums for culinary use, which he calls "*Cayan butter* or *Pepper-Pot,*" that "is by some of the English people mightily esteem'd."[39] As a case in point, "India pepper" and "Cayan pepper" appeared in Hannah Glasse's popular mid-eighteenth century *The Art of Cookery Made Plain and Easy* as seasonings in a few of the more exotic recipes, such as a cod and pork chowder, a stewed turtle dressed "the West-India Way," and stewed "green pease the Jews way."[40]

Instructions for cultivating several varieties of capsicums for both ornament and consumption appeared in editions of the *Gardeners Dictionary* published during Miller's lifetime, as well as in the revised and enlarged 1807 edition by the Cambridge botanist (and Rousseau's translator) Thomas Martyn;[41] but it was John Abercrombie who deserves the most credit for securing a place for capsicums in the English kitchen garden. Abercrombie's highly popular book *Every Man His Own Gardener,* first published in 1767 and going through many editions into the nineteenth century, followed the time-honored method of organizing gardening information according to a monthly "Kalendar" of operations; but his detailed, literally down-to-earth directions were written with the novice in mind rather than the experienced gardener, who already knew how to prepare a hot bed or construct a cucumber frame. Described once again as "much esteemed for pickling," capsicums appear in the kitchen-garden section as esculents to be sown in a hot bed in March and

BACK TO THE GARDEN 259

ultimately transplanted in late May "into beds of rich earth in the common ground."[42] Abercrombie's influence can be measured by the fact that his instructions for growing capsicums appear in the culinary section under "Plants Used in Preserves and Pickles" in John Loudon's *Encyclopedia of Gardening*, the massive and definitive source of British horticultural information through the first half of the nineteenth century.[43] It is probably no coincidence that the children's tongue-twister "Peter Piper picked a peck of pickled peppers" dates in print from 1813—further evidence of capsicums' rising popularity in England not just as a spice to be purchased, but as an esculent grown in the kitchen garden as well.[44] According to Thackeray in *Vanity Fair*, 1813 was the same year that Becky Sharpe, in her campaign to engage herself to the nabob Jos Sedley, suffered agonies after sampling a "horrid pepper-dish" of Indian curry fired by cayenne, only to make matters worse by consuming an even hotter green chili, whose name, which sounded like "something cool," led her to think it would douse rather than fuel her gustatory flames. Capsicums found a better place in George Spratt's *Flora Medica* as "one of the most simple and powerful stimulants we possess, its action not being followed by any narcotic effects," a sure sign of the growing appreciation for the "sharp and biting" quality found suspect by Gerard—and despite being deemed so "abominable" by Becky.[45] Spratt's accompanying illustration certainly conveys an equal appreciation for the species' beauty (see fig. 41).[46]

Nevertheless, it was the tomato (*Solanum lycopersicum*) that first won the hearts of the Victorians as a favorite condiment before finally earning a distinguished place in the kitchen gardens of all classes by the last quarter of the nineteenth century—a long delay that began in its early association with its closely related Old World cousin the eggplant (*Solanum melongena*), whose European history serves as an appropriate starting point for the tomato's own story. As David Gentilcore explains in his fascinating *Pomodoro! A History of the Tomato in Italy*, when the tomato arrived from Spain in the mid-sixteenth century, the Italian herbalist Mattioli identified it as "another species of eggplant," a nightshade indigenous to either India or Africa and introduced some time earlier to Italy where it quickly earned a bad reputation, in part because of its recognizable affinity to its other Old World solanaceous kin, including the notorious mandrake. For example, Mattioli's friend and fellow botanist Pietro Antonio Michiel claimed that eating eggplants could lead to everything from headaches and melancholy to leprosy and fevers.[47] As a badge of its dishonor, the eggplant's common name in Italian, "melanzana," thus became associated with (if not derived from) the Latin pejorative "mala insana," which in turn gave birth to the English "madde or raging Apple" by which it was known to Gerard.[48]

FIGURE 41. *Capsicum annum*, George Spratt, *Flora Medica*, vol. 1: 91. (Courtesy of the Wellcome Collection)

Although he never mentions the term "eggplant," Gerard explains its derivation when he describes the fruit as "great and somewhat long, of the bignesse of a swans egge, and sometimes much greater, of a white colour, sometimes yellowe, and often browne." Gerard reports that mad apples were eaten in Spain and northern Africa, adding that in Toledo the preserved fruit is consumed "all winter to procure lust," a belief likely originating in the mandrake connection.[49] In fact, in his sixteenth-century English translation of Dodoens's herbal (by way of Clusius's French version), Henry Lyte claimed that "Raging or mad apples" were also called in English "Amorous Apples, and Apples of love"—terms that Gerard, writing about twenty years later, would reserve just for the tomato.[50] Lyte offers no explanation for the eggplant's sexier English name; but as the botanist Charles Heiser wryly comments in his classic book on the nightshades, "Everyone knows that love and insanity are closely allied."[51] In any case, Gerard energetically discouraged the mad apple's consumption in England, "for doubtlesse these apples have a mischeevous qualitie, the use thereof is utterly to be forsaken." Instead, he recommended they be grown for their "pleasure" and "rarenesse"—even though his own plants had disappointed by not producing ripe fruit in England's chilly climate.[52]

Linnaeus established *Solanum melongena* as the eggplant's scientific name in 1735, the species designation taken from another Italian term known to and mentioned by Gerard.[53] In his 1754 *Gardeners Dictionary*, however, Philip Miller preferred to consider *Melongena* as the genus and still opted for the pejorative common name "mad-apple," even though he adds "egg-plant" as a recent, but "confusing," English alternative. Miller also mentions that the Turks' term *Badanjan* had given rise to several colorful English corruptions— Brown Johns, Brown Jolly, and Baron Jelly, the first two obviously responding to the plant's striking dark-purple skin (and the third surely an inspired improvisation on the second). Names aside, Miller's entry reveals that by the mid-eighteenth century, acceptance of the plant as an esculent was increasing on the Continent, where it was "greatly cultivated" for food—if also preserved for winter use "to provoke a venereal Appetite." English tastes had not yet changed, however: the plants were still grown only "as Curiosities" and the fruit never consumed—"except by some *Italians* or *Spaniards*, who have been accustom'd to eat of them in their own Countries."[54]

During the first quarter of the nineteenth century, *Solanum melongena* finally cast off its negative image in England as information about its culinary uses from other parts of the world became more available. For example, Loudon's *Gardener's Magazine* reported in 1826 on the Royal Navy officer Captain Peter Rainier's "directions for cultivating and cooking the brinjall," an eggplant variety that had "dark-coloured, elongated fruit" popular for curries and "made-dishes" in India. The plant was "also established as an esculent in the French gardens, under the name of aubergine," which had evolved from the aforementioned Indian term. Although Rainier doesn't say so, the French were probably responsible for the simple method of preparation he preferred: slicing the parboiled fruit lengthwise, seasoning it with butter, salt, and pepper, and broiling it "on a gridiron."[55] Rainier was undoubtedly one of the earliest proponents of the edible eggplant in England, for Loudon's contemporaneous *Encyclopedia of Gardening* describes only the white-colored variety of the fruit and doesn't mention the term "aubergine," which was destined to become its most common name in England by the end of the century when, as the *OED* records, it also entered the language as a distinctive shade

262 VICTORIAN NIGHTSHADES

of deep purple. Loudon does, however, omit the mad-apple moniker, calling it "egg-plant" instead, and listing it along with its solanaceous relatives peppers and tomatoes as "Plants used as Preserves and Pickles." Loudon gives no details about the eggplant's culinary applications in England, but notes that "in French and Italian cookery, it is used in stews and soups, and for the general purposes of the love apple"—a comparison that attests to England's familiarity with and increasing appreciation of the tomato by the early nineteenth century, while indicating once again how entangled the fate of these two nightshades had been for the past three hundred years.[56]

The history of the tomato in England rightly begins with Gerard, who knew "apples of love" from personal experience, his commentary informed by the plants he grew from seed in his own garden, "where," he says, "they do increase and prosper." Yet despite his plants' fecundity, Gerard didn't know quite what to make of them; and this resistance to easy categorization would plague the tomato's reception in England as it had in Italy. For example, Gerard was obviously intrigued with the tomato's unusual habits and "bright shining" red fruit, which his predecessor Lyte, following Clusius, had insistently called "strange."[57] Gerard describes in detail the plant's weak stalks, "trailing upon the grounde," unable to support its soft but heavy "faire and goodly apples" filled with moisture and reddish pulp—the fruit it produced in such abundance. Like most observers before and after, he was also struck by the strong—and to him offensive—smell: "The whole plant is of a ranke and stinking savour." The plant's name already presented problems, but Gerard simply followed recent European botanical tradition in identifying the plants as "Apples of Love" in English and *Poma Amoris* in Latin, saying nothing about the aphrodisiacal qualities these terms would seem to imply.[58] In fact, David Gentilcore claims that the tomato had no such propensities and surmises that confusion with the tomatillo (*Physalis ixocarpa*), another New World nightshade whose encased fruit reminded the Spanish naturalist Francisco Hernandez of female genitalia, likely gave rise to the Latin *Poma Amoris* and eventually to the French *pommes d'amours*, from whence came the English "love apple." The Spanish name *tomates*, taken from the Aztec *tomatl* by which the tomatillo was also known in Mexico where both species were first domesticated, merely reinforced the confusion.[59] Gerard doesn't mention the term "tomato" or its variations, but includes "golden apple" as the alternate English name, one derived from the Italian *pomo d'oro*, which described the color of some tomatoes, including a variety grown by Gerard.[60] However, "*pomo d'oro*" in Italy was also commonly used for edible citrus fruits or even figs and melons,[61] certainly in response to their color, but undoubtedly to suggest their desirability as well, by allusion to the precious golden apples growing in the Garden of the Hesperides—a mythic association that Gerard mentions, only to reject.[62]

Indeed, rather than color or mythical associations, Gerard was most concerned with the plant's "temperature," a matter of importance to Renaissance herbalists still influenced by medieval humoral theory that based the nutritional and/or medicinal value of almost everything on its elemental qualities, its relative degree of heat and moisture, which in turn determined its effect on the human body's "humors," the essential fluids that governed one's health and 'temperament.'[63] Since achieving a proper humoral balance was the requisite goal for a healthy body, anything ingested whose nature pushed the limits of cold or heat, wetness or dryness, was considered suspect and probably toxic or caustic—like the infamous mandrake, deemed to be extremely cold, or like capsicums, known to be extremely hot. Of course, fitting a newly introduced plant like the tomato into existing humoral categories presented a challenge, but in this case it was one that Gerard was prepared to meet in light of his own observations. Disputing Dodoens's opinion that the golden apple is "not fully so colde as Mandrake," Gerard counters, "but to my judgement it is very colde, yea perhaps even to the highest degree of coldnesse." As evidence, Gerard describes "carelessly" tossing out some cut branches in "the allies of my garden," where despite that summer's excessive heat, lack of rain, and the hardness of the ground, they took root and grew "as fresh where I cast them, as before I cut them off; which argueth the great coldnesse conteined therein." Gerard then adds, "True it is that it doth argue also a great moisture, wherewith the plant is possessed, but as I have said not without great cold, which I leave to every mans censure."[64] Given this experiment, neither a romantic name nor presumptive classical credentials could allay Gerard's suspicions about eating this strange fruit; and after reporting that tomatoes, boiled in oil and seasoned with salt and pepper, were served as a dish "in Spain and those hot regions," he voiced the earlier herbalists' verdict that "they yeelde very little nourishment to the bodie, and the same naught and corrupt." Gerard concludes, "Likewise they do eate the apples with oile, vineger and pepper mixed together for sauce to their meate, even as we in these cold countries do mustarde"[65]—a comment that inadvertently foreshadows the tomato's importance in Victorian times as the main ingredient in a popular sauce and in ketchup, destined ultimately to become mustard's major rival among the world's favorite condiments.

Gerard's opinion apparently held sway in England for over a hundred years, with the tomato gradually beginning to gain some acceptance as a foodstuff by the mid-eighteenth century, its progress documented in succeeding editions of Miller's *Gardeners Dictionary*. Following the French botanist Joseph de Tournefort, Miller listed "love-apples" under the genus *Lycopersicon,* another alias used since the sixteenth century translated as "wolf peach," no doubt to describe the unsavory, probably poisonous, but not identical species that Galen, with whom the term originated, had in mind.[66] Miller's contemporary and friend Linnaeus would adopt the same pejorative as the species name instead, classifying the tomato as *Solanum lycopersicum*—and thus setting off a debate about scientific nomenclature that continues today. But whether as genus or species, the term registered the prevailing suspicions about the plant; and Miller's verdict in 1735 closely echoed Gerard by noting that despite the fruit being eaten in Italy and Spain, "considering their great Moisture and Coldness, the Nourishment they afford must be bad."[67] By 1754, however, he was reporting that love apples were "now much used in England" in soups, to which they provide "an agreeable Acid," before softening his previous caveat thusly: "There are some Persons who think them not wholesome."[68] Finally, by 1768 Miller was giving one *Lycopersicon* species—the "Apple-bearing Nightshade ... commonly called tomatas by the Spaniards"—the name "*Esculentum*"—an acknowledgment of the love apple's identity henceforth as primarily an edible rather than an ornamental plant. Miller now recommended planting this species "in the borders of the kitchen garden" rather than as a pleasure-garden curiosity where their odor had diminished their appeal: "The Plants emit so strong an Effluvium, as renders them unfit to stand near an Habitation, or any Place that is frequented; for upon their being brush'd by the Cloaths, they send forth a very strong disagreeable Scent."[69] The previous year John Abercrombie still considered love apples (as well as capsicums and eggplants) to be appropriate additions to the pleasure garden; but he, too, gives priority to their place in the kitchen garden, providing detailed instructions for planting out "tomatos, or love-apples" in May after raising them from seed in a hot-bed.[70]

By the turn of the century the tomato was finding its way to becoming a British culinary staple: the 1797 *Encyclopaedia Britannica* reported it was in "daily use" for soups or "garnishes to flesh-meats," even though by all indications the fruit served only as an accompaniment to other dishes and was not eaten raw.[71] The tomato scholar Andrew F. Smith explains that Jews in England were the exception, since many were of Portuguese or Spanish descent—that is, from "those hot regions" mentioned by Gerard, where the fruit had originally found favor after its introduction to Europe.[72] However, cultivating tomatoes in England was almost solely confined to the wealthy or to enterprising market gardeners, for whom growing subtropical, labor-intensive plants was either affordable or profitable. In a letter to her sister Cassandra in the fall of 1813, Jane Austen's question and comment, "Have you any tomatas? Fanny and I regale on them every day," implies that she and her favorite niece were eating fruit raised at Godmersham Park, her brother's large estate in Kent where she was visiting (and the possible model for the eponymous *Mansfield Park*, the novel whose heroine, probably not by chance, is named "Fanny").[73] The question also suggests that Cassandra might be enjoying homegrown tomatoes from their much humbler kitchen garden at Chawton Cottage where Jane lived with her sister and mother, who were known to be avid gardeners. If so, the Austen ladies' plants would have been a novelty in the neighborhood.

The best evidence that the tomato had recently gained a firm foothold in England comes in 1819, when it debuted at the London Horticultural Society in the most comprehensive report on the species to date presented by the secretary, Joseph Sabine, who begins by remarking on the tomato's "great use . . . of late years for culinary purposes," for which it was being "regularly grown" in both private and commercial gardens.[74] The paper makes clear that what had been considered in England a strange plant and certainly a suspect esculent was now being appreciated in some quarters for its "agreeable acid," described by Sabine as "a quality very unusual in the ripe fruit of vegetables," one "quite distinct from any other product of the kitchen garden." Sabine probably provides the first official record of what would become the tomato's most universal application when he adds that "its juice is preserved for winter use, in the manner of ketchup," the condiment that Andrew Smith claims had recently become "the rage of the English-speaking world on both sides of the Atlantic"[75]—a concentrated sauce often made from walnuts or mushrooms until tomatoes ultimately eclipsed their use. Sabine presents tomato's natural history; discusses the familiar large red, cherry, and pear-shaped varieties; and includes a detailed account of the market gardener John Wilmot's astounding success growing six hundred plants, which produced on average twenty pounds of fruit apiece.[76]

In 1819 the tomato also made what seems to have been its only appearance in floriography, with *pomme d'amour* featured as the "Emblem of Discord" in Alexis Lucot's early *Emblèmes de flore et des végétaux*. Lucot's definition deftly draws together the French and Italian common names by alluding to the Hesperidean golden apple that the Goddess of Discord, in revenge for not being invited to the wedding of Peleas and Thetis, used to incite a quarrel among Juno, Athena, and Venus by promising the fruit to the one judged "most beautiful" of the three goddesses. When the arbiter Paris chose Venus because she could reward him with Helen, the most beautiful woman among mortals, Discord's prize metaphorically transformed into an apple of love.[77] Lucot's cleverness notwithstanding, his definition didn't catch on, even though "discord" had in fact worked to spread the fruit's popularity in France—and appropriately to Paris's namesake city. As Clarissa Hyman explains in her global history of the tomato, *pomme d'amour* lived up to its name during the French Revolution when it won the hearts of Parisians, as "comrades" from Provence brought their culinary tastes and their produce to the capital.[78]

In 1822 Loudon's *Encyclopedia of Gardening* summarized the culinary conquest of Europe by the "love-apple" as follows: "Though a good deal used in England in soups, and as a principal ingredient in a well-known sauce for mutton; yet, our estimation and uses of the fruit are nothing to those of the French and Italians, and especially the latter. Near Rome and Naples, whole fields are covered with it, and scarcely a dinner is served up in which it does not in some way or other form a part." Loudon added the tomato's role "in confectionary, as a preserve; and when green, as a pickle" to the now-familiar list of its culinary uses; and he printed copious instructions for cultivation, including Wilmot's innovative methods taken from the recent Horticultural Society report.[79] In the 1835 edition of the *Encyclopedia*, Loudon was also providing information from several articles on culinary use and cultivation that had appeared in his *Gardener's Magazine*, giving further evidence that the tomato was coming into its own as an extremely desirable esculent.[80] And in 1837 Dickens confirmed in his first novel the popularity of that aforementioned "well-known sauce for mutton"—while introducing a comic twist to the tomato's amorous associations in the process. When Pickwick goes to trial on trumped-up charges of breaking a promise of marriage to his landlady, the prosecutor misrepresents Pickwick's dinner request as evidence of the defendant's covert bid for a romantic assignation. After reading aloud the message, "'Garraways, twelve o'clock. Dear Mrs. B.—Chops and tomata sauce. Yours, *Pickwick*,'" the prosecutor demands of the jury, "Gentlemen, what does this mean? Chops and tomata sauce. Yours, Pickwick! Chops! Gracious heavens! and tomata sauce! Gentlemen, is the happiness of a sensitive and confiding female to be trifled away, by such shallow artifices as these?"[81]

Dickens describes Pickwick's landlady Mrs. Bardell as having "a natural genius for cooking, improved by study and long practice, into an exquisite talent"[82]—thus a woman fictively capable of making her own tomato sauce, for which a recipe had been readily available in England since 1808 in Maria Rundell's bestselling *A New System of Domestic Cookery*, written for the middle-class housewife and in its latest edition by 1833. To be sure, Rundell's "Tomata Sauce, for hot or cold Meats" needed the hand of a skillful cook since her instructions are short on details, with no information about how much of the requisite capsicum-vinegar, garlic, and salt to add to an unspecified

BACK TO THE GARDEN

number of ripe tomatoes rendered "quite soft" in an oven still hot from baking bread. Presumably, the recipe was to be made in quantity, given that Rundell directs her readers to "keep the mixture in small wide-mouthed bottles, well-corked, and in a dry cool place"—an indication, along with the preservative vinegar, that her sauce was indistinguishable from tomato ketchup, as was often the case for much of the nineteenth century.[83] Fortunately, the less culinarily gifted could purchase this popular condiment ready-made, for at least by mid-century there were numerous tomato sauces being sold by the bottle in London alone. For instance, an 1852 Sanitary Commission Report entitled "On Sauces and Their Adulterations" appearing in the *Lancet* revealed that ten of the thirty-three commercial sauces analyzed were made primarily from tomatoes, including two of the chef Alexis Soyer's recipes marketed by the venerable firm Crosse and Blackwell. Since the analyses showed that red dyes had been detected in many of the samples, the author promoted at-home production, for which cause the *Lancet* printed a few recipes.[84]

Although sauces and ketchups were by far the most popular uses for tomatoes, the recipes for stewed, pureed, and baked tomatoes that began appearing in cookbooks published around mid-century record their gradual acceptance as side dishes, usually to serve with meats. Of ten tomato recipes in Eliza Acton's *Modern Cookery* (1845), the two most elaborate were for "forced tomatas"—that is, stuffed with ingredients like ham, mushrooms, and seasonings, then topped with buttered bread crumbs before baking—what Queen Victoria's onetime chef Charles Francatelli referred to as "tomatas au gratin" or "Tomatas, a la Provencale."[85] Unusually, Acton also provided a recipe for raw "Tomatas en Salade," commenting, "These are now often served in England in the American fashion, merely sliced, and dressed like cucumbers, with salt, pepper, oil, and vinegar."[86] In that mid-century classic *The Book of Household Management* (1861), Isabella Beeton's use of tomatoes was less adventurous, but they regularly appear as accompaniments for meat in her recipes and bills of fare while also serving as indispensable flavorings for numerous soups and gravies.[87] As usual, Mrs. Beeton interspersed her recipes with information about the tomato lifted from other sources; but the most curious of these items comes in a section headed "Medical Memoranda," quoting the first part of an article from the *Gardeners' Chronicle* about the health benefits of the tomato touted by the American doctor John Bennett, who vigorously promoted it in treating dyspepsia, diarrhea, and "all those affections of the liver and other organs where calomel is indicated." Titled "The Tomato Medicinal," Beeton's entry apparently supports Dr. Bennett's claims; but it rather tellingly fails to include the doctor's endorsement of eating tomatoes raw and omits his directions for serving sliced tomatoes "with salt, pepper, and vinegar, as you do Cucumbers"[88]—Eliza Acton's "Tomatas en Salade"—even though the article calls this the most common method of preparation.[89]

Despite whatever reservations Mrs. Beeton had about eating tomatoes raw, their importance as an ingredient in her recipes, in addition to her apparent espousal of Dr. Bennett's extravagant claims, helped to bolster the public's confidence in tomato's salutary effects on the human body. Writing in *The Book of Garden Management* just the year after the release of Isabella's block-buster cookbook, Samuel Beeton—her husband and the mastermind behind their publishing business—continued to promote the tomato as a versatile esculent, one "singularly wholesome, and very useful, especially in cases of bad digestion," thereby adding another influential voice to the growing chorus of enthusiasts who worked to transform the tomato into a Victorian health food during the last third of the century. Beeton lamented that the fruit was "not appreciated or cultivated as it ought to be," and urged his readers to grow their own plants even though he acknowledged, "There is obviously some little difficulty in our climate in fruiting and ripening tomatoes to perfection." Stripped to its essentials, Beeton's recommended directions consisted of starting the seeds in a cucumber frame, transplanting the crop against a south wall in the spring, and hoping for a stretch of hot, sunny weather lasting into fall. He warns that "even with the greatest care," cold weather would prevent ripening outdoors, in which case the fruit could be cut from the main stem and dried in a greenhouse or an oven.[90] Clearly, growing tomatoes in England could be a challenge for the impecunious or the faint of heart.

In any case, Victorian cookbooks and the horticultural press record that the popularity of the tomato went well beyond the initial favor it found as a staple ingredient in soups and sauces; but it entered the households of the middle and lower classes as a foodstuff some time before being planted in their gardens. When Loudon noted in early editions of the *Encyclopedia of Gardening* that tomatoes, eggplants, and capsicums—the three nightshades then primarily in use for preserves and pickles—together "occupy but a few square yards of the largest kitchen-garden . . . and few of them are seen in that of the cottager," his observation would hold true until the latter third of the century, for a couple of reasons.[91] First, given the time, expense, labor, and risk required to bring these three sun-loving nightshades to fruition in a private garden, only the wealthy could afford to grow them as esculents; for the "cottager" who depended on the kitchen garden for sustenance, they quite simply were not worth the trouble. Second, fresh as well as preserved tomatoes and capsicums began to be available for purchase in urban England from commercial sources at least by the mid-1820s so that the supply at first

BACK TO THE GARDEN

matched and—in the case of tomatoes—sometimes exceeded the demand. According to reports in Loudon's magazine and the *Gardeners' Chronicle*, both nightshades were regularly sold in London at Covent Garden Market, where tomatoes were supplied by local market gardeners when in season, but also shipped from such places abroad as Lisbon, Marseilles, and even Algiers.[92]

Tomatoes were the only one of the three nightshades that would eventually be grown for more than their limited use in "preserves and pickles" during the Victoria era; but there is plenty of evidence to show that despite their meteoric rise to popularity in the latter part of the century, for most people they were an acquired taste. Many references in the *Gardeners' Chronicle* attest to the sentiment expressed by a writer in 1876: "Few people like them on the first trial; but after the taste for them is once acquired, nothing is more palatable."[93] Toward the end of the century, commentators looked back with amazement at the change that had taken place. As one writer put it in 1895, "It is strange that the present generation should crave for tomatos, although fifty years ago few people liked them." And another remarked the following year, "I remember my first bite at a Tomato, and why I ever tried another I am at a loss to imagine—yet they have become both pleasant and desirable."[94] As was characteristic of the class divide in culinary habits and preferences, the bias against tomatoes was particularly pronounced among the lower classes. In an 1876 article entitled "Little Known Dainties," the author, attributing the reluctance even to try tomatoes to the lamentable, if understandable conservatism of the "uneducated classes," explained that whereas "the rich vary their fare according to their fancy, without regard to their pockets—the poor must first consider their means, and then please their appetites according to their purse." Consequently, the idea that the alien and pricey tomato was "a very wholesome article of diet, like the time-honoured Cabbage and popular Onion, never enters into their thoughts."[95] Resistance to eating tomatoes raw, however, was not class specific; and as late as 1881 a commentator writing about the tomato's purported health benefits continued to doubt "that they will ever become acceptable as ordinary uncooked fruit."[96]

In any case, by the eighties most accounts indicate that the tomato's popularity had become universal. In 1882 the *Gardeners' Chronicle* reported on the extensive glass houses—thirty in all—recently opened for growing tomatoes by a market gardener in Kent to take advantage of the huge demand for this "now favourite vegetable." Marveling over the revolution that had taken place "within recent times" in the country's appetite for tomatoes, the writer once again recalls the typical first reaction not only to their unique—and thus peculiar—flavor, but also to the odor that had so offended John Gerard and Philip Miller: "On first acquaintance taste and smell are alike to most people disgusting, yet there is nothing that we can think of so different to other articles of food that people so soon take a liking to; added to which, the unanimous verdict of medical men as to their wholesomeness has gone far to popularize them."[97] Over the next decade the market for tomatoes exploded: an 1892 article cited a report from a United States Commercial Agent urging US growers to consider exporting some of their crop to England to meet the sky-rocketing demand. Agent Smyth noted that presently "over one million square feet of glass surface in Great Britain [was] exclusively devoted" to the tomato's commercial production—certainly evidence for his claim that "the epicurean taste prefers the hothouse article." Nevertheless, in the off-season tomatoes were coming in from the climate-friendly Channel Islands where the plants could be "grown in the open air," as well as from the Azores, the Canary Islands, France, and Spain. By the nineties there seemed to be no limit to the number of tomatoes the British could consume. The incredible revolution in the "Englishman's taste" they had effected moved Agent Smyth to end with a poetic flourish, one that eloquently expressed how far the tomato had come: "What was once the Love Apple of his forefathers, a mere object of romance and curiosity mixed up with all sorts of mythological traditions, is now served on his table and eaten 'blood raw'—no longer 'Like to the Apples on the dead sea's shore, all ashes to the taste.'"[98] Two years later, in an article promoting "Market Gardening" as "A Profession for Our Sons," the author put the tomato first as England's most profitable commercial plant for cultivation under glass: "Of all the crops peculiarly suited to the market-grower, none is more deservedly recommended than Tomato-growing."[99]

In 1897 the tomato's phenomenal success was verified by none other than Arthur Sutton, scion of the renowned house of Sutton and Sons of Reading, famous, among other things, for developing the Magnum Bonum potato and for experiments with *Solanum* hybrids beginning in the mid-eighties. In the keynote address entitled "The Progress of Vegetable Cultivation during Queen Victoria's Reign," presented at the Royal Horticultural Society's conference at the Crystal Palace to mark Her Majesty's sixty years on the throne, Sutton claimed that "the tomato has increased in popularity to a greater extent and more rapidly" in those years than any other of the country's most important vegetables. Appropriately for the occasion, Sutton praised Owen Thomas, the Queen's gardener at Frogmore, for assisting the tomato's advance by developing one of the best of the new varieties whose numbers were now "legion," whereas at mid-century only one variety, the "common Red," seemed to have been available. Moreover, now there were several cultivars especially adapted "for earliest out-door use," an improvement that no doubt helped to bring about the "enormous impetus" to growing tomatoes that had occurred

FIGURE 42. Sutton's Royal Dwarf Cluster tomato, *Gardeners' Chronicle*, n.s., vol. 10 (December 14, 1878): 772. (Image from Biodiversity Heritage Library; contributed by UMass Amherst Libraries)

in the "last quarter of a century." Sutton, however, attributes this change to "the more cultivated taste of the masses of the people," adding that the tomato "is now found in almost every garden from the cottager's upwards."[100]

The development of many new varieties and a change in taste were no doubt the major factors that accelerated the tomato's popularity so dramatically over the Victorian period; but other forces were at work as well to quicken the pace of its widespread cultivation. After mid-century, cheaper glass and prefab technology led to the manufacture and sale of small, portable greenhouses affordable by the middle classes, making it possible to start plants from seed early and easily; and these developments also meant that for gardeners of more modest means, young plants were readily available for purchase at reasonable prices. Seedsmen's advertisements for tomatoes took an upswing in the seventies, with smaller-fruited cultivars evidently

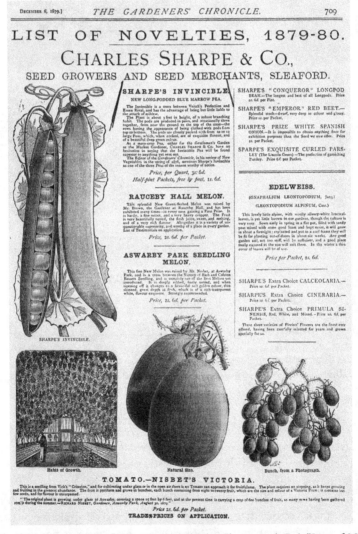

FIGURE 43. Tomato—Nisbet's Victoria, in Charles Sharpe and Co.'s "List of Novelties, 1879–80," *Gardeners' Chronicle*, n.s., vol. 12 (December 6, 1870): 709. (Image from Biodiversity Heritage Library; contributed by UMass Amherst Libraries)

BACK TO THE GARDEN 273

becoming especially popular. For example, the Carters' firm—"The Queen's Seedsmen"—heavily advertised its plum-sized Green Gage tomato, and Sutton's brought out a cherry variety, the Royal Dwarf Cluster (see fig. 42).[101] Most extraordinary was an ad from Charles Sharpe and Company promoting Nisbet's Victoria, pictured as filling an entire greenhouse with a canopy of plum tomatoes (fig. 43).[102] Finally, Shirley Hibberd, the foremost advocate for amateur gardening, should be given credit for his role in providing the "enormous impetus" to the tomato's cultivation among all social classes that Sutton mentioned. As editor of the *Floral World and Garden Guide* beginning in 1858 and the *Gardener's Magazine* from the 1860s, Hibberd regularly printed detailed instructions on growing tomatoes as well as capsicums and eggplants.[103] By the time he published *The Amateur's Kitchen Garden* in 1877, he already considered the tomato to be a universal Victorian favorite, thanks in great part to the inspiration he had provided. Hibberd expresses his enthusiasm for this "noble fruit" in my chapter's epigraph, whose final words are worth repeating here. By the end of the nineteenth century the tomato had indeed become "a citizen of the world, the welcome guest of every household, and the subject of the poor man's special care. And no wonder, for like its cousin, the potato, it is so thoroughly useful and so accommodating in respect of the cultivation it requires, that it suits every garden and every table, and, strange to say, gratifies every taste, for it was never heard that, when fairly presented, this noble fruit displeased anyone."[104] The nightshades' conquest of Britain was complete.

The Garden Solanaceous

When the Victorian era officially came to a close in January 1901, the Queen's sixty-four-year reign had witnessed radical changes in English horticulture. England's primacy as the foremost industrial country in the world had made possible the British Empire's expansion to every climate on the globe, where its botanists and plant hunters reaped the spoils. They had brought back exotic species of every imaginable shape and habit, to be kept alive and healthy by British technological prowess once these vegetable immigrants settled in their new home. Fifty years earlier Joseph Paxton's design for a giant greenhouse, inspired both by the climatic requirements and plant anatomy—that is, the rigid leaf structure—of the tropical *Victoria regia* water lily, had created what became the century's most acclaimed monument to the marriage of technology and botany: the original Crystal Palace in Hyde Park that housed the Great Exhibition of 1851.[105] The construction of that miracle in sheet glass and cast iron had been one of the great engineering feats of the Queen's reign, and in its reincarnation three years later at Sydenham, the Palace remained a potent symbol for the extraordinary advances in gardening that characterized the era. The Crystal Palace stood for England's triumph over the vegetable kingdom, an achievement in no small measure made possible by the technology that had given birth to the structure itself. In its many permutations—from the greenhouse, stove, and conservatory to its versions in miniature like the cold frame, glazed window box, and even the diminutive Wardian case—the Victorian glass house modernized gardening. Under cover and indoors, no longer at the mercy of climate or season, plants flourished as their gardeners forced, bred, studied, and trained thousands of non-native species as diverse as towering palms and pines to low-growing pelargoniums and petunias, raising endless new varieties and creating astonishing hybrids so at ease in this manufactured environment they could have been to the manner born.

In the meantime, advances in science and technology reconfigured England's topography. By 1901 its population had more than doubled since Victoria came to the throne in 1837; but as the country transformed from an agrarian to an industrial society, the working classes crowded into towns and cities, losing their connection to the land and leaving most of the open countryside to the wealthy few who controlled England's economy and more specifically, its agriculture. Congested conditions in urban areas led civic-minded leaders like Paxton to sponsor the creation of parks and even cemeteries as public grounds open for all to enjoy; but the decrease of green space available for private use made its value more palpable, only to increase the desire among England's humblest citizens for a garden of one's own. And as the Sydenham Crystal Palace and Park gave proof, the abundance of technological and horticultural riches on display furthered the rage for gardening among all classes. Periodicals devoted to botany and plant cultivation proliferated during the nineteenth century, providing detailed information about growing new species while advertising everything from seeds and starts to a full range of gardening supplies, including glass houses scaled down for the smallest garden. Paxton himself patented a design for prefab structures that went on sale in January 1860, marketed by Samuel Hereman in the *Gardeners' Chronicle* as "Hothouses for the Million." The advertisement promoted "these Buildings of unparalleled cheapness" as adaptable to many configurations, affording "persons of limited means a luxury hitherto confined to the wealthy," and making year-round gardening a universal possibility (see fig. 44).[106]

From mid-century on, local horticultural societies and florists' clubs sprang up throughout Britain, sponsoring competitions and lectures to support

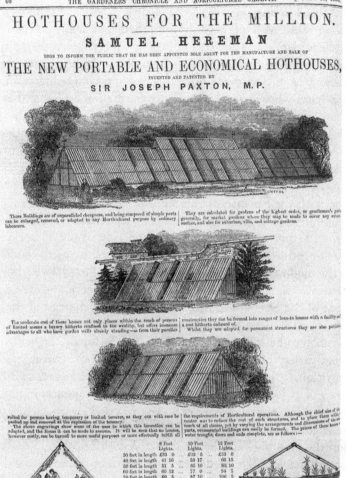

Figure 44. "Hothouses for the Million," *Gardeners' Chronicle*, vol. 21 (January 21, 1860): 60. (Image from Biodiversity Heritage Library; contributed by the Missouri Botanical Garden, Peter H. Raven Library)

the increasing number of amateurs who had taken up gardening.[107] As a result, botanical literacy spread to the farthest reaches of the kingdom, for the working-class "Million" began raising their own fruit, vegetables, trees, shrubs, foliage plants, and flowers not only for sustenance and ornament; the deluge of correspondence and queries appearing in the horticultural journals was testimony to the fact that among this new wave of gardeners, there were many hobbyists whose eagerness to cultivate the unfamiliar and exotic led them to a deeper investigation and appreciation of plant life. By 1909 the garden historian Alicia Amherst was describing horticulture in England as having evolved from the "passion of the few" in 1890 to its current status as "the craze of the many," an indication of the accelerated pace at which the Victorians had fallen in love with growing plants by the end of the era.[108]

And so with the Solanaceae: the same scientific, technological, and demographic forces that revolutionized horticulture laid the groundwork, as it were, for the nightshades' wholesale occupation of the Victorian garden during the final third of the century, permanently recasting the family's character for the modern British imagination. Margaret Plues, the author of popular botanies in the sixties, articulates the more balanced and even laudatory assessment of the Solanaceae that would prevail at the end of the century, thanks to the increasing popularity of its edible species. Plues asserts that "there is much to be said in favour" of the family, for the nightshades "possess a great variety of qualities; some poisonous, some harmless, some actually nutritious." Moreover, even the dangerous species serve as "beneficent medicines": "The Deadly Nightshade and Henbane are cases in point. Rash meddlers with these get illness, or even death, as the reward of their temerity, whilst the agonised sufferer blesses the same Henbane for his much-needed sleep, and the painful headache yields before homæopathic doses of Belladonna as to a magic charm. Truly the Nightshades have no need to blush for their members."[109] Leading the way to this new acceptance was the common potato *Solanum tuberosum*, a member of the family's largest and signature genus, as the first edible nightshade to gain a sizable foothold in British gardens and in most of its citizens' hearts. As early as 1819, Rees's *Cyclopedia* had suggested that the potato represented the type of the comforting quality indicated by its genus name;[110] but the real work of redemption began at mid-century when the blight raised the tuber's reputation from humble vegetable to martyr to noble survivor—enforcing an acute awareness of its eminence as the nation's most important non-grain foodstuff. Like the potato, the other important edible nightshades—capsicums, tomatoes, and eggplants—had been in Gerard's Holburn garden during Elizabethan times; but it was not until the mid-nineteenth century that the requisite infrastructure was in place

BACK TO THE GARDEN

for successful widespread cultivation of these latter three esculents. When in 1851 the botanist Edward Forbes noted the nightshades' split personality—its "strange mixture in one family of man's deadly enemies, with several of his valued friends"—his use of "deadly" underscored the facility with which the word still served as an epithet for indigenous nightshades; but his singling out tomatoes and capsicums as "valued friends" signaled that change was already taking place thanks to New World immigrants, marking mid-century as the turning point for the family reputation.[111]

Moreover, the recently established scientific name "Solanaceae," which replaced Linnaeus's ominous "Luridae" for these historically "suspect" plants, reinforced the shifting emphasis away from evil and toward the plants' consolatory qualities. Whereas the Old World nightshades, steeped in superstitious lore about their magical and poisonous powers, conjured apparitions of witches and whores, the New World nightshades evoked the wholesomeness of Victorian housewives, those paragons of domestic virtue and providers of home comforts—the very images of consolation. And although the Old World species belladonna, henbane, bittersweet, and black nightshade were still considered by many to be scourges—invaders of gardens, outcasts haunting ruins and waste places, or wildings climbing banks and hedgerows—their New World relatives were settling in comfortably as welcome tenants in every imaginable sort of cultivated space. By 1901 nightshades were everywhere: potatoes, tomatoes, and capsicums filled farms, fields, allotments, plots, kitchen gardens, and the vast greenhouses of marketers. Popular ornamental nightshades like petunias and nicotianas emerged from glass houses to grace parterres, borders, cottage flowerbeds, trellises, pots, and boxes. There were a host of new ornamental solanums: in *The Subtropical Garden* (1879), William Robinson recommended cultivating thirteen different solanums for the beauty of their foliage;[112] and in *Home and Garden* (1901), the famed garden designer Gertrude Jekyll particularly praised the gorgeous *Solanum jasminoides* and *Solanum crispum*, which had become established as hardy garden climbers in the south of England.[113] Conservatories sported stunning daturas, brugmansias, and solandras. Representatives of two nightshade genera, *Physalis* and *Lycium*,[114] had been around since the eighteenth century, but had recently become better known and more visible: the two *Physalis* species bearing fruit with papery husks, popularly known respectively as winter cherries and cape gooseberries, were grown as curiosities and the latter for preserves. Cultivated for hedging, the prickly and woody *Lycium* or box-thorn, also known wryly as the Duke of Argyll's teaplant in commemoration of his lordship's legendary (but possibly apocryphal) misidentification, had naturalized in Suffolk hedgerows.[115] But for whatever reason they were planted, all these nightshades were grown for the pleasure of the gardener, fulfilling the purpose at the root of the Solanaceae name: the Latin *solor*, to comfort and soothe, which became in Middle English *solas*, the great Chaucer's word for pleasure, joy, or solace of the most deeply satisfying nature.[116] Indeed, these nightshades gave the kind of comfort needed to endure the trials of a modern age.

The family's horticultural success was not confined to the New World immigrants, however. The second half of the Victorian era saw as well a major change in the garden prospects of the two most notorious Old World Solanaceae indigenous to England: belladonna, the nightshade primarily responsible for the family's deadly name, the species so vilified by Gerard and his successors; and henbane, its equally poisonous kin. Both species had maintained a modest place for centuries in physic gardens; but as medical science learned how to extract and thus control the potent tropane alkaloids that made these narcotic nightshades highly dangerous but effective drugs, acres of the plants were cultivated "on an agricultural scale" in walled and guarded farms operated by the English pharmaceutical houses founded by William Ransom, Peter Squire, and Stafford Allen, their firms supplying the home drug market and earning England acclaim for its high-quality products.[117] Thus, after being driven from English gardens centuries earlier and eradicated almost to extinction, belladonna, the darkest of nightshades, was finally being pampered and protected like the beautiful lady invoked by her name.

Of course, the nightshade that would ultimately prove to be most dangerous of all was another New World immigrant, one that had arrived about the same time as the potato, to be quickly adopted by the British as the nation's favorite recreational drug: *Nicotiana tabacum*. From the sixteenth century on, as millions of Britain's citizens discovered tobacco's extraordinary power both to soothe and to stimulate, tobacco became the herbal epitome of *solas*: as Oscar Wilde said of the cigarette, "the perfect type of a perfect pleasure."[118] Ironically, however, its very popularity caused tobacco's cultivation in England for commercial purposes to be prohibited shortly after its introduction, since plants imported from Britain's colonies proved to be a lucrative source of revenue for the government. And like the other spoils of empire that became essential to British life such as cotton and tea, the manufacture of finished tobacco products proved immensely profitable. Connoisseurs could still grow tobacco in private gardens for personal use; and thanks to its extreme toxicity, tobacco maintained an important place in English horticulture as a most effective insecticide. Tobacco smoke was extensively employed to fumigate greenhouses and frames, while a liquor obtained from tobacconists could be sprinkled on outdoor plants to rid them of aphids.[119] Nevertheless, this seductive nightshade's most important effect was the mastery it wielded over countless numbers of Victorian men, women, and children who became addicted to snuffing, to chewing, and most of all, to smoking tobacco. Indeed, no other plant could be said to have captured a nation's minds and hearts—not to mention its lungs—so effectively. By the dawn of the twentieth century, tobacco was, as W. A. Penn claimed, truly Britain's sovereign herb.[120]

In short, the revelations about the importance and scope of the nightshade family that occurred during Victoria's reign were nothing short of miraculous, and not least among these discoveries was the firsthand information they provided about how species evolve, in this case by artificial selection. Scores of potato, petunia, and tomato varieties were produced over the course of the nineteenth century, giving visible evidence of the many possibilities inherent in creating new and better species. Just as Arthur Sutton had reported on his company's work to hybridize solanums and create disease-resistant potato varieties, so Liberty Hyde Bailey, the acknowledged "Father of American Horticulture,"[121] chose the hybridization of the petunia as a prime example of evolution taking place from two recognized species in record time—"within little more than half a century."[122] And in an essay wonderfully titled "The Amelioration of the Tomato," Bailey recounts how the species had achieved universal popularity in America within only thirty years thanks to the introduction of the highly successful Trophy variety, which marked "a great landmark in American vegetable growing." Bailey notes that in 1870, just when "the time was ripe for a tomato of a new type" to supplant the small, rough-skinned, irregular-shaped, and watery specimens then available, the "scientific farmer" Colonel George Waring produced what was needed: the large, early, smooth, apple-shaped, and flavorful Trophy. Bailey shows how the history of the Trophy's development through shrewd selection and crossing "illustrates the great plasticity of the tomato, and how quickly it responds to good or bad or modified treatment."[123] Indeed, the species' astonishing adaptability—its capacity to "improve" according to the desires and standards of another species—forged a bond between plant and gardener that ensured the tomato's survival and "amelioration" under domestication, as it had with its relatives the potato, petunia, and capsicum.

It was therefore in every way fitting that at the Royal Horticultural Society's Fruit Show held at the Crystal Palace in 1897 as part of the Queen's Diamond Jubilee festivities, the tomato was honored as the stellar example of horticultural progress in England over the past sixty years. Magnificent tomato cultivars from Her Majesty's Gardens at Frogmore and the Reading nurseries of Sutton and Sons rivaled all the other fruits on display, outdoing perennial English favorites like apples and pears, grapes and strawberries.[124]

A market gardener in attendance named Roupell noted that "there was practically an unlimited demand" for tomatoes nowadays; and even though Arthur Sutton had extolled the tomato's triumph as a vegetable in his lecture during the celebration, Roupell said that most of his crop were sold "for dessert purposes or for salads"—that is, "not so much used as a vegetable as a fruit" and presumably eaten raw—thereby underscoring the tomato's success on both sides of the Atlantic as a versatile, delicious, and favorite esculent.[125] After a horticultural apprenticeship in England lasting three hundred years, the ascendancy of the tomato toward the end of the nineteenth century marked the nightshades' final, decisive victory over the curse that had plagued the family ever since Gerard expelled belladonna from the English garden and the fate of its family members took such widely divergent and dramatic turns.

Arthur Rackham's delightful illustration "Mr. and Mrs. Vinegar at Home," the frontispiece for Flora Annie Steel's 1918 collection *English Fairy Tales*, visualizes the essential place tomatoes and peppers had attained in British culinary culture by the end of the era, providing a charming coda to the history of the Victorian nightshades (see fig. 45).[126] The tale "Mr. and Mrs. Vinegar" itself, which originates from oral tradition in the west of England and dates from the mid-nineteenth century, recounts the misadventures of a hapless couple after the hotheaded wife accidentally shatters the vinegar bottle they live in with her broom.[127] Steel altered the narrative by changing the dwelling to a pickle-jar; but Rackham probably selected the horticultural details for the illustration, which depicts the tale's back story: the happier time described in the opening paragraph, when "Mr. Vinegar tilled his garden with a pickle-fork and grew vegetables for pickling," before his wife "lost her temper with a cobweb" and brought down the "pickle-jar house about her ears."

Considering the book's date of publication at the end of the First World War, Rackham's decision to illustrate the Vinegars "at home" as a portrait of peaceful domesticity—representing the pre-disaster Pickalilli Cottage situated within a bountiful working garden—poignantly emphasizes the timeliness of the scene as a nostalgic, retrospective commentary on an idyllic past: in this case, a return to an imaginary Victorian garden with real vegetables in it. Despite the inclusion of a few elves picking mushrooms and loading the wagon for market, the picture is less fairy-tale than wishful fancy. And that Steel chose to substitute a uniformly cylindrical pickle-jar for a narrow-necked and thus presumably less habitable vinegar bottle as the cottagers' domicile was especially fortunate, for the jar elicits a comparison to actual dwellings connected with gardens, the greenhouses and conservatories that became so integral to cultivating plants during the mid-Victorian period. On a fanciful, miniature scale, it is also reminiscent of that most iconic of glass houses, the Crystal Palace, which was designed to accommodate and cultivate people rather more than their vegetable charges. In short, the illustration's quaint, touching charm derives from the way the scene conveys a deeply satisfying sense of "at-home-ness" by surrounding a fanciful glass dwelling with the familiar bounty of a cottager's kitchen garden—here, the actual crops needed to flavor vinegar and make various sorts of pickles and ketchups: cabbages, cauliflower, raspberries, walnuts, cucumbers, garlic, and of course, peppers and tomatoes.

The vigorously growing nightshades surpass all the other produce. With their robust luxuriance, they dominate the right side of the illustration while partially framing Pickalilli Cottage and its sour-faced tenant Mrs. Vinegar, whose acid expression seems less menacing thanks to her colorful costume

FIGURE 45. "Mr. and Mrs. Vinegar at Home," Arthur Rackham, frontispiece, Flora Annie Steel, *English Fairy Tales*. (Image from Wikipedia Commons; digitized by Internet Archive)

that visually echoes the rich red of the capsicums and cheerful orange of the tomatoes so close by. Their vibrant presence suffuses the illustration with the deep pleasure rooted in their solanaceous name, cheerfully belying any adverse association with night or shade, and promising consolation for the bitterness of the time to come.

NOTES

1. A Family Plot

1. Lindley, *Vegetable Kingdom*, xxiv.
2. Ibid., fig. 227, 618. In the figure, the "ripe fruit of Solanum dulcamara" (2) is mislabeled. They are *Solanum nigrum* (black nightshade) berries.
3. Ibid., 622, 685.
4. Heiser, *Fascinating Nightshades*, 33. Heiser defines the curse as "the belief that all plants in this family were poisonous."
5. For an intriguing study of how the global movements of plants, because of their "transcendent referentiality"—their living "both within and outside the novel's narrative world"—informs the transition from realism to modernism in Victorian fiction, see Chang, *Novel Cultivations*, 5.
6. Endersby, *Orchid*, 77–78.
7. "The Private History of the Palace of Glass," *Household Words*, January 18, 1851: 385–91. See also Holway, *Flower of Empire*, for the story of *Victoria regia*; Keogh, *Wardian Case*; and Armstrong, *Glassworlds*, esp. chaps. 6 and 7. For an account of how Britain's vegetable traffic included both imports and exports, see Herbert, *Flora's Empire*, 61.
8. Forbes, "Vegetable World," I, II.
9. Ibid., I.
10. Ibid.
11. Ibid., IV–VI.
12. Ibid., VIII.
13. Ibid., I.
14. Tennyson, *In Memoriam*, section 56: 36.
15. Forbes, VIII.
16. Thompson, *Mystic Mandrake*, 168.
17. Ruck, "Gods and Plants," 140–41.
18. Grieve, *Modern Herbal*, vol. 2: 583.
19. Schultes, Hofmann, and Ratsch, eds., "Hexing Herbs," 86–91.
20. Gerard, *Herbal* [1633], 37.
21. Ibid., 340–41.
22. Shteir, *Cultivating Women, Cultivating Science*, 13–20.
23. Linnaeus, *Philosophia Botanica*, 300.
24. Beaumont and Fletcher, *Coxcomb*, act 2, scene 1, 329.
25. J. C. Loudon, *Encyclopedia of Plants*, 154–55.
26. Friend, *Flowers and Flower Lore*, 531.

284 NOTES TO PAGES 11–21

27. Skeat, ed., *Etymological Dictionary*, s.v., "nightshade."
28. *Official Catalogue*, vol. 3: 1336, 1337.
29. Salaman, *History of the Potato*, 243.
30. Ibid., 427, 438.
31. For example, Greenaway, *Language of Flowers*, 33.
32. *Official Catalogue*, vol. 1: 207, 208; vol. 2: 642, 650, 651, 966; vol. 3: 1043, 1048, 1069, 1070, 1088, 1153, 1242, 1381; vol. 3: 1126–27.
33. Ibid., vol. 2: 874, 952, 972, 973, 976, 978, 981, 993; vol. 2: 1312, 1319, 1334, 1386; vol. 1: 981.
34. Allen and Hatfield, *Medicinal Plants*, 198–99.
35. Burton, *Anatomy of Melancholy*, 228.
36. Gerard, *Herbal* [1633], 356; 357–58.
37. Ibid., 359–61.
38. Goodman, *Tobacco in History*, 59.
39. Gately, *Tobacco*, 157–58, 185.
40. Hilton, *Smoking*, 2, 11.
41. Ibid., 140–44.
42. Fairholt, *Tobacco*, 1–2.
43. Cooke, *Seven Sisters of Sleep*, 243.
44. Hilton, 63.
45. Penn, *Soverane Herbe*, preface; Ruskin, *Queen of the Air*, 19: 368.
46. *Official Catalogue*, vol. 2: 813.
47. Ibid., vol. 2: 873, 938, 950, 963, 974, 993; vol 2: 592, 605, 606, 627, 639, 801. Vol. 3: 1040, 1087, 1126, 1132, 1153, 1259, 1261, 1314, 1369, 1371, 1386, 1402, 1409, 1416, 1453. Vol. 1: 303; supplement, 198.
48. "Tobacco," *The Illustrated London News*, August 9, 1851, 191.
49. "The Most Popular Plant in the World," *Chambers's Journal*, vol. 22, no. 50 (December 16, 1854): 393–95; 393.

2. Bittersweet

1. Thoreau, *Wild Fruits*, 95.
2. Blackmore, *Maid of Sker*, 173.
3. Reid, *Origin of British Flora*, 63, 134, 175.
4. Watson, *Cybele Britannica*, vol. 2, 185–86.
5. Thoreau, 95.
6. Darwin, *Origin of Species*, 484–85.
7. Hyman, *Tangled Bank*, 33.
8. See Ayres, *Aliveness of Plants*; Darwin, *Power of Movement*, 1.
9. Darwin, *Movements and Habits*, 43.
10. Ibid., 43, 72, 111–12.
11. Ibid., 43.
12. Cooke, *Freaks and Marvels*, 191, 193.

NOTES TO PAGES 21–33

13. Eliot, *Adam Bede*, chap. 18: 233–34.
14. Eliot, *Felix Holt*, "Author's Introduction," 76–77.
15. Bromfield, "Catalogue of Plants," 595.
16. Bentham, *British Flora*, 1st ed., 384–85; for example, 7th ed., 318–19.
17. To J. D. Hooker, July 21, [1858]. Darwin Correspondence Project.
18. Morris, "Unknown Church," 29.
19. See Jill, Duchess of Hamilton, Hart, and Simmons, *Gardens of William Morris*, 147.
20. Mackail, *William Morris*, vol. 1: 200; Keats, *Complete Poems*, 348.
21. MacCarthy, *William Morris*, 221–23.
22. Morris, *Life and Death of Jason*, book 7: 132.
23. Chaucer, "Canon's Yeoman's Tale," line 878: "For unto hem it is a bitter sweete"; Wright, ed., *Canterbury Tales*, 313.
24. Anderson, *Herbals*, 152–53.
25. Turner, *New Herball*, vol. 2: 638.
26. Allen, and Hatfield, eds. *Medicinal Plants*, 198.
27. Sumner, *Medicinal Plants*, 98–99.
28. Watts, *Dictionary of Chemistry*, vol. 5: 345–46; vol. 4: 642.
29. Lindley, *Flora Medica*, 512.
30. Thoreau, 95.
31. For example, Marder, *Plant-Thinking*, and Hall, *Plants as Persons*.
32. Taylor, *Sagacity and Morality*, 104, 115–18.
33. Mabey, *Flora Britannica*, 302.
34. Scourse, *Victorians and Their Flowers*, 31–32.
35. Ibid., 30.
36. Hey, *Moral of Flowers*, 5, note.
37. Ibid., vii.
38. Ibid., 75.
39. Ibid.; Stephenson and Churchill. *Medical Botany*, vol. 1: plate 17.
40. Hey, 76.
41. Ibid.
42. Ibid., 77.
43. Ibid.
44. Ibid.
45. Ibid., 78.
46. Ibid., 75.
47. Coleridge, "To a Friend," in *Poetical Works*, part 1, vol. 2: 133.
48. Twamley, "A November Stroll," in *Romance of Nature*, 217.
49. Ibid., 218–19.
50. Ibid., plate 28.
51. Ann Radcliffe, *Mysteries of Udolpho*, vol. 2, chap. 5. Project Gutenberg eBook. https://www.gutenberg.org/files/3268/3268-h/3268-h.htm.
52. Burke, *Sublime and Beautiful*, 73.

286 NOTES TO PAGES 33–44

53. Ibid., 51.

54. Radcliffe, *Journey of 1794,* 487.

55. Burke, 142–43; 157–58.

56. Gilpin, *Three Essays,* 3–5.

57. Gilpin, *Observations,* 166.

58. Ibid., v, before page 165.

59. Beckford, *Memoirs,* vol. 1: 150–51.

60. Wordsworth, *Prelude,* book 2: lines 102–7, 209.

61. Ibid., lines 110–14, 209.

62. Ibid., lines 116–17, 209–10.

63. Ibid., lines 125–28, 210.

64. Walpole, *Correspondence,* vol. 39, To Henry Seymour Conway, December 31, 1774: 237, and note 18.

65. West, *Antiquities of Furness,* 66.

66. Ibid., 191.

67. "Meaning of Bekan," vol. 6: 56–57.

68. Ibid., 153, 210, 355–56.

69. T. West, *Antiquities of Furness,* 191.

70. Ibid., William Close, appendix, 376–77.

71. Radcliffe, *Journey of 1794,* 489.

72. Explore Low Furness: "Vale of Nightshade Walk," https://www.explore lowfurness.co.uk/guidedwalks.htm.

73. See Wood, "Visualizations of Furness Abbey."

74. Landow, *Theories of John Ruskin,* chap. 3: 237–38.

75. Ruskin, *Works,* vol. 5: *Modern Painters,* vol. 3, chap. 3, 67–68.

76. Ruskin, *Works,* vol. 6: *Modern Painters,* vol. 4, chap. 3, 68.

77. Ibid.

78. Ibid., vol. 19, *Queen of the Air,* chap. 2: 368–69.

79. Keats, *Complete Poems,* 348.

80. de Almeida, *Medicine and Keats,* chap. 12: 170–71.

81. Shelley, *Rosalind and Helen,* lines 207–10.

82. Shelley, *Poetry and Prose, Epipsychidion,* 380, lines 256–66.

83. Crook and Guiton, *Shelley's Venomed Melody,* 137, 147–48, 152.

84. Bagehot, "Percy Bysshe Shelley" (1856), in *Literary Studies,* vol. 1: 110.

3. Dulcamara

1. Seaton, *Language of Flowers,* 68.

2. Ibid., 61–62.

3. Goody, *Culture of Flowers,* 230–31.

4. Seaton, 68.

5. Delachénaye, *Abécédaire,* 149.

6. de Latour, *Le langage des fleurs,* 299–300.

NOTES TO PAGES 44–56

7. Ibid., plate following 238.

8. Shoberl, *Language of Flowers*, 254.

9. Ibid., plate following 254.

10. Seaton, 143–45.

11. Shoberl, 299–300.

12. Eliot, *Felix Holt*, "Author's Introduction," 76–77.

13. Shelley, *Poetry and Prose*, "Epipsychidion," 380, line 257.

14. Phillips, *Floral Emblems*, 306.

15. Tyas, *Sentiment of Flowers* (1842), vi.

16. Ibid., 351–60.

17. Ibid., 320.

18. Ibid., 321.

19. Ibid.

20. Tyas, *Sentiment of Flowers* (1869), xii; plate 7: 278.

21. Shteir, *Cultivating Women, Cultivating Science*, chaps. 2–5.

22. Ibid., 153–57.

23. Rousseau, *Letters on Botany*, 19.

24. Ibid., 71.

25. Ibid., Martyn, Translator's Preface, v.

26. Ibid., x–xi.

27. Ibid., Letter 16: 190–91.

28. Roberts and Wink, eds. *Alkaloids*, 31; Thiselton-Dyer, *Folk-Lore of Plants*, 48.

29. Linnaeus, *Philosophica Botanica*, 300.

30. Rousseau (Martyn), 193–94.

31. Ibid., 197–98.

32. Ibid., 200–201.

33. Wakefield, *Introduction to Botany*, 2.

34. Ibid., 61.

35. Ibid., 62.

36. Ibid., 62–63.

37. Ibid.

38. Ibid., 63.

39. Ibid.

40. Ibid., 64.

41. Lindley, *Ladies' Botany*, vol. 1: 180.

42. Ibid., iv.

43. Ibid., 4.

44. Ibid., 184–85.

45. Ibid., 186–87.

46. Ibid., 186.

47. Ibid.

48. Ruskin, *Works*, vol. 19: 368.

49. Ibid., 369.

50. Lindley, vol. 1: 187–88.
51. Ibid., 77.
52. Ruskin, *Works*, vol. 19: 369.
53. Lindley, plate 15.
54. Jane Loudon, *Botany for Ladies*, vi.
55. Ibid., 142.
56. Ibid., 143.
57. Pratt, *Flowering Plants* (1855), vol. 4: 62.
58. Ibid., vol. 2 (1905), plate 151.
59. Ibid., 69.
60. Pratt, *Wild Flowers*, vol. 1: 57.
61. Hall, *Elves in Anglo-Saxon England*, 98, 155.
62. Gerard, *Herbal* [1633], 350; Thoreau, *Wild Fruits*, 96.
63. Gerard, 350.
64. Culpeper, *Complete Herbal* [1826], 11–12.
65. Coles, *Art of Simpling*, 76.
66. Davis, *Popular Magic*, 110, 119, 133–37.
67. Culpeper, 11.
68. Coles, 106.
69. Mabberley, *Plant-Book*, 669.
70. Vickery, *Dictionary of Plant Lore*, 35.
71. Culpepper, 11; Coles, 54–55.
72. Allen and Hatfield, eds., *Medicinal Plants*, 198–99.
73. Vickery, 35; Allen and Hatfield, 199.
74. Pulteney, *Writings of Linnaeus*, 476; *Medical and Philosophical Commentaries*, vol. 3 (1775): 15.
75. Gataker, *Observations on Solanum*, 6, 25–26; Bromfeild, *English Nightshades*, iii–iv; 67–77.
76. Woodville, *Medical Botany*, vol. 2: 241–42.
77. Spratt, *Flora Medica*, vol. 1: 9–10.
78. Ibid., vol. 1: 11–12.
79. Woodville, vol. 5: 151–52.
80. Hahnemann, *Organon of Medicine*, 55.
81. Nicholls, *Homeopathy*, 73.
82. Hahnemann, *Organon*, 330–33.
83. Ibid.
84. Hahnemann, *Materia medica pura*, vol. 1: 18.
85. Ibid., 583.
86. Ruddock, *Stepping-Stone to Homeopathy*, 3, 8.
87. *Punch* frequently leveled attacks on patent medicines and homeopathy, as in "The Finest Balsams of Arabia," in which homeopathy is mentioned as an example of "quack medicines" like the product named in the article's title; see *Punch*, vol. 26 (1854): 40.

NOTES TO PAGES 67–78

88. Ibid., 214–15.
89. Ashbrook, *Donizetti*, 73.
90. Ibid., 553.
91. Donizetti, *L'elisir d'amore*, act 1, scene 6, page 83.
92. Ashbrook, *Donizetti*, 175.
93. Cheer, *Great LaBlache*, 217–18; 144.
94. Henry Morley, "Latest Intelligence from Spirits," *Household Words*, vol. 11 (June 30, 1855): 513–15.
95. Wilkie Collins, "Dr. Dulcamara, M.P.," *Household Words*, vol. 18 (December 18, 1858): 49–52.
96. Dickens, *Uncollected Writings*, vol. 2: 619–25.
97. "Bright View of Reform," 111.
98. "Dr. Dulcamara in Dublin," *Punch*, vol. 51 (November 10, 1866): 193, 232.
99. Stedman, *W. S. Gilbert*, 34.
100. Gilbert, *Dulcamara*, scene 3; Rappaport, *Beautiful for Ever*.
101. "Dr. Dulcamara Up to Date," *Punch*, vol. 105 (November 11, 1893): 218–19.
102. Donizetti, *L'elisir d'amore*, act 1, scene 5, page 64.
103. Yonge, *Herb of the Field*, 125.
104. "Nightshade and Nightshade," *Punch*, vol. 56 (June 5, 1869): 236.

4. Belladonna

1. Schedule A in "A Bill to Regulate the Sale of Poisons and Alter and Amend the Pharmacy Act," *Pharmaceutical Journal*, 2nd ser., vol. 9, no. 12 (1867–68): 576.
2. Tanner, *Woodland Plants*, 188.
3. Syme, *English Botany*, illustrated by Sowerby, vol. 6 (1866), plate 934.
4. Christison, *Treatise on Poisons*, 720.
5. Squire, *Companion to the British Pharmacopoeia*, 43.
6. Gerard, *Herbal* [1633], 340–41.
7. Culpeper, *Complete Herbal*, 104–5.
8. Ibid.
9. Buchanan, *History of Scotland*, vol. 1, book 7: 264.
10. Skeat, *Etymological Dictionary*, s.v. "dwale."
11. Chaucer, "Reeve's Tale," line 241, in *Complete Works*, vol. 4: 121.
12. *Promptorium parvulorum*, s.v. "dwale," 134.
13. For example, Johns, *Flowers of the Field*, 64. Shakespeare, however, supposedly got his information from another contemporary history, Holinshed's *Chronicles*, where the soporific agent is described as "the juice of mekilwoort berries," which produced essentially the same results (Holinshed, *Chronicles*, vol. 5: 267).
14. Coles, *Art of Simpling*, 90.
15. Miller, *Gardeners Dictionary* (1735), vol. 1, s.v. "belladona."
16. For example, Miller, *Gardeners Dictionary*, vol. 1 (1754), s.v. "belladona"; and 8th ed. (1767), s.v. "atropa."

NOTES TO PAGES 78–88

17. Martyn, ed., Rousseau's *Letters on Botany*, 196–97.
18. Wakefield, *Introduction to Botany*, 63.
19. Johnson, ed., *Cottage Gardeners' Dictionary*, s.v. "atropa," 94.
20. "Poisoning with the Berries of *Atropa Belladonna*, or Deadly Nightshade," *Pharmaceutical Journal*, vol. 6, no. 4 (October 1846): 174–77.
21. Taylor, *On Poisons*, 617.
22. *Old Bailey Proceedings Online* (www.oldbaileyonline.org, version 8.0), September 1846, trial of JAMES HILLARD (t18460921–1863).
23. "Poisoning with the Berries."
24. Stearn, *John Lindley, 1799–1865*, 51.
25. Lindley, *Medical and Oeconomical Botany*, 204, illustration 276.
26. "Familiar Botany: Deadly Nightshade," *The Gardeners' Chronicle*, September 12, 1846, 612.
27. Johns, *Flowers of the Field*, 64; compare to Britton and Holland, *English Plant-Names*, s.v. "dwale," 162.
28. "Familiar Botany: Deadly Nightshade."
29. Ibid.
30. "Home Correspondence: Atropa Belladonna," *The Gardeners' Chronicle*, October 17, 1846, 694.
31. "Poisonous Plants of England Portrayed and Described," *The Lady's Newspaper*, issue 158 (Saturday, January 5, 1850): 3–4.
32. Johns, *Flowers of the Field*, 64.
33. Yonge, *Herb of the Field*, v.
34. Ibid., 123.
35. Ibid.
36. Ibid., 124.
37. Ibid., 124–25.
38. Pratt, *Flowering Plants* [1855], vol. 4: 71–72.
39. Bromfield, "Catalogue of Plants," 594.
40. "Nightshade and Nightshade," *Punch*, vol. 56 (June 5, 1869): 236.
41. Bromfield, 594.
42. "Netley Abbey," 543.
43. Ibid., 545.
44. Bromfield, 594.
45. Ibid., 205.
46. Ibid., 595.
47. Ibid., 596; Lees, *Botanical Looker-Out*, 299–300.
48. Ingrouille, *Historical Ecology*, 265.
49. Bromfield, 594, note.
50. "Evidence of Rabbits in Roman Times, Say Academics," *BBC News*, April 18, 2019; Ingrouille, 241, 263.
51. "Alleged Poisoning by Atropine," *Pharmaceutical Journal*, vol. 7, no. 3 (September 1, 1865): 127–29.

52. Bromfield, 595; Lees, 300.

53. Bromfield, 596.

54. Watson, *Cybele Britannica*, vol. 1: 2.

55. Ibid., vol. 1: 63–64.

56. Ibid., vol. 1: 63; vol. 2: 186.

57. Ibid., vol. 2: 186.

58. T. Ashby, "Excavations at Caerwent," 401–5.

59. Ingrouille, 241.

60. Allen, "British and Foreign," 123.

61. Allen, 132.

62. Ibid., 129.

63. Ibid., 132.

64. Dickens, *Our Mutual Friend*, book 1, chap. 11: 128, 140.

65. "Poisoning by Belladonna," *Pharmaceutical Journal*, 2nd ser., vol. 11, no. 1 (July 1869): 45–46.

66. Bell and Redwood, *Progress of Pharmacy in Great Britain*, 277–78; Holloway, *Royal Pharmaceutical Society*, 221–30.

67. "Prevention of the Misuse of Poisons," *Pharmaceutical Journal*, 2nd ser., vol. 11, no. I (July 1869): 1–3.

68. Dickens, *David Copperfield*, chap. 28: 413.

69. Bell and Redwood, *Progress of Pharmacy*, 369–381; Schedule A in "A Bill to Regulate the Sale of Poisons and Alter and Amend the Pharmacy Act."

70. *British Pharmacopoeia, 1867,* 49–51; 54–55; 105, 114, 172, 182, 320–21; 352–53.

71. J. H. Whelan, "Some Medical and Surgical Uses of Belladonna or Its Alkaloid," *The Lancet*, vol. 2 (September 2, 1882): 348–49; 348.

72. "Poisoning by Belladonna."

73. "Dudgeon, Robert Ellis," *Oxford Dictionary of National Biography*, 1912 Supplement. https://en.wikisource.org/wiki/Dictionary_of_National_Biography,_1912_supplement/Dudgeon,_Robert_Ellis.

74. Hahnemann, *Materia medica pura*, vol. 1: 198–255.

75. Lai, "Legacy of Atropos," 1794–95. Quoting H. G. Morton, "Atropine Intoxication: Its Manifestations in Infants and Children," *Journal of Pediatrics* 14 (1939): 755–60.

76. Hahnemann, vol. 1: 198.

77. Dudgeon, *Lectures*, xxv–xxvi.

78. Alcott, *Little Women*, chap. 17: 142.

79. Dudgeon, 540.

80. John Harley, "Lectures on the Physiological Action and Therapeutic Uses of Conium, Belladonna, and Hyoscyamus Alone and in Combination with Opium," *Pharmaceutical Journal*, ser. 2, vol. 9 (April 1868): 471–75.

81. Harley, *Vegetable Neurotics*, preface, v.

82. Ibid., vi.

83. Hahnemann, vol. 2, 285.

NOTES TO PAGES 97–104

84. Harley, *Vegetable Neurotics*, v.
85. Ibid., 309.
86. Ibid., 290.
87. Ibid., 244.
88. Ibid.
89. Ibid.
90. Whelan, 348.
91. Holzman, "Legacy of Atropos," 248.
92. Harley, *Vegetable Neurotics*, 244.
93. Ibid., 239, 247, 250, 254, 266, 268.
94. Ibid., 245.
95. Ibid., 244.
96. Ibid., 239.
97. Burnett, *Outlines of Botany*, vol. 2, 1995.
98. Robert Christison's *Treatise on Poisons* of 1829; and Pereira's *Elements of Materia Medica*, first published 1842.
99. Pereira, *Elements of Materia Medica*, vol. 2: 472. De Claubry's account first appeared in the December 1813 issue of the French medical periodical *Sedillot's Journal*; Pereira's source is Mathieu Orfila's *Treatise on Poisons*.
100. Forbes, "Note on Belladonna," 405.
101. Pereira, vol. 2: 470; 473.
102. Hahnemann, *Materia medica pura*, vol. 1: 254–55.
103. "Sheffield Pharmaceutical and Chemical Association," *Pharmaceutical Journal*, 3rd ser., vol. 1 (November 26, 1870): 430–31.
104. Pereira, *Elements of Materia Medica*, vol. 2: 473.
105. Beverley, *History of Virginia*, chap. 4: 110.
106. Harley, 339.
107. Ibid., 344.
108. Pereira, vol. 2, 467, 468.
109. Charles Dickens to Catherine Dickens, November 1, 1838, *Letters*, vol. 1: 1833–56, 2nd. ed. (London: Chapman & Hall, 1880).
110. Dickens, *Our Mutual Friend*, book 1, chap. 10: 114.
111. Harley, 337.
112. Ibid., 330–32.
113. Bentley and Redwood, eds., *Dr. Pereira's Elements of Materia Medica*, 1872. For example, references to Harley, 605, 606, 607.
114. Harley, ed., *Royle's Manual*, 6th ed., 1876. For example, see pages on Atropa Belladonna, 488–96.
115. Holmes, "Horticulture," 56.
116. Jonathan Pereira, "Mitcham: Its Physic Gardeners and Plants," *Pharmaceutical Journal*, vol. 10, no. 3 (September 1, 1850): 115–16.
117. Cited by Holmes, "Horticulture," 60.
118. "Garden Memoranda," *Gardeners' Chronicle*, October 19, 1861, 929.

NOTES TO PAGES 104–114

119. Stafford Allen, *Romance of Empire Drugs*, 20.
120. "The Cultivation of Medicinal Plants at Hitchin, Herts," *Gardeners' Chronicle*, May 19, 1860: 456–57.
121. Wallace, *Children of the Labouring Poor*, 12.
122. Kilmer, "Lands Where Drugs Grow," 160.
123. Holmes, "Horticulture," 60.

5. Victoria's Secrets

1. Ticknor, *Miss Belladonna*, chap. 1: 2, 5–6.
2. King, *Bloom*, 3–8.
3. Markham, *English Housewife*, chap. 1: 40.
4. Ruskin, *Queen of the Air, Works*, vol. 19, chap. 2: 368.
5. Forbes, "Note on Belladonna," 403.
6. Ibid., 404.
7. Gerard, *Herbal* [1633], 341; Beaumont and Fletcher, *Coxcomb*, act 2, scene 1, page 329.
8. Martyn, ed., Rousseau's *Letters on Botany*, 196–97.
9. Lindley, *Loudon's Encyclopedia of Plants*, 154–55.
10. Ibid., 155.
11. "On the Legends and History of Certain Plants," 149–53.
12. Friend, *Flowers and Flower Lore*, 531.
13. "Antropa—Nightshade," Bodleian Libraries, University of Oxford, John Johnson Collection: Trade in Prints and Scraps 18 (33b).
14. Lucot, *Emblèmes*, 21.
15. Martyn, 197; Milton, *Lycidas*, 99: lines 75–76.
16. For example, Burke, *Miniature Language of Flowers*,10; *The Language of Flowers: An Alphabet of Floral Emblems*, London: T. Nelson and Sons, 10; *The Language and Poetry of Flowers*, London: Routledge and Sons, 11.
17. For example, *The Language of Flowers*, London: Ernest Nister [18—]; Ingram, *Flora Symbolica* [1869], 355.
18. Greenaway, *Language of Flowers*, 9.
19. For example, Ingram, *Flora Symbolica*, 355, 357.
20. Pratt and Miller, *Language of Flowers*, 17, 20, 27, 26.
21. Campbell, "Don't Say It with Nightshades," 607–15.
22. Thackeray, *Mrs. Perkins's Ball*, 22.
23. Seaton, *Language of Flowers*, 68.
24. Thackeray, *Vanity Fair*, chap. 4: 75.
25. Ibid., chap. 29: 343.
26. Ibid., chap. 67: 789.
27. Ibid., chap. 64: 752.
28. Ibid., chap. 64: 752–773.
29. Ibid., chap. 64: 753, 754.

294   NOTES TO PAGES 115–125

30. Ibid., chap. 67: 795, 796.
31. Meredith, *Richard Feverel*, chap. 36: 384–85; chap. 37: 408–9, 416.
32. Ibid., chap. 38: 426, 430, 422, 440, 441, 435.
33. Gilfillan, preface, *Nightshade*.
34. "Nightshade," *The Saturday Review*, vol. 4, no. 90 (July 18, 1857): 6–67.
35. Johnston, *Nightshade*, chap. 5: 25; chap. 24: 182.
36. Ibid., chap. 22: 165.
37. Bulwer-Lytton, *Strange Story*, chap. 65: 329.
38. Fitzgerald, *Bella Donna*, vol. 1, book 1, chap. 1: 15.
39. Ibid., vol. 1, book 1, chap. 16: 179; vol. 2, book 3, chap. 1: 157, 158.
40. Ibid., vol. 1, book 2, chap. 11: 297; vol. 2, book 3, chap. 12: 273.
41. Fitzgerald, *Seventy-Five Brooke Street*, "Advertisement," vol. 1.
42. "Jenny Bell," *The Saturday Review*, vol. 21, no. 540 (March 3, 1866): 267.
43. Dickens, *Pictures from Italy*, 352; *Little Dorrit*, chap. 1: 4.
44. William Michael Rossetti, *Dante Gabriel Rossetti*, vol. 1, chap. 34: 315.
45. Hake, *Parables and Tales, The Deadly Nightshade*, stanza 29: 87; stanza 4: 78; stanza 15: 82.
46. Ibid., *The Lily of the Valley*, stanza 22: 67.
47. Ibid., *The Deadly Nightshade*, stanza 1: 77; stanza 2: 78.
48. Hake, *Memoirs*, 236; Binding for Hake's *Parables and Tales*, https://www .victorianweb.org/art/design/books/cooke2.html.
49. To Alfred Hake, ca. September 29, 1872, 72.85, in Dante Gabriel Rossetti, *Correspondence*, vol. 5: 292.
50. Christina Rossetti, *Sing-Song*, 116–17.
51. Tarr, "Covent Goblin Market."
52. Brown, *Memoir*, in *Dwale Bluth*, vol. 1: 13.
53. *Dwale Bluth*, vol. 1:166, 130, 120.
54. Ibid., vol. 1: 130.
55. Ibid., vol. 1: 142, 144.
56. Ibid., vol. 1: 177.
57. Ibid., vol. 1:144.
58. Ingram, *Oliver Madox Brown*, 226, 171.
59. *Dwale Bluth*, vol. 1: 178–79.
60. Ibid., vol. 1: 161; 110, 114; 198.
61. Ibid., vol. 1: 201, 202.
62. Ibid., vol. 1: 207–8.
63. Ibid., vol. 1: 232.
64. Ibid., vol. 1: 234.
65. Hartley, *Go-Between*, chap. 21: 280.
66. Hardy, *Jude the Obscure*, part 1, chap. 6: 32.
67. Ibid., part 6, chap. 6: 297, 298; part 6, chap. 7: 303–5.
68. Crane, *Floral Fantasy*, [15–16].
69. Crane, *Flower Wedding*, 37.

NOTES TO PAGES 125–133

70. Lankester, *Flowers Worth Notice*, 95.
71. "Englishwoman's Conversazione," in *The Englishwoman's Domestic Magazine*, issue 31 (July 1, 1867): 390.
72. "Fine Eyes for Foolish Girls," *Punch*,3 vol. 31 (August 16, 1856): 70.
73. "False Fine Eyes," *Punch*, vol. 47 (July 16, 1864): 23.
74. "Song of the Passee Belle," *Punch*, vol. 57 (July 17, 1869): 20.
75. Hibbert, *Queen Victoria*, 464.
76. Braddon, *Phantom Fortune*, chap. 30: 254.
77. Bailey, *Use of Belladonna*, 15, 19, 29–65.
78. Ibid., 22–23.
79. Ibid., 12, 13–14.
80. Tilt, *Uterine Therapeutics*, 139, 264–265; 366.
81. Ibid., 150.
82. Ad in *Boston Post*, August 30, 1896: 19. Retrieved from the U.S. Forestry Service site: www.fs.fed.us/wildflowers/ethnobotany/Mind_and_Spirit/images/solanaceae/PlastersAdvertisement.
83. J. Bower Harrison, "The Effects of a Belladonna Plaster," *The British Medical Journal*, vol. 1, no. 594 (May 18, 1872): 520–21.
84. Tilt, *Uterine Therapeutics*, 150.
85. Harley, *Vegetable Neurotics*, 231; Tilt, *Uterine Therapeutics*, 150, 151, 152, 153, 372, 150.
86. Tilt, *Change of Life*, v.
87. Ibid., 103.
88. Ibid., 153.
89. Ibid., 38–39; 102; 156, 103–4.
90. Tilt, *Uterine Therapeutics*, 3.
91. Meigs, *Woman: Diseases and Remedies*, Letter 2: 34.
92. Thoms, "Anesthésie à la reine," 340.
93. Velpeau, *Treatise on Midwifery*, 361.
94. Meigs, *Woman: Diseases and Remedies*, Letter 44: 681.
95. Conquest, *Outlines of Midwifery*, 122–23.
96. Ramsbotham, *Obstetric Medicine*, 240.
97. Simpson, *Obstetrical and Gynaecological Works*, 10.
98. Simpson, *Clinical Lectures*, 55, 104, 183, 215, 250, 501. West, *Diseases of Women*, 87, 106, 214, 645–646, 403. Ashwell, *Diseases Peculiar to Women*, 98, 151, 181, 280, 283.
99. Simpson, *Obstetrical and Gynaecological Works*, 770; 664–666. See, for example, Churchill, *Diseases of Women*, 37; Smith, *Practical Gynaecology*, 86, 167.
100. Gill, *Florence Nightingale*, 458–59.
101. "Witchcraft in the 19th Century," *The Times*, issue 16338 (Monday, February 13, 1837): 6, col. F; issue 17461 (Saturday, September 12, 1840): 6, col. B; issue 23106 (Thursday, September 23, 1858): 11, col. B. "Witchcraft in the Nineteenth Century," *All the Year Round*, n.s., vol. 2 (November 6, 1869): 541–44.

102. Davis, *Popular Magic*, 20–21.
103. Moran, "Light No Smithfield Fires," 123–24.
104. "Witchcraft in the 19th Century," *The Times*, issue 23106 (Thursday, September 23, 1858): 11, col. B.
105. Friend, *Flowers and Flower Lore*, 61; Thiselton, Folklore of Plants, 73.
106. Folkard, *Plant Lore*, 95.
107. Ibid., 91.
108. Ibid., 425, 426, 428.
109. Ibid., 460, 93.
110. Jonson, *The Masque of Queens*, in *Works*, vol. 7: 131, note 9; Folkard, 85.
111. See note 9.
112. Friend, 529.
113. Friend, 538–39, chap. 18, note 2, 664–65.
114. Brand, *Popular Antiquities*, vol. 3: 9.
115. Bacon, *Silva Silvarum*, 198–99, paragraph 902.
116. Mann, *Murder, Magic, and Medicine*, 87.
117. Scot, *Discoverie of Witchcraft*, chap. 8: 105.
118. Middleton, *The Witch*, act 1, scene 2, lines 37–41: 368; Shadwell, *Lancashire-Witches*, act 3: 41.
119. Ainsworth, *Lancashire Witches*, book 2, chap. 10: 216; book 2, chap. 10: 216–17, book 2, chap. 13: 369, book 3, chap. 4: 469; book 1, chap. 10: 221.
120. Ibid., book 1, chap. 9: 205.
121. Michelet, *La sorcière* [trans. *Satanism and Witchcraft* by A. R. Allinson], "Notes and Elucidations," 317.
122. Ibid., 82.
123. Ibid., "Introduction," x.
124. Ibid., x–xi.
125. Linnaeus, *Philosophia Botanica*, 300.
126. Michelet, chap. 9: 87, 81, 81 note 3.
127. Ibid., chap. 9: 81, 81–82, 82–84.
128. Ibid., chap. 9: 82, 83, 84.
129. Ibid., chap. 9: 84.
130. Ibid., chap. 9: 84 note 7.
131. Ibid., chap. 9: 85–86; Hecker, *Epidemics*, 96.
132. Hecker, 100.
133. Michelet, chap. 9: 85–86.
134. Gauthier, "Why Witches?," 199–203.
135. For excellent discussions of the late nineteenth-century glamorization of the witch in art and theater, see Purkiss, *Witch in History*, 39; and Wygant, *Medea, Magic, and Modernity*, 21.
136. Elzea, *Frederick Sandys*, color plate 26, page 53; text, pages 184–85.
137. Swinburne, *Complete Works*, vol. 15: 209.
138. Michelet, chap. 12: 177.

6. The Triumph of the Potato

1. Browne, *Darwin: Voyaging*, 348.
2. Eiseley, *Darwin's Century*, esp. 330–31.
3. Zuckerman, *Potato*, 1998; McNeill, "How the Potato Changed the World's History," 1999; Reader, *Potato*, 2008.
4. Reader, 27; Smith, *Potato: A Global History*, 9.
5. Allan, *Darwin and His Flowers*, 91, 111, 113.
6. Keynes, ed., *Darwin's Zoology Notes*.
7. Browne, 133,176.
8. Darwin, *Voyage of the* Beagle, 224.
9. Browne, 135.
10. Darwin's *Beagle* Library; Caldcleugh's *Travels*, vol. 2: 46. Darwin Online.
11. Darwin, *Voyage of the* Beagle, 242.
12. Porter, "Darwin's Plant Specimens," 157; Ristaino and Pfister, "'Painfully Interesting Subject,'" 1038–39.
13. Baker, "Tuber-Bearing Species," 492–93, 507.
14. Darwin, *Voyage of the* Beagle, 225 and note.
15. Sabine, "Native Country of the Wild Potatoe," 254.
16. Ibid., plate 11, following 254.
17. Darwin, *Origin of Species*, 327; Browne, 361–65.
18. See note 13.
19. Hawkes, "Chilean Wild Potato Species," 671–73.
20. Reader, 82.
21. Zuckerman, 11.
22. Salaman, *History of the Potato*, 113; Reader, 21–23.
23. Salaman, 24.
24. Reader, 24.
25. Darwin, *Voyage of the* Beagle, 225.
26. Ames and Spooner, "DNA," 252–57.
27. Zuckerman, 12–13.
28. Reader, 27, 29–30.
29. Pollan, *Botany of Desire*, xiv–xv, 230–31.
30. Darwin to J. S. Henslow, October 28, 1845, Darwin Correspondence Project, "Letter no. 921."
31. *The Gardeners' Chronicle and Agricultural Gazette*, no. 34 (August, 23, 1845): 575.
32. Reader, 198–212.
33. *The Gardeners' Chronicle and Agricultural Gazette*, no. 40 (Saturday, October 3, 1846): 661.
34. Darwin to W. D. Fox [before October 3, 1846], Darwin Correspondence Project, "Letter no. 13809."
35. Browne, *Darwin: Power of Place*, 220–11.

36. Darwin to W. D. Fox [February 13, 1845], Darwin Correspondence Project, "Letter no. 827."
37. Darwin, *Variation*, vol. 1:330–31.
38. Ibid., vol. 1: 384.
39. Browne, *Darwin: Power of Place*, 202, 204, 275–76.
40. Darwin, *Variation*, vol. 2: 314.
41. Ibid., vol. 2:169.
42. Ibid., vol. 2: 146.
43. Cathcart, "Cultivated Potato," 281–300.
44. Ibid., 294, 297.
45. Hooker, "Solanum maglia," vol. III (1884): plate 6756; Hooker, *Antarctic Voyage*, 329–32.
46. "Hybridising Potatos at Reading," *The Gardeners' Chronicle*, n.s., vol. 22 (November 8, 1884): 585–86.
47. *The Gardeners' Magazine*, vol. 27 (December 6, 1884): 689.
48. Ibid., vol. 28 (September 12, 1885): 516–17.
49. *The Gardeners' Chronicle*, n.s., vol. 24 (October 24, 1885): 528, 530.
50. Ibid., n.s., vol. 26 (November 20, 1886): 656.
51. *The Indian Forester; A Monthly Magazine of Forestry, Agriculture, Shikar & Travel*, vol. 13 (1887): 190–92.
52. "The Potato Experiments at Reading," *The Gardeners' Chronicle*, n.s., vol. 6 (December 4, 1886): 723–24.
53. "The Potato Centenary Celebration," Ibid., 720.
54. "On the Wild Forms of Tuberous Solanum," *The Gardeners' Chronicle*, n.s., vol. 6 (December 11, 1886): 746.
55. Salaman, 176.
56. Porter, "Darwin's Plant Specimens," 93: 157.
57. "New Potatos," *The Gardeners' Chronicle*, ser. 3, vol. 14 (November 25, 1893): 656–57; "Experiments on Cross-Breeding Potatos, etc.," *The Gardeners' Chronicle*, ser. 3, vol. 25 (February 4, 1899): 78.
58. Sutton, "Potatos," 416–22.
59. Salaman, 167.
60. "Cross-bred Potatos," *The Gardeners' Chronicle*, n.s., vol. 26 (November 20, 1886): 656.
61. Fenn, "Looking Back," 99–100.
62. Salaman, 169; Reader, 218–21.
63. Reader, 160–61; 218.
64. Salaman, 427; Reader, 114–15.
65. Ibid., 82.
66. Gerard, *Herbal* [1633], 596–98; Zuckerman, 14.
67. Salaman, 424–25; Zuckerman, 8–9; Reader, 78.
68. Dickens, *Bleak House*, chap. 1: 4.
69. Salaman, 238, 447.

NOTES TO PAGES 164–173

70. Salaman, 188–90, 204, 214–15; 343; Woodham-Smith, *Great Hunger*, 18–32.
71. Woodham-Smith, 146.
72. Salaman, 232–33; Woodham-Smith, 35.
73. Pollan, 202.
74. Cobbett, 44, 41, 43.
75. "Peasant Cottage Interior," *Pictorial Times*, February 7, 1846, from "Views of the Famine."
76. Thackeray, *Irish Sketch Book*, 108.
77. Anderson, *Predicting the Weather*, 65–81.
78. Engels, *Condition of the Working Class*, 91, 61, 77, 73.
79. Ibid., 272.
80. Marx, *Eighteenth Brumaire*, 124.
81. Reader, 130.
82. Zuckerman, 140.
83. Salaman, 343.
84. Woodham-Smith, 82.
85. Adam Smith, in Malthus, *Principles of Population* (1798), 43.
86. Malthus, *Principles of Population* (1890), 260.
87. *Gardeners' Chronicle*, no. 37 (September 13, 1845): 623.
88. Woodham-Smith, 204–6.
89. Ibid., 101, 136–41, 145.
90. Ibid., 195–96.
91. Ibid., 188–204.
92. Ibid., 361–65.
93. Ibid., 44–46; 51–52.
94. Ibid., 49–54.
95. Ibid., 86–87, 307.
96. Trevelyan, *Irish Crisis*, 108.
97. Woodham-Smith, 37.
98. Trevelyan, 2.
99. Ibid., 8.
100. Ibid., 1.
101. Ibid., 6.
102. Ibid., 5, 11, 22, 201.
103. Woodham-Smith, 408–9, 24; John Stuart Mill, *England and Ireland*, 16.
104. Woodham-Smith, 415.
105. Williams, *Daniel O'Connell*, 2–3.
106. Ibid., 27.
107. Woodham-Smith, 16–18.
108. Williams, 37.
109. "The Real Potato Blight of Ireland," *Punch*, vol. 9 (1845): 255; Gray, "Punch and the Great Famine"; Williams, 87–88.
110. Heath, "Great Agi-Tater."

NOTES TO PAGES 173–186

111. Williams, 34.
112. Salaman, 120.
113. "'Save Me from My Friends!,'" *Punch*, vol. 69 (August 28, 1875): 81.
114. "The Irish Cinderella and Her Haughty Sisters, Britannia and Caledonia," *Punch*, vol. 10 (1846): 181; Gray, "Punch and the Great Famine."
115. "Justice to Punch and Ireland," *Punch*, vol. 83 (September 23, 1882): 142.
116. Wiseman, review, 417.
117. Ibid.
118. "Consolation for the Million.—The Loaf and the Potato," *Punch*, vol. 13 (1847): 95; Zuckerman, 211–12.
119. "Grand Vegetable Banquet to the Potato on His Late Recovery," *Punch*, vol. 17 (1849): 204.
120. Neill, *Account of British Horticulture*, 254.
121. J. C. Loudon, *Encyclopedia of Gardening* (1822), 697–98.
122. Stearn, *John Lindley*, 1799–1865, 31.
123. J. C. Loudon, ed., *Encyclopedia of Plants*, 1078.
124. Ibid., 158–60, 157.
125. Lindley, *Introduction to the Natural System of Botany*, 232.
126. Lindley, *Natural System of Botany*, xiii, 293.
127. Lindley, *Ladies Botany*, vol. 1: 286.
128. Acton, *Modern Cookery, in all its Branches*, 300–305.
129. Acton, *Modern Cookery for Private Families*, 309–10.
130. Ibid., xi.
131. Francatelli, *Modern Cook*, 73, 93.
132. Soyer, *Gastronomic Regenerator*, 470–71.
133. Soyer, *Modern Housewife*, 332–33.
134. Ibid., 333.
135. Cowan, *Relish*, 125–32.
136. Francatelli, *Plain Cookery*, 70–73, 14–15.
137. Isabella Beeton, *Household Management*, 583–84, 582–83.
138. Colquhoun, *Taste*, 294.
139. Hogg, *Vegetable Kingdom*, 548–549; Beeton, 582, 589; J. C. Loudon, *Encyclopedia of Agriculture*, 778.
140. *Private Life of the Queen*, 140, 141, 229.
141. *Paterson's National Benefit*, 35.
142. Salaman, 166, 168.
143. *Paterson's National Benefit*, 28.
144. Lucot, *Emblèmes*, 33.
145. Shoberl, *Language of Flowers*, 238.
146. Zuckerman, 103.
147. Ibid., 82–85.
148. Tyas, *Sentiment of Flowers* [1842], 61.
149. Tyas, *Language of Flowers*, 166; Greenaway, *Language of Flowers*, 33.

NOTES TO PAGES 186–196

150. Phillips, *Cultivated Vegetables*, vol. 2, 78–92; *Floral Emblems*, 77.
151. Dickens, *Pickwick Papers*, chap. 2: 16; chap. 6: 68; chap. 7: 90; chap. 8: 99; chap. 10: 124; chap. 11: 137; chap. 12: 155; chap. 16: 210; chap. 39: 552; chap. 45: 639.
152. Ibid., chap. 45: 641–642.
153. Dickens, *Oliver Twist*, chap. 23: 127; *Nicholas Nickleby*, chap. 8: 92.
154. Salaman, 240.
155. Catherine Dickens ["Lady Maria Clutterbuck"], *What Shall We Have for Dinner?*, 45–46, 53.
156. Dickens, *Old Curiosity Shop*, chap. 36: 272–73.
157. Dickens, *Christmas Carol*, Stave 1, 18.
158. Mayhew, *London Labour*, vol. 1: 173–75, 167.
159. Ibid., 174.
160. Dickens, *Sketches by Boz*, "Scenes," chap. 2, "The Streets—Night," 53.
161. Panayi, *Fish and Chips*, 28.
162. Dickens, *Tale of Two Cities*, chap. 5: 29.
163. Panayi, 27.
164. Warren, *Cookery for Maids*, 88.
165. Mayhew, vol. 1: 165–166.
166. Panayi, 15, 28; Dickens, *Oliver Twist*, chap. 26: 184.
167. Panayi, 36.
168. Dickens, *Pickwick Papers*, chap. 19: 260; *Little Dorrit*, book 1, chap. 9: 100; book 2, chap. 5: 476.
169. Dickens, *David Copperfield*, chap. 11: 155.
170. Trollope, *Framley Parsonage*, chap. 14: 168.
171. Hardy, *Tess of the d'Urbervilles*, chap. 50: 429.
172. Hardy, *Under the Greenwood Tree*, part 4, chap. 3: 125–27.
173. Ruskin, *Works*, vol. XIX, *Queen of the Air*, chap. 2, 368–69.
174. Ibid., vol. XXV, *Proserpina*, 227.
175. Ibid., vol. XIX, chap. 2, 367.

7. SUBLIME TOBACCO

1. Mackenzie, *Sublime Tobacco*, 250–52.
2. Stirling, *Life's Little Day*, 264.
3. Mackenzie, 252.
4. Ibid., 86–87.
5. Gately, *Tobacco*, 188–89; 151–52; 157–58.
6. Apperson, *Social History of Smoking*, 99–109.
7. Hilton, *Smoking*, 20–22.
8. Fairholt, *Tobacco*, 56–57.
9. Apperson, 148–149.
10. Mayhew, *London Labour*, vol. 1: 12.
11. Ibid., vol. 2: 176.

302   NOTES TO PAGES 196–208

12. Ibid., vol. 1: 109, 110.
13. Ibid., vol. 1: 65–66.
14. Kingsley, *Westward Ho!* chap. 7: 96. Hilton, 29–30.
15. Gately, 40–41.
16. Gerard, *Herball* [1597], chap. 63: 287–88.
17. Ibid., 286–87.
18. Goodman, *Tobacco in History,* 6.
19. Gerard, chap. 63: 288.
20. Ibid., 286.
21. Ibid., 288.
22. Mackenzie, 103, 105–6, 108.
23. Goodman, 143–44.
24. Ibid., 59.
25. Lord Byron, *The Island, or Christian and His Comrades,* canto 1, XIX, lines 448–59; 438, in *Complete Poetical Works,* 425.
26. Hilton, 20–22.
27. Gately, 154–55.
28. Mackenzie, 72. Thevet's account was translated by Thomas Hacket as *The New Found World* in 1568.
29. Mackenzie, 4, 6.
30. Gately, 286.
31. Steinmetz, *Tobacco,* iii–iv, xv–xvi.
32. Samuel Solly, Lecture 1, "Clinical Lectures on Paralysis." *The Lancet,* vol. 2 (December 13, 1856): 641–643; 641.
33. "Tobacco Controversy—Is Smoking Injurious?" *The Lancet,* vol. 1 (February 7, 1857): 152–54.
34. Steinmetz, *Tobacco,* 11; Hilton, 63.
35. Steinmetz, *Tobacco,* iv.
36. Ibid., 1–6.
37. Ibid., 8.
38. Ibid., 16–17, 18–23.
39. Ibid., 11–12.
40. Ibid., 69–70, 73.
41. Ibid., 76–77, 87.
42. Ibid., 103, 105, 107, 106.
43. Ibid., 107, 108, 109.
44. Ibid., frontispiece.
45. Ibid., 144, 160.
46. Review, *The Lancet* (April 4, 1857), 1: 353.
47. "The Lancet Verdict on the Tobacco Controversy Slightly Di-versified," by Quid, *The Lancet* vol. 1 (April 18, 1857): 418.
48. See Umberger, "In Praise of Lady Nicotine," 236–37; Penn, *Soverane Herbe,* vi; Apperson, 7.

NOTES TO PAGES 208–217

49. Fairholt, *Tobacco*, v–vi, 9, 322.
50. Ibid., 327.
51. Ibid., 1–2.
52. Ibid., 2, frontispiece; Goodspeed, *Genus Nicotiana*, 372–75; 134–35.
53. Ibid; J. C. Loudon, *Encyclopedia of Plants*, 137.
54. Fairholt, 3; Gerard [1633], chap. 62: 284; Mackenzie, 67: "The famous latakia of Syria is *N. tabacum.*"
55. Lindley, *Medical and Oeconomical Botany*, 205.
56. Fairholt, 3; Lindley, *Medical and Oeconomical Botany*, 205; Goodspeed, 393.
57. Goodman, 2.
58. Steinmetz [aka "A Veteran of Smokedom"], *Smoker's Guide*, "To the Reader"; Bragge, *Bibliotheca Nicotiana*, 9; Fairholt, 9.
59. Fairholt, 13–16, 45–46.
60. Ibid., 45, 47.
61. Ibid., 50–52, 51, 53.
62. Ibid., 53–55, 59, 60, 65; 58–59.
63. Ibid., 87–94.
64. Ibid., 107, 116, 330, 119–20, 125.
65. Schama, *Embarrassment of Riches*, 189.
66. Ibid., 127–28. Fairholt quotes "the excellent little *Paper of Tobacco*" (a history written in 1839 by William Andrew Chatto, alias "Joseph Fume," from whom he draws considerable information).
67. Apperson, 108–9; Fairholt, 130.
68. Ibid., 134–35.
69. Ibid., 147, 148, 150, 151.
70. Ibid., 147–48.
71. Ibid., 149.
72. Ibid., 152, 161, 159.
73. Ibid., 172, 173, 176, 185–86.
74. Ibid., 194–195, 196 note.
75. Ibid., 180–181; Mackenzie, 254.
76. Goodman, 91.
77. Fairholt, 220–21.
78. Mayhew, vol. 2: 145–46.
79. Fairholt, 231–32.
80. Ibid., 237–38.
81. Ibid., 239, 234, 252, 268; Mackenzie, 165–72.
82. Ibid., 252, 254, 259, 293.
83. Mayhew, vol. 1: 440.
84. Fairholt, 264, 278–279, 273–274, 293.
85. Browne, *Darwin: Power of Place*, 234.
86. Fairholt, 273.
87. Ibid., 278, 251.

88. Ibid., 296–304.

89. Ibid., 306–12.

90. Ibid., 325.

91. Ibid., 331.

92. Ibid., 214, 217.

93. Penn, vi.

94. Ibid., 197.

95. Ibid., 203.

96. Goodman, 99.

97. Hilton, 102, 84; Mackenzie, 280.

98. Goodman, 98–99.

99. Penn, 200.

100. Goodman, 206–9.

101. Gately, 177–79.

102. Penn, 204–5.

103. Wilde, *Dorian Gray*, chap. 6: 116; Gately, 204–205; Hilton, 55.

104. Penn, 205.

105. Klein, *Cigarettes Are Sublime*, xi.

106. Ibid., 62, xi.

107. Ibid., 63.

108. Ibid., 64.

109. Steinmetz, *Smoker's Guide, Philosopher and Friend*, "To the Reader."

110. Bragge, 38–39.

111. Steinmetz, *Smoker's Guide*, 95, 111.

112. Penn, chap. XVI: 278–95.

113. Seaton, "Cope's and Tobacco," 5, 21–22.

114. Altick, "Cope's Tobacco Plant," 333, 339; Hilton, 44.

115. "John Fraser and His Collection," in "Smokescreen: the Victorian Vogue for Tobacco" Exhibition arranged by Dr. Maureen Watry, Head of Special Collections and Archives, University of Liverpool Library, May 2002. https://libguides .liverpool.ac.uk/library/sca/smokescreenexhib.

116. Seaton, 10; Hilton, 44.

117. Quoted in Altick, "Cope's Tobacco Plant," 337.

118. Ibid.

119. "Cope's Smoke Room Booklets," *Smokescreen*.

120. Altick, "Cope's Tobacco Plant," 335–36.

121. See Carmen Abroad. https://carmenabroad.org/performance-run/5445 and https://carmenabroad.org/performance-run/5450.

122. Thomson, "Tobacco at the Opera," 293–94.

123. Ibid., 294.

124. For example, *The Smoker's Garland*, part 3, *Cope's Smoke-Room Booklets*, no. 10.

125. Hatton, "A Day in a Tobacco Factory," *Cope's Smoke Room Booklets*, no. 14: 41, 43.

NOTES TO PAGES 226–233

126. Crawford, *Cigarette-Maker's Romance*, 10–15; Hatton, 53, 55.
127. Hatton, 46, 48.
128. Ibid., 45, 56, 58.
129. Gately, 179.
130. Hatton, 59–60.
131. Ibid., 61.
132. Mackenzie, 29–233; Cockerell, "Tobacco in Victorian Literature," 89–99; Altick, "The Favorite Vice of the Nineteenth Century," in *Presence of the Present*, 240–74; Hilton, 41–59; Gately, 185–203.
133. Hilton, 17–20; Doyle, "The Red-Headed League," in *Sherlock Holmes*, 43.
134. Thackeray, *Vanity Fair*, chap. 11: 145; chap. 13: 154; chap. 25: 291.
135. Brontë, *Jane Eyre*, chap. 23: 278–79.
136. Dickens, *Pickwick Papers*, chap. 30: 407–8.
137. Ibid., chap. 37: 518; chap. 40: 563; chap. 42: 588.
138. Ibid., chap. 20: 277.
139. Ibid., chap. 2: 6; chap. 14: 177; chap. 20: 270.
140. Ibid., chap. 33, Illustration facing p. 457.
141. For example, Ibid., chap. 31: 425–26.
142. Mackenzie, 229.
143. Altick, "The Favorite Vice of the Nineteenth Century," 244–48.
144. Dickens, *Martin Chuzzlewit*, chap. 19: 310, 313; *Bleak House*, chap. 14: 188, 190, 193.
145. *Nicholas Nickleby*, chap. 26: 331; *David Copperfield*, chap. 28: 427; *Bleak House*, chap. 20: 278, 273.
146. *Old Curiosity Shop*, chap. 4: 37, chap. 5: 38.
147. *Little Dorrit*, chap. 18: 211, 213, 214.
148. *Our Mutual Friend*, book 1, chap. 3: 19; chap. 12: 147, 150; book 2, chap. 1: 229; chap. 2: 234, 240; chap. 6: 286–89, 291–92, 294–95; chap. 15: 407; book 3, chap. 10: 534, 541; chap. 12: 628.
149. Meller, *Nicotiana*, 116–32.
150. Trollope, *Warden*, chap. 16: 148–49.
151. *Oliver Twist*, chap. 8: 56–57; chap. 18: 129–32; chap. 25: 178–79. *Our Mutual Friend*, book 1, chap. 5: 57–59.
152. *David Copperfield*, chap. 3: 32, 37; chap. 10: 141–42; chap. 30: 438–42; chap. 31: 448; chap. 50: 714–15; chap. 51: 733, 735; chap. 10: 146.
153. *Dombey and Son*, chap. 17: 231; chap. 25: 362; chap. 32: 456; chap. 39: 542, 551–52, 554; chap. 49: 683–85; chap. 50: 713; chap. 56: 788; chap. 62: 875. chap. 6: 72; chap. 27: 381–82.
154. *Bleak House*, chap. 27: 386–87; chap. 49: 669–75.
155. Ibid., chap. 8: 106–109.
156. *Great Expectations*, chap. 10: 69; chap. 16: 113; chap. 27: 211. chap. 25: 197. chap. 40: 313.
157. *Hard Times*, chap. 3: 132; chap. 8: 179. *Little Dorrit*, book 2, chap. 13: 581–85.

306 NOTES TO PAGES 233–245

158. *Edwin Drood*, chap. 5: 48; chap. 23: 264–67; chap. 1: 2.
159. *American Notes*, chap. 8: 112.
160. *Martin Chuzzlewit*, chap. 16: 261, 266.
161. *Bleak House*, chap. 43: 593.
162. *Little Dorrit*, book 1, chap. 1: 7–8; chap. 11: 128, 129; book 2, chap. 6: 495; chap. 28: 748–49, 750, 752; chap. 30: 785.
163. Forster, *Life of Dickens*, vol. 2: 292.
164. Ibid., 277; 292; 293.
165. Kaplan, *Dickens*, 536.
166. "Woman's Emancipation," *Punch*, vol. 21: 3.
167. John Leech, "Bloomerism—An American Custom," *Punch*, vol. 21: 141; "Something More Apropos of Bloomerism," vol. 21: 196.
168. Ouida, *Under Two Flags*, vol. 1, chap. 6: 65; vol. 1, chap. 3: 42.
169. Ibid., vol. 1, chap. 1: 14.
170. Ibid., vol. 1, chap. 1: 14; vol. 1, chap. 6: 64; vol. 1, chap. 9: 104.
171. Ibid., vol. 2, chap. 1: 180, 184.
172. Ibid., vol. 2, chap. 1: 184; "Punch's Fancy Portraits.—No. 45. Ouida." *Punch*, vol. 81: 83.
173. "Our New Novel, Entitled STRAPMORE! A Romance by Weeder," *Punch*, vol. 74 (March 9–May 11, 1878): 105, 107, 117–18, 129, 131, 142–43, 153, 177, 213–15.
174. Steinmetz, *Smoker's Guide*, 63, 48.
175. Ibid., 147.
176. Barrie, *My Lady Nicotine*, chap. 1: 4, 10.
177. Ibid., chap. 3: 19.
178. Ibid., chap. 32: 265.
179. Ibid.
180. *The Smoker's Garland*, part 3, *Cope's Smoke Room Booklets*, no. 10.
181. Ibid., "Why Women Hate the Weed," 157–59.
182. Ibid., 159.
183. Altick, *Presence of the Present*, 204; Richard Peto, Sarah Darby, Harz Deo, Paul Silcocks, Elise Whitley, and Richard Doll, "Smoking, Smoking Cessation, and Lung Cancer in the UK since 1950: Combination of National Statistics with Two Case-Control Studies," *BMJ*, vol. 321 (August 5, 2000): 323–29; 323.

8. Back to the Garden

1. Paxton, *Crystal Palace*, 9; Piggott, *Palace of the People*, 53, 67–122.
2. Wilkinson, *Victorian Gardener*, 55–56; Stuart, *Garden Triumphant*, 43–45; Elliott, *Victorian Gardens*, 134.
3. Willes, *Gardens of the Working Class*, 246–47; Wilkinson, 56; Elliott, *Victorian Gardens*, 89.
4. Beaton, "Crystal Palace," 39.
5. Elliott, *Flora*, 216; Stuart, *Plants That Shaped Our Gardens*, 72.

NOTES TO PAGES 246–255

6. Beaton, "The Flower-Garden," 18; "Flower-Garden Plan.—No. 10," 153–54; "Crystal Palace," 38.

7. Beaton, "Bedding Geraniums," 3; "Crystal Palace," 40.

8. Castel, Kusters, Koes, "Inflorescence in Petunia," 2235–46. *"Petunia Nyctaginiflora," Curtis's Botanical Magazine* 52 (1825): plate 2552; *"Petunia nyctaginiflora violacea," Paxton's Magazine of Botany*, vol. 2 (1836): 173; Maund, *Botanic Garden*, vol. 3: no. 208.

9. *"Petunia Violacea," Paxton's Magazine*, vol. 1 (1834): 7.

10. *"Petunia nyctaginiflora violacea," Paxton's Magazine of Botany*, vol. 2 (1836): 173.

11. J. C. Loudon, *Encyclopedia of Gardening* (1835), book 2, chap. 8, subsection 27: 1051.

12. Jane Loudon, *Instructions in Gardening for Ladies*, 363, 365.

13. Jane Loudon, *Gardening for Ladies and Companion to the Flower Garden*, 227.

14. Pratt and Miller. *Language of Flowers*, 27.

15. Jane Loudon, *Gardening for Ladies and Companion to the Flower Garden*, 227.

16. For example, advertisements, *The Gardeners' Chronicle and Agricultural Gazette*, vol. 10 (April 30, 1859): 383.

17. Glenny, "The Petunia," 247.

18. Willes, chap. 4, "A Passion for Flowers," *Gardens of the Working Class*, 90–112; Wilkinson, *Victorian Gardener*, 36–40.

19. Glenny, "The Petunia," 248; Glenny, *Properties of Flowers and Plants*, 69.

20. Wilkinson, 36; Glenny, *Properties of Flowers and Plants*, 70.

21. Holland, "Properties and Culture of the Petunia," 149, 151.

22. Ibid., 150.

23. First-Class Certificate of Merit to Mssrs. G. Henderson and Son for Petunia Queen, "Royal Botanic Gardens List of Awards, Wednesday, July 4, 1860," *The Gardeners' Chronicle and Agricultural Gazette*, vol. 20 (July 7, 1860): 619; Advertisement, "New Plants for 1863, B. S. Williams's Paradise and Victoria Nurseries, London, featuring 3 of Holland's petunias, including 2 awarded Royal Botanic Society First-Class Certificates from 1862, Crimson Gem and Flower of the Day; First-Class Certificate to Holland for Petunia Royalty, Royal Botanic Society Exhibition, Wednesday, May 13, 1863," *Gardeners' Chronicle*, vol. 23 (May 16, 1863): 458; First class for seedling Petunia Duchess of Northumberland, "Central Society of Horticulture," *Gardener's Weekly Magazine, and Floricultural Cabinet*, vol. 5 (June 6, 1863): 179.

24. Hibberd, *Garden Favourites*, 341, 361, 362, 365, 366.

25. Hibberd, "Reaction against the Bedding Mania," 193–94.

26. Hibberd, *Amateur's Flower Garden*, 86.

27. Hibberd, *Familiar Garden Flowers*, 111, 110, 111, 109.

28. Willes, 208–9, 257–58; Wilkinson, 27.

29. Robinson, *English Flower Garden*, 4th ed., v, 193, 12.

30. Ibid., 64, 66, 68.

31. Bailey, *Survival of the Unlike*, 472.

308   NOTES TO PAGES 255–264

32. Howell, *Flora Mirabilis*, 67.
33. Andrews, *Peppers*, 1–10; Naj, *Peppers*, 11–14.
34. Gerard, *Herball* [1597], book 2, chap. 66: 292–93.
35. Ibid.
36. Gerard, *Herbal* [1633], book 2, chap. 71: 364–65.
37. Ibid., book 2, chap. 71: 364.
38. Evelyn, *Acetaria*, 161, 183.
39. Miller. *Gardeners Dictionary* [1735], vol. 1: s.v. "capsicum."
40. Glasse, *Art of Cookery* [1774], 272, 332, 368.
41. Miller, *Gardeners Dictionary* [1768], vol. 1, s.v. "capsicum"; Martyn, ed., *Gardener's and Botanist's Dictionary* [1807], vol. 1, part 1, s.v. "capsicum."
42. Abercrombie, *Every Man His Own Gardener*, 135.
43. Loudon, *Encyclopedia of Gardening* [1822], book 1, part 3, subsection 3: 764. The same information attributed to Abercrombie in the 1835 edition, book 1, part 3, subsection 3: 880.
44. Harris, ed., *Peter Piper's Principles of Pronunciation*, s.v. "P."
45. Thackeray, *Vanity Fair*, chap. 3: 61–62, chap. 3: 64.
46. Spratt, *Flora Medica*, vol. 1: 91.
47. Gentilcore, *Pomodoro!*, 1–2, 5, 9, 11–12.
48. Gerard, *Herball* [1597], book 2, chap. 54: 274.
49. Ibid.
50. Lyte, *New Herball*, chap. 85: 507.
51. Heiser, *Fascinating Nightshades*, 47.
52. Gerard, *Herball* [1597], book 2, chap. 54: 274.
53. Ibid.
54. Miller. *Gardeners Dictionary* [1754], vol. 2, s.v. "melongena."
55. Peter Rainier, quoted in "Notices of Communications to the Horticultural Society, between Jan. 1, 1822, and Jan. 1, 1823, of which Separate Accounts Have Not Been Published in the Transactions," *The Gardener's Magazine and Register of Rural and Domestic Improvement*, vol. 1 (1826): 306, 307.
56. Loudon, *Encyclopedia of Gardening* [1822], book 1, part 3, subsection 2: 764.
57. Gerard, *Herball* [1597], book 2, chap. 55: 275; Lyte, *New Herball*, chap. 86: 508.
58. Gerard, ibid.
59. Gentilcore, 9–11.
60. Gerard, ibid.
61. Gentilcore, 4.
62. Gerard, ibid.
63. For an excellent discussion of humoral theory, see Albala, *Eating Right*, 48–52.
64. Gerard, ibid.
65. Ibid., 276.
66. Miller, *Gardeners Dictionary* [1735], vol. 2, s.v. "lycopersicon"; Smith, *Tomato in America*, 13.

NOTES TO PAGES 264–269

67. Miller, *Gardeners Dictionary* [1735], vol. 2, s.v. "lycopersicon."
68. Miller, *Gardeners Dictionary* [1754], vol. 2, s.v. "lycopersicon."
69. Miller, *Gardeners Dictionary* [1768], vol. 2, s.v. "lycopersicon."
70. Abercrombie, *Every Man His Own Gardener*, 183, 213, 157.
71. *Encyclopaedia Britannica*, 3rd ed., s.v. "solanum," 597–98.
72. Smith, 19.
73. Letter 48, Jane Austen to Cassandra, Monday, October 11, 1813, in Woolsey, *Letters of Jane Austen*, 215.
74. Sabine, "*Love Apple* or *Tomato*," 342.
75. Smith, *Pure Ketchup*, 19.
76. Sabine, 342–54.
77. Lucot, *Emblèmes*, "Pomme d'Amour," 132.
78. Hyman, *Tomato*, 47.
79. Loudon, *Encyclopedia of Gardening* [1822], book 1, part 3, subsection 1: 765, 764.
80. Loudon, *Encyclopedia of Gardening* [1835], book 1, part 3, subsection 1: 879.
81. Dickens, *Pickwick Papers*, chap. 34: 473, 482.
82. Ibid., chap. 12: 150.
83. Rundell, *Domestic Cookery*, 153–54.
84. Analytical Sanitary Commission, "On Sauces and Their Adulterations," *The Lancet*, vol. 2, no. 23 (December 4, 1852): 527–30.
85. Acton, *Modern Cookery*, 319–320; Francatelli, *Modern Cook*, 350.
86. Acton, 318.
87. Isabella Beeton, *Household Management*, 251–53, 595–97.
88. Ibid., 1096.
89. "The Tomato," *The Gardeners' Chronicle*, 19 (November 19, 1859): 932.
90. Samuel Beeton, *Garden Management*, 362–63.
91. Loudon, *Encyclopedia of Gardening* [1822], book 1, part 3, section 11: 763; *Encyclopedia of Gardening* [1835], book 1, part 3, section 11: 879.
92. For example, tomatoes sold by the sieve and capsicums by the hundred in the following Covent Garden Market reports from *The Gardener's Magazine*, vol. 5 (1829): 616, 741; 6 (1830): 119; 7 (1831): 624; 8 (1832): 624; 9 (1833): 634, 724. Reports of tomatoes and capsicums sold at Covent Garden in *The Gardeners' Chronicle*, vol. 1 (September 11, 1841): 600; tomatoes from Lisbon and Marseilles, *The Gardeners' Chronicle*, vol. 12 (January 31, 1852): 70; a Horticultural Society exhibit of tomatoes from Algiers, *The Gardeners' Chronicle*, vol. 13 (April 23, 1853): 262; tomatoes in Covent Garden coming from Algiers, by way of Paris, *The Gardeners' Chronicle*, vol. 13 (May 1853): 339.
93. *The Gardeners' Chronicle*, n.s., vol. 5 (March 18, 1876): 370.
94. "Tomato Culture," *The Gardeners' Chronicle*, ser. 3, vol. 17 (January 19, 1895): 80; *The Gardeners' Chronicle*, ser. 3, vol. 20 (July 11, 1896): 46.
95. "Little Known Dainties. — II. 'Tomatos,'" *The Gardeners' Chronicle*, n.s., vol. 6 (September 16, 1876): 358.

NOTES TO PAGES 269–279

96. "Tomatos," *The Gardeners' Chronicle*, n.s., vol. 16 (August 6, 1881): 177.

97. "Tomato Growing for Market," *The Gardeners' Chronicle*, n.s., vol. 18 (December 30, 1882): 847.

98. "The Tomato in Great Britain," *The Gardeners' Chronicle*, ser. 3, vol. 11 (June 11, 1892): 759.

99. "Market Gardening. A Profession for Our Sons," *The Gardeners' Chronicle*, ser. 3, vol. 15 (April 14, 1894): 462.

100. Sutton, "Progress of Vegetable Cultivation," 386–88.

101. Green Gage Tomato, Carters' advertisement, *The Gardeners' Chronicle*, n.s., vol. 9 (February 16, 1878): 200; Sutton's Royal Dwarf Cluster, *The Gardeners' Chronicle*, n.s., vol. 10 (December 14, 1878): 772.

102. Nisbet's Victoria, "List of Novelties, 1879–80," Charles Sharpe & Co., *The Gardeners' Chronicle*, n.s., vol. 12 (December 6, 1879): 709.

103. An excellent example is the calendar of "Work for the Week Commencing January 19th" beginning with "Capsicums, Tomatoes, and Egg Plants to be sown at once, and placed in heat. Use light rich soil, sow thin, and prick the plants out to strengthen as soon as they are large enough" (*The Gardeners' Magazine*, vol. 10 [January 19, 1867]: 28). *The Floral World and Garden Guide* also featured regular tasks for the garden year, as well as special articles, such as "Culinary Uses for Tomatoes," vol. 3 (1860): 229–30; and "The Tomato, and Its Uses," by James Cuthill, of Camberwell, n.s., vol. 9 (1866): 111–12.

104. Hibberd, *Amateur's Kitchen Garden*, 225.

105. Armstrong, *Victorian Glassworlds*, 173.

106. Advertisement, "Hothouses for the Million," *The Gardeners' Chronicle*, vol. 20 (January 21, 1860): 60.

107. Willes, chap. 9, 224–45.

108. Amherst. *History of Gardening*, 306.

109. Plues, *Rambles in Search of Wild Flowers*, 206, 207.

110. Rees. *Cyclopedia*, vol. 33, s.v. "solanum."

111. Forbes, "Vegetable World," II, VI.

112. Robinson, *Subtropical Garden*, 190–95.

113. Jekyll, *Home and Garden*, 252–53.

114. Burbridge, *Cultivated Plants*, 538.

115. Nicholson, ed., *Dictionary of Gardening*, vol. 5: 115; vol. 4: 306; Beaton, "The Cape Gooseberry," 251–53; Mabey, *Flora Britannica*, 300.

116. "Solas," Middle English Dictionary online, University of Michigan. https://quod.lib.umich.edu/m/middle-english-dictionary/dictionary?utf8=%E2%9C%93&search_field=hnf&q=solas.

117. Kilmer, "Lands Where Drugs Grow," 157.

118. Wilde, *Dorian Gray*, chap. 6, 116.

119. Johnson, ed, *Cottage Gardeners' Dictionary*, s.v. "tobacco," 876–77.

120. Penn, *Soverane Herbe*, 120.

121. Seeley, "Liberty Hyde Bailey," 1204.

NOTES TO PAGES 279–280

122. Bailey, *Survival of the Unlike*, 472.
123. Ibid., 481, 480, 482.
124. "Societies. Royal Horticultural," *The Gardeners' Chronicle*, ser. 3, vol. 12 (October 9, 1897): 257–58.
125. "Discussion" at Conference, "Fruit Culture in Her Majesty's Reign, 1837–1897," *The Gardeners' Chronicle*, ser. 3, vol. 12 (October 9, 1897): 259.
126. Arthur Rackham, "Mr. and Mrs. Vinegar At Home," frontispiece, Flora Annie Steel, *English Fairy Tales*.
127. Halliwell-Phillipps, *Rhymes and Nursery Tales*, 26–29.

BIBLIOGRAPHY

Abercrombie, John. *Every Man His Own Gardener.* J. F. and C. Rivington, T. Longman, B. Law, et al., 1787.

Acton, Eliza. *Modern Cookery for Private Families.* London: Longman, Green, Longmans & Roberts, 1860.

———. *Modern Cookery, in All Its Branches.* London: Longman, Brown, Green & Longmans, 1845.

Ainsworth, William Harrison. *The Lancashire Witches* [1848]. Manchester, UK: E. J. Morten, 1976.

Albala, Ken. *Eating Right in the Renaissance.* Berkeley and Los Angeles: U of California P, 2002.

Alcott, Louisa May. *Little Women* [1869]. Boston: Little, Brown, 1922.

All the Year Round. Dickens Journals Online. University of Buckingham. https://www.djo.org.uk.

Allan, Mia. *Darwin and His Flowers: The Key to Natural Selection.* New York: Taplinger, 1977.

Allen, David E., and Gabrielle Hatfield, eds. *Medicinal Plants in Folk Tradition: An Ethnobotany of Britain and Ireland.* Portland, OR: Timber Press, 2004.

Allen, Grant. "British and Foreign." *Falling in Love, and Other Essays on More Exact Branches of Science.* London: Smith, Elder, 1889: 123–36.

Almeida, Hermione de. *Romantic Medicine and John Keats.* New York: Oxford UP, 1991.

Altick, Richard. "Cope's Tobacco Plant: An Episode in Victorian Journalism." *Papers of the Bibliographical Society of America,* vol. 45 (1951): 33–50.

———. *The Presence of the Present.* Columbus: Ohio State UP, 1991.

Ames, Mercedes, and David M. Spooner. "DNA from Herbarium Specimens Settles a Controversy about Origins of the European Potato." *American Journal of Botany,* vol. 95, no. 2 (2008): 252–57.

Amherst, Alicia. *A History of Gardening in England.* New York: E. P. Dutton, 1910.

Anderson, Frank J. *An Illustrated History of the Herbals.* New York: Columbia UP, 1977.

Anderson, Katherine. *Predicting the Weather: Victorians and the Science of Meteorology.* Chicago: U of Chicago P, 2005.

Andrews, Jean. *Peppers: The Domesticated Capsicums.* Austin: U of Texas P, 1984.

Apperson, G. L. *The Social History of Smoking.* London: Martin Secker, 1914.

Armstrong, Isobel. *Victorian Glassworlds: Glass Culture and the Imagination, 1830–1880.* Oxford: Oxford UP, 2008.

Ashbrook, William. *Donizetti and His Operas*. Cambridge: Cambridge UP, 1983.

Ashby, T. "Appendix A: Excavations at Caerwent, Monmouth, 1904–1905." *Report of the Seventy-Sixth Meeting of the British Association for the Advancement of Science, York, 1906.* London: John Murray, 1907.

Ashwell, Samuel. *A Practical Treatise on the Diseases Peculiar to Women.* 3rd ed. Philadelphia: Blanchard & Lea, 1855.

Ayres, Peter. *The Aliveness of Plants: The Darwins at the Dawn of Plant Science.* London: Pickering & Chatto, 2008.

Bacon, Sir Francis. *Silva Silvarum, or A Natural History in Ten Centuries.* 9th ed. London: William Lee, 1670.

Bagehot, Walter. *Literary Studies.* 2 vols. London: Dent, 1911.

Baker, J. G. "A Review of the Tuber-Bearing Species of *Solanum*." *Journal of the Linnean Society, Botany,* vol. 20 (1884): 489–507.

Bailey, John. *Observations Relative to the Use of Belladonna in Painful Disorders of the Head and Face.* London: W. Thorne, 1818.

Bailey, Liberty Hyde. *The Survival of the Unlike: A Collection of Evolution Essays Suggested by the Study of Domestic Plants.* New York: Macmillan, 1901.

Barrie, J. M. *My Lady Nicotine.* Illustrated M. B. Prendergast. Boston: Joseph Knight, 1896.

Beaton, Donald. "Bedding Geraniums." *The Cottage Gardener and Country Gentleman's Companion,* vol. 16 (1856): 2–4.

———. "The Cape Gooseberry." *The Cottage Gardener,* vol. 20 (July 27, 1858): 251–53.

———. "Crystal Palace." *The Cottage Gardener and Country Gentleman's Companion,* vol. 13 (1855): 38–40.

———. "The Flower-Garden." *The Cottage Gardener,* vol. 4 (1850): 18–19.

———. "Flower-Garden Plan.—No. 10." *The Cottage Gardener and Country Gentleman's Companion,* vol. 11 (1854): 153–54.

Beaumont, Francis, and John Fletcher. *The Coxcomb.* Ed. A. R. Waller. Cambridge: Cambridge UP, 1910.

Beckford, William, *Memoirs of William Beckford of Fonthill, Author of Vathek.* Ed. Cyrus Redding. 2 vols. London: Charles J. Skeet, 1859.

Beeton, Isabella. *The Book of Household Management.* London: S. O. Beeton, 1861.

Beeton, Samuel. *The Book of Garden Management.* London: Ward, Lock & Tyler [1862].

Bell, Jacob, and Theophilus Redwood. *Historical Sketch of the Progress of Pharmacy in Great Britain.* London: Pharmaceutical Society of Great Britain, 1880.

Bentham, George. *Handbook of the British Flora.* London: Lovell Reeve, 1858.

———. *Handbook of the British Flora.* 7th ed. Revised J. D. Hooker. London: L. Reeve, 1892.

Bentley, Robert, and Henry Trimen. *Medicinal Plants.* London: J. & A. Churchill, 1875.

Bentley, Robert, and Theophilus Redwood, eds. *Dr. Pereira's Elements of Materia Medica and Therapeutics.* London: Longmans, Green, 1872.

BIBLIOGRAPHY

Beverley, Robert. *The History of Virginia, in Four Parts.* Reprinted from the author's 2nd revised edition, London, 1722. Introduction by Charles Campbell. Richmond, VA: J. W. Randolph, 1855.

Blackmore, R. D. *The Maid of Sker* [1872]. Edinburgh and London: William Blackwood & Sons, 1893.

Braddon, Mary Elizabeth. *Lady Audley's Secret* [1861–62]. Ed. Jenny Bourne Taylor. London: Penguin Books, 1998.

———. *Phantom Fortune* [1883]. [Chicago: Belford, Clarks, 188–?].

Bragge, William. *Bibliotheca Nicotiana: A Catalogue of Books about Tobacco.* Birmingham: Privately printed, 1880.

Brand, John. *Observations on the Popular Antiquities of Great Britain.* Arranged, Revised, and Greatly Enlarged by Sir Henry Ellis. 3 vols. London: Henry G. Bohn, 1849.

"A Bright View of Reform." *Bentley's Miscellany,* vol. 45 (1859): 111.

British Pharmacopoeia, 1867. Published under the direction of the General Council of Medical Education and Registration of the United Kingdom, pursuant to the Medical Act, 1858. London: Spottiswoode, 1867.

Britton, James, and Robert Holland. *A Dictionary of Plant Names.* Part 2. London: For the English Dialect Society, Trübner, 1882.

Bromfeild, William. *An Account of the English Nightshades and Their Effects.* London: R. Baldwin & G. Woodfall, 1757.

Bromfield, William A. "A Catalogue of the Plants Growing Wild in Hampshire, with Occasional Notes and Observations on Some of the More Remarkable Species." *The Phytologist: A Popular Botanical Miscellany,* vol. 3 (1847–50): 579–80; 593–609.

Brontë, Charlotte. *Jane Eyre* [1847]. Ed. Michael Mason. London: Penguin Classics, 1996.

Brown, Oliver Madox. *The Dwale Bluth, Hebditch's Legacy, and Other Literary Remains.* Ed. William M. Rossetti and F. Hueffer. 2 vols. London: Tinsley Brothers, 1876.

Browne, Janet. *Charles Darwin: The Power of Place.* Princeton, NJ: Princeton UP, 2002.

———. *Charles Darwin: Voyaging.* Princeton, NJ: Princeton UP, 1995.

Buchanan, George. *History of Scotland.* 3rd ed. 2 vols. London: J. Bettenham, 1733.

Bulwer-Lytton, Edward. *A Strange Story* [1862]. New York: Harper & Brothers, 1873.

Burbridge, F. W. *Cultivated Plants: Their Propagation and Improvement.* London: William Blackwood & Sons, 1877.

Burke, Edmund. *A Philosophical Enquiry into the Origin of Our Ideas of the Sublime and Beautiful.* 2nd ed. New York: Harper & Brothers, 1844.

Burke, Mrs. L. *The Miniature Language of Flowers.* London: George Routledge & Sons, 1865.

Burnett, Gilbert. *Outlines of Botany.* 2 vols. London: John Churchill, 1835.

BIBLIOGRAPHY

Burton, Robert. *The Anatomy of Melancholy* [1621]. Ed. Holbrook Jackson with introduction by William H. Gass. New York: New York Review of Books, 2001.

Bynum, W. F., and Roy Porter. *Medical Fringe and Medical Orthodoxy, 1750–1850.* London: Croom Helm, 1987.

Byron, Lord [George Gordon]. *The Complete Works of Byron.* Cambridge Edition. Revised Robert F. Gleckner. Boston: Houghton Mifflin, 1975.

Caldcleugh, Alexander. *Travels in South America during the Years 1819-20-21: Containing an Account of the Present State of Brazil, Buenos Ayres, and Chile.* 2 vols. London: John Murray, 1825. Darwin Online. https://darwin-online.org.uk.

Campbell, Elizabeth. "Don't Say It with Nightshades: Sentimental Botany and the Natural History of *Atropa Belladonna.*" *Victorian Literature and Culture,* vol. 35, no. 1 (2007): 607–15.

Candolle, Alphonse de. *Origin of Cultivated Plants.* New York: D. Appleton, 1886.

Carter, Tom. *The Victorian Garden.* Salem, NH: Salem House, 1985.

Castel, Rob, Elske Kusters, and Ronald Koes. "Inflorescence Development in Petunia: Through the Maze of Botanical Terminology." *Journal of Experimental Botany,* vol. 61, issue 9 (May 2010): 2235–46.

Cathcart, Earl. "On the Cultivated Potato." *Journal of the Royal Agricultural Society of England,* vol. 20 (1884): 266–300.

Chang, Elizabeth Hope. *Novel Cultivations: Plants in British Literature of the Global Nineteenth Century.* Charlottesville: U of Virginia P, 2019.

Chaucer, Geoffrey. *The Canterbury Tales: A Prose Version in Modern English.* Ed. David Wright. New York: Vintage Books, 1964.

———. *The Complete Works.* Ed. Walter Skeat. 2nd ed. 7 vols. Oxford: Clarendon Press, 1894–1900.

Cheer, Clarissa Lablache. *The Great LaBlache: Nineteenth-Century Operatic Superstar, His Life and His Times.* Bloomington, IN: Xlibris Corporation, 2009.

Christison, Robert. *A Treatise on Poisons, in Relation to Medical Jurisprudence, Toxicology, and the Practice of Physic.* 2nd ed. Edinburgh: Black, 1832.

Churchill, Fleetwood. *On the Diseases of Women: Including Those of Pregnancy and Childbed.* Philadelphia: Blanchard & Lea, 1857.

Cobbett, William. *Cottage Economy.* New York: John Doyle, 1833.

Cockerell, Hugh. "Tobacco and Victorian Literature." *Ashes to Ashes: The History of Smoking and Health.* Ed. S. Lock, L. A. Reynolds, and E. M. Tansey. Atlanta, GA: Rodopi, 1998: 89–99.

Coleridge, Samuel Taylor. *Poetical Works.* Part 1. Vol. 2. New Brunswick, NJ: Princeton UP, 2001.

Coles, William. *The Art of Simpling: An Introduction to the Knowledge and Gathering of Plants* [1655]. St. Catharines, ON: Provoker Press, 1968.

Colquhoun, Kate. *Taste: The Story of Britain through Its Cooking.* London: Bloomsbury, 2008.

Conquest, John Tricker. *Outlines of Midwifery: Intended as a Textbook for Students and*

BIBLIOGRAPHY 317

a Book of Reference for Junior Practitioners [1820]. Ed. James M. Winn. London: Longman, Brown, Green & Longmans, 1854.

Cooke, Mordecai Cubitt. *Freaks and Marvels of Plant Life, or Curiosities in Vegetation.* London: Society for Promoting Christian Knowledge, 1882.

———. *The Seven Sisters of Sleep* [1860]. Rochester, VT: Park Street Press, 1997.

Cowan, Ruth. *Relish: The Extraordinary Life of Alexis Soyer, Victorian Celebrity Chef.* London: Weidenfeld & Nicolson, 2006.

Crane, Walter. *A Floral Fantasy in an Old English Garden.* London: Harper & Brothers, 1899.

———. *A Flower Wedding* [1905]. London: V & A Publishing, 2011.

Crawford, Marion. *The Cigarette-Maker's Romance* [1890]. New York: MacMillan, 1894.

Crook, Nora, and Derek Guiton. *Shelley's Venomed Melody.* Cambridge: Cambridge UP, 1986.

The Crystal Palace Exhibition Illustrated Catalogue. Special issue, *The Art-Journal* [1851]. New York: Dover, 1970.

Culpeper, Nicholas. *Culpeper's Complete Herbal, and English Physician* [1826]. Leicester, UK: Magna Books, 1992.

D'Arcy, W. G. "The Classification of the *Solanaceae.*" *The Biology and Taxonomy of the* Solanaceae. Ed. J. G. Hawkes, R. N. Lester, and A. D. Skelding. New York: Academic Press, 1979.

Darwin, Charles. *The Movements and Habits of Climbing Plants.* 2nd ed. London: John Murray, 1875.

———. *The Origin of Species.* New York: Collier Books, 1962.

———. *The Power of Movement in Plants.* London: John Murray, 1880.

———. *The Variation of Animals and Plants under Domestication.* 2 vols. London: John Murray, 1868.

———. *The Voyage of the* Beagle [1839]. London: Penguin Books, 1989.

Darwin Correspondence Project, University of Cambridge. https://www .darwinproject.ac.uk.

Davis, Owen. *Popular Magic: Cunning-Folk in English History.* London: Hambledon Continuum, 2003.

De Almeida, Hermione. *Romantic Medicine and John Keats.* New York: Oxford UP, 1991.

Delachénaye, B. *Abécédaire de flore ou langage des fleurs.* Paris: P. Didot L'Ainé, 1811.

Dickens, Charles. *American Notes* [1842] *and Pictures from Italy* [1846]. Oxford: Oxford UP, 1994.

———. *Bleak House* [1853]. Oxford: Oxford UP, 1994.

———. *A Christmas Carol* [1843]. *Christmas Books.* Oxford: Oxford UP, 1994.

———. *David Copperfield* [1850]. Oxford: Oxford UP, 1994.

———. *Dombey and Son* [1848]. Oxford: Oxford UP, 1987.

———. *Great Expectations* [1862]. Oxford: Oxford UP, 1994.

———. *Hard Times* [1859]. Oxford: Oxford UP, 1994.

318   BIBLIOGRAPHY

———. *The Letters of Charles Dickens*. Vol. 1, 1833–58. London: Chapman & Hall, 1880.

———. *Little Dorrit* [1857]. Oxford: Oxford UP, 1994.

———. *Martin Chuzzlewit* [1844]. Oxford: Oxford UP, 1994.

———. *The Mystery of Edwin Drood* [1870]. Oxford: Oxford UP, 1987.

———. *Nicholas Nickleby* [1839]. Oxford: Oxford UP, 1994.

———. *The Old Curiosity Shop* [1841]. Oxford: Oxford UP, 1994.

———. *Oliver Twist* [1838]. Oxford: Oxford UP, 1994.

———. *Our Mutual Friend* [1865]. Oxford: Oxford UP, 1992.

———. *Pickwick Papers* [1837]. Oxford: Oxford UP, 1994.

———. *Sketches by Boz* [1839]. Oxford: Oxford UP, 1994.

———. *A Tale of Two Cities* [1859]. Oxford: Oxford UP, 1994.

———. *The Uncollected Writings of Charles Dickens: Household Words, 1850–1859*. Ed. Harry Stone. 2 vols. London: Allen Lane, 1969.

Dickens, Catherine [Lady Maria Clutterbuck]. *What Shall We Have for Dinner?* London: Bradbury & Evans, 1852.

Donizetti, Gaetano. *L'elisir d'amore*. Vocal score. [New York]: Kalmus [198?]. http://www.dlib.indiana.edu/variations/scores/caw8891/large/index.html.

Doyle, Arthur Conan, Sir. "The Red-Headed League." *The Adventures of Sherlock Holmes*. New York: Grosset & Dunlap [1892].

Dunal, Michel Felix. *Histoire naturelle, médicale et économique des Solanum, et des genres qui ont été confondus avec eux*. Paris: Amand Koenig, 1813.

Dudgeon, Robert Ellis. *Lectures on the Theory and Practice of Homeopathy*. Manchester, UK: Henry Turner, 1854.

Eiseley, Loren. *Darwin's Century: Evolution and the Men Who Discovered It*. London: Scientific Book Guild, 1959.

Eliot, George, *Adam Bede* [1859]. Ed. Stephen Gill. Harmondsworth, Middlesex, UK: Penguin Books, 1985.

———. *Felix Holt* [1866]. Ed. Peter Coveney. Harmondsworth, Middlesex, UK: Penguin Books, 1972.

Elliott, Brent. *Flora: An Illustrated History of the Garden Flower*. Buffalo, NY: Firefly Books, 2001.

———. *Victorian Gardens*. Portland, OR: Timber Press, 1986.

Elzea, Betty. *Frederick Sandys: A Catalogue Raisonné*. Woodbridge, Suffolk, UK: Antique Collectors' Club, 2001.

Emboden, William A., Jr. *Bizarre Plants: Magical, Monstrous, Mythical*. New York: Macmillan Press, 1974.

Encyclopaedia Britannica. 3rd ed. Edinburgh: A. Bell & C. Macfarqhar, 1797.

Endersby, Jim. *Orchid: A Cultural History*. Chicago: U of Chicago P, 2016.

Engels, Friedrich. *The Condition of the Working Class in England in 1844*. Trans. Florence Kelley Wischnewetzky. London: George Allen & Unwin, 1943.

Evelyn, John. *Acetaria: A Discourse of Sallets*. London: Rob. Scot, Ric. Chiswell, George Sawbridge & Benj. Tooke, 1706.

BIBLIOGRAPHY 319

Fairholt, F. W. *Tobacco: Its History and Associations*. London: Chapman & Hall, 1859. Reissue, Detroit, MI: Singing Tree Press, 1968.

Fenn, Robert. "Looking Back." *Journal of Horticulture and Cottage Gardener*, 3rd ser., vol. 8 (February 7, 1884): 99–100.

Fitzgerald, Percy [Gilbert Dyce]. *Bella Donna, or The Cross before the Name*. 2 vols. London: Richard Bentley, 1864.

———. *Jenny Bell*. 3 vols. London: Richard Bentley, 1866.

———. *Seventy-Five Brooke Street*. 3 vols. London: Tinsley Brothers, 1867.

The Floral World and Garden Guide. London: Groombridge & Sons, 1858–80.

Folkard, Richard. *Plant Lore, Legends, and Lyrics*. London: Sampson Low, Marston, Searle & Rivington, 1884.

Forbes, Edward. "On the Vegetable World as Contributing to the Great Exhibition." *The Crystal Palace Exhibition Illustrated Catalogue*. Special issue, *The Art-Journal* [1851]. New York: Dover, 1970.

Forbes, T. R. "Why Is It Called 'Beautiful Lady'? A Note on Belladonna." *Bulletin of the New York Academy of Medicine* 53 (1977): 403–6.

Forster, John. *The Life of Charles Dickens*. 2 vols. Philadelphia: J. D. Lippincott, 1874.

Francatelli, Charles. *The Modern Cook*. London: R. Bentley, 1846.

———. *Plain Cookery Book for the Working Classes*. London: Routledge, Warne & Routledge, 1853. Rpt. Scolar Press, 1977.

Freer, S., trans. *Linnaeus' "Philosophia botanica."* Oxford: Oxford UP, 2003.

Friend, Hilderic. *Flowers and Flower Lore*. 2nd ed. London: W. Swan Sonnenschein, 1884.

The Gardeners' Chronicle and Agricultural Gazette. London: Published for the Proprietors, 1841–.

The Gardeners' Magazine. Conducted by Shirley Hibberd. London: Gardeners' Magazine Office, 1862–82.

Gataker, Thomas. *Observations on the Internal Use of the Solanum*. London: R. & J. Dodsley, 1757.

Gately, Iain. *Tobacco: A Cultural History of How an Exotic Plant Seduced Civilization*. New York: Grove Press, 2001.

Gauthier, Xavière. "Why Witches?" Trans. Erica M. Eisinger. *The New French Feminisms*. Ed. Elaine Marks and Isabelle de Courtivon. New York: Schocken Books, 1981: 199–203.

Gentilcore, David. *Pomodoro! A History of the Tomato in Italy*. New York: Columbia UP, 2010.

Gerard, John. *The Herbal or General History of Plants*. Revised Thomas Johnson. [1633]. New York: Dover Publications, 1975.

———. *The Herball or General Historie of Plants*. London: John Norton, 1597.

Gibbs-Smith, C. H. *The Great Exhibition of 1851: A Commemorative Album*. London: Victorian and Albert Museum, Her Majesty's Stationery Office, 1964.

Gilbert, W. S. *Dulcamara, or The Little Duck and the Great Quack*. London: Strand Printing & Publishing, 1866.

Gill, Gillian. *The Extraordinary Upbringing and Curious Life of Miss Florence Nightingale*. New York: Ballantine Books, 2004.

Gilpin, William. *Observations, Relative Chiefly to Picturesque Beauty, Made in the Year 1772, on Several Parts of England; Particularly the Mountains and Lakes of Cumberland and Westmoreland*. London: Printed for R. Blamire, 1792.

———. *Three Essays: On Picturesque Beauty; On Picturesque Travel; and On Sketching*. London: Printed for R. Blamire, 1794.

Glasse, Hannah ["By a Lady"]. *The Art of Cookery Made Plain and Easy*. New ed. London: W. Strahan, J. and F. Rivington, et al., 1774.

Glenny, George. "The Petunia." *The Gardener and Practical Florist*, vol. 1 (1843): 247–48.

———. *The Properties of Flowers and Plants*. 2nd ed. London: Houlston & Wright, 1859.

Goodman, Jordan. *Tobacco in History: The Cultures of Dependence*. London: Routledge, 1993.

Goodspeed, Thomas H. *The Genus* Nicotiana [1954]. New Delhi: A. J. Reprints Agency, 1982.

Goody, Jack. *The Culture of Flowers*. Cambridge: Cambridge UP, 1993.

Graves, Robert James. *A System of Clinical Medicine*. Dublin: Fannin, 1843.

Gray, Peter. "Punch and the Great Famine." *History Ireland*, issue 2 (Summer 1993): 26–33.

Green, Thomas. *The Universal Herbal, or Botanical, Medical, and Agricultural Dictionary*. 2nd ed. Vol. 1. London: Caxton Press, 1824.

Greenaway, Kate. *Language of Flowers* [1884]. New York: Dover Publications, 1992.

Grieve, Maud. *A Modern Herbal* [1931]. 2 vols. New York: Dover Publications, 1971.

Griggs, Barbara. *Green Pharmacy: The History and Evolution of Western Medicine*. Rochester, VT: Healing Arts Press, 1981.

Hahnemann, Samuel. *Materia medica pura*. 2 vols. Trans. R. E. Dudgeon. Liverpool: Hahnemann Pub. Society; London: E. Gould and Son, 1880–81.

———. *Organon of Medicine*. Trans. R. E. Dudgeon. 5th ed. London: W. Headland, 1849.

Hake, Gordon. Binding for *Parables and Tales*. Photograph and text by Simon Cooke. https://victorianweb.org/art/design/books/33.html.

———. *Memoirs of Eighty Years* [1892]. London: Richard Bentley & Son, 1892.

———. *Parables and Tales*. London: Chapman & Hall, 1872.

Hall, Alaric. *Elves in Anglo-Saxon England: Matters of Belief, Health, Gender and Identity*. London: Boydell Press, 2007.

Hall, Matthew. *Plants as Persons: A Philosophical Botany*. Albany: State University of New York Press, 2011.

Halliwell-Phillipps, James Orchard. *Popular Rhymes and Nursery Tales*. London: John Russell Smith, 1849.

Hansen, Harold A. *The Witch's Garden*. Trans. Muriel Crofts. Santa Cruz: Unity Press / Michael Kesend, 1978.

BIBLIOGRAPHY 321

Hardy, Thomas. *Jude the Obscure* [1895]. Ed. Irving Howe. Boston: Houghton Mifflin, 1965.

———. *Tess of the d'Urbervilles.* Ed. David Skilton. Harmondsworth, Middlesex, UK: Penguin Books, 1978.

———. *Under the Greenwood Tree.* Ed. Tim Dolin. London: Penguin Books, 1998.

Harley, John. *The Old Vegetable Neurotics, Hemlock, Opium, Belladonna, and Henbane: Their Physiological Action and Therapeutical Use Alone and in Combination.* London: Macmillan, 1869.

Harris, John, ed. *Peter Piper's Practical Principles of Plain and Perfect Pronunciation* [1813]. Rpt. Philadelphia: Willard Johnson, 1836.

Hartley, L. P. *The Go-Between* [1953]. New York: New York Review of Books, 2002.

Hatton, Joseph. "A Day in a Tobacco Factory." *Cope's Smoke Room Booklets*, No. 14. Liverpool: At the Office of "Cope's Tobacco Plant," 1893: 40–61.

Hawkes, John G. "A Chilean Wild Potato Species and the Publication of Its Name, *Solanum maglia* (*Solanaceae*)." *Taxon*, vol. 42, no. 3 (August 1993): 671–73.

Heath, William. "A Sketch of the Great Agi-Tater." London: Thomas McLean, 1829.

Hecker, Justus. *The Epidemics of the Middle Ages.* Trans. B. G. Babington. London: Sherwood, Gilbert & Piper, 1835.

Heiser, Charles B., Jr. *The Fascinating World of the Nightshades.* New York: Dover Publications, 1987.

Herbert, Eugenia W. *Flora's Empire: British Gardens in India.* Philadelphia: U of Pennsylvania P, 2011.

Hey, Rebecca. *The Moral of Flowers.* London: Longman, Rees, Orme, Brown, Green & Longman, and J. Hatchard, 1833. Rpt. Kessinger Publishing, 2010.

Hibberd, Shirley. *The Amateur's Flower Garden: A Handy Guide to the Formation and Management of the Flower Garden* [1871]. London: Groombridge & Sons, 1884.

———. *The Amateur's Kitchen Garden.* London: Groombridge & Sons, 1877.

———. *Familiar Garden Flowers.* London, Paris & New York: Cassell, Petter, Galpin [n.d.].

———. *Garden Favourites; Their History, Properties, Cultivation, Propagation, and General Management in All Seasons.* London: Groombridge & Sons, 1858.

———. "Reaction against the Bedding Mania." *The Floral World and Garden Guide*, n.s., vol. 9 (1866): 193–94.

———. *The Town Garden: A Manual for the Management of City and Suburban Gardens.* London: Groombridge & Sons, 1855.

Hibbert, Christopher. *Queen Victoria: A Personal History.* Cambridge, MA: Da Capo Press, 2001.

Hill, Arthur. *The History and Functions of Botanic Gardens.* Annals of the Missouri Botanical Garden, vol. 2, 1915.

Hilton, Matthew. *Smoking in British Popular Culture, 1800–2000: Perfect Pleasures.* New York: Manchester UP, 2000.

Hobhouse, Henry. *Seeds of Change: Five Plants That Transformed Mankind.* New York: Harper & Row, 1987.

Hobhouse, Hermione. *The Crystal Palace and the Great Exhibition*. London: Athlone, 2002.

Hodgson, William. *Flora of Cumberland*. Carlisle, UK: W. Meals, 1898.

Hogg, Robert. *The Vegetable Kingdom and Its Products*. London: W. Kent, 1858.

Holinshed, Raphael. *Holinshed's Chronicles of England, Scotland, and Ireland*. 6 vols. London: J. Johnson, et al., 1808.

Holland, James. "Properties and Culture of the Petunia." *The Floral World and Garden Guide*, vol. 4 (July 1861): 149, 151.

Holloway, S. W. F. *Royal Pharmaceutical Society of Great Britain, 1841–1991: A Political and Social History*. London: Pharmaceutical Press, 1991.

Holmes, E. M. "The Cultivation of Medicinal Plants in Lincolnshire." Supplement, *Scientific American*, vol. 12, no. 304 (October 29, 1881): 4849.

———. "Horticulture in Relation to Medicine." *Journal of the Royal Horticultural Society*, vol. 31 (1906): 42–61.

Holway, Tatiana. *The Flower of Empire: An Amazonian Water Lily, The Quest to Make It Bloom, and the World It Created*. New York: Oxford UP, 2013.

Holzman, Robert. "The Legacy of Atropos, the Fate Who Cut the Thread of Life." *Anesthesiology*, vol. 89 (1998): 241–49.

Hooker, Joseph D. *The Botany of the Antarctic Voyage of the H.M. Discovery Ships Erebus and Terror in the Years 1839–1843*. London: Reeve Brothers, 1844.

———. "Solanum maglia." *Curtis's Botanical Magazine*, vol. 111 (1884): plate 6756.

———. *The Student's Flora of the British Islands*. London: Macmillan, 1870.

Hoppen, K. Theodore. *The Mid-Victorian Generation, 1846–1886*. Oxford: Oxford UP, 1998.

Household Words. Dickens Journals Online. University of Buckingham. https://www.djo.org.uk.

Howell, Catherine Herbert. *Flora Mirabilis: How Plants Shaped World Knowledge, Health, Wealth, and Beauty*. Washington, DC: National Geographic, 2009.

Hyman, Clarissa. *Tomato: A Global History*. London: Reaktion Books, 2019.

Hyman, Stanley Edgar. *The Tangled Bank: Darwin, Marx, Frazer and Freud as Imaginative Writers*. New York: Athenaeum, 1962.

Ingram, John. *Flora Symbolica: or, The Language and Sentiment of Flowers*. London: Frederick Warne, [1869].

———. *Oliver Madox Brown: A Biographical Sketch*. London: Elliot Stock, 1883.

Ingrouille, Martin. *Historical Ecology of the British Flora*. London: Chapman & Hall, 1995.

Jekyll, Gertrude. *Home and Garden: Note and Thoughts, Practical and Critical, of a Worker in Both*. London: Longmans, Green, 1900.

Jill, Duchess of Hamilton, Penny Hart, and John Simmons. *The Gardens of William Morris*. New York: Stewart, Tabori & Chang, 1998.

Johns, Charles Alexander. *Flowers of the Field*. London: Society for Promoting Christian Knowledge, 1853.

Johnson, George W., ed. *The Cottage Gardeners' Dictionary*. London: W. S. Orr, 1852.

BIBLIOGRAPHY 323

Johnston, William. *Nightshade*. London: Simpkin, Marshall, 1858. Rpt. Nabu Press, 2011.

Jonson, Ben. *The Works*. Ed. W. Gifford. 9 vols. Vol. 7. London: W. Bulmer, 1816.

Kaplan, Fred. *Dickens: A Biography*. New York: William Morrow, 1988.

Keats, John. *The Complete Poems*. Ed. John Barnard. Harmondsworth, Middlesex, UK: Penguin Books, 1988.

Keogh, Luke. *The Wardian Case: How a Simple Box Moved Plants and Changed the World*. Chicago: U of Chicago P, 2020.

Keynes, Richard, ed. *Charles Darwin's Zoology Notes and Specimen Lists from H.M.S. Beagle*. Cambridge: Cambridge UP, 2000. Darwin Online. https://darwin-online.org.uk.

Kilmer, F. B. "In Lands Where Drugs Grow." *The American Journal of Pharmacy*, vol. 72 (April 1900): 155–70.

King, Amy M. *Bloom: The Botanical Vernacular in the English Novel*. Oxford: Oxford UP, 2003.

Kingsley, Charles. *Westward Ho! or The Voyages and Adventures of Sir Amyas Leigh, Knight, of Burrough, in the County of Devon. In the Reign of Her Most Glorious Majesty Queen Elizabeth* [1855]. New York: Charles Scribner's Sons, 1948.

Klein, Richard. *Cigarettes Are Sublime*. Durham, NC: Duke UP, 1993.

Lai, D. C. "More on Legacy of Atropos, with Special Reference to Datura stramonium." *Anesthesiology*, vol. 90 (June 1999): 1794–95.

Landau, George P. *The Aesthetic and Critical Theories of John Ruskin*. Princeton, NJ: Princeton UP, 1971. Hypertext, *The Victorian Web*. https://www.victorianweb.org/authors/ruskin/secondarymaterials.html.

The Language of Flowers. London: Ernest Nister [18–].

The Language of Flowers: An Alphabet of Floral Emblems. London: T. Nelson & Sons, 1858.

The Language and Poetry of Flowers. London: Routledge & Sons, 186–.

Lankester, Phebe. *Wild Flowers Worth Notice*. London: R. Hardwicke, 1861.

Latour, Charlotte de (pseudonym). M. Aimé Martin. *Le langage des fleurs*. Brussels: Louis Hauman, 1830. Nabu Public Domain Reprint.

Lees, Edwin. *The Botanical Looker-Out among the Wild Flowers of England and Wales*. London: Hamilton, Adams, 1851.

Lindley, John. *Flora Medica*. London: Longman, Orme, Brown, Green & Longmans, 1838.

———. *An Introduction to the Natural System of Botany*. London, Longman, Rees, Orme, Brown & Green, 1830.

———. *Ladies' Botany, or A Familiar Introduction to the Natural System of Botany*. 3rd ed. 2 vols. London: James Ridgeway & Sons, 1834–37.

———. *Medical and Oeconomical Botany*. London: Bradbury & Evans, 1856.

———. *The Natural System of Botany*. London: Longman, Rees, Orme, Brown, Green, & Longman, 1836.

324 BIBLIOGRAPHY

———. *The Vegetable Kingdom, or The Structure, Classification, and Uses of Plants.* London: Bradbury & Evans, 1846.

Linnaeus, Carl. *Philosophia Botanica* [1751]. Trans. Stephen Freer. Oxford: Oxford UP, 2003.

Linton, Eliza Lynn. *Witch Stories.* London: Chapman & Hall, 1861.

Lizars, John. *The Use and Abuse of Tobacco.* 8th ed. Philadelphia: P. Blakiston Son, 1883.

Loudon, Jane. *Botany for Ladies, or A Popular Introduction to the Natural System of Plants.* London: John Murray, 1842.

———. *Gardening for Ladies and Companion to the Flower Garden.* London: Wiley & Putnam, 1843.

———. *Instructions in Gardening for Ladies.* London: John Murray, 1840.

Loudon, John Claudius, ed. *An Encyclopedia of Agriculture.* Longman, Hurst, Rees, Orme, Brown & Green, 1826.

———. *An Encyclopedia of Gardening.* London: Longman, Hurst, Rees, Orme & Brown, 1822.

———. *An Encyclopedia of Gardening.* London: Longman, Rees, Orme, Brown, Green & Longman, 1835.

———. *An Encyclopedia of Plants.* London: Longman, Rees, Orme, Brown & Green, 1829.

Lucot, Alexis. *Emblèmes de flore et des végétaux.* Paris: L. Janet, 1819.

Lyte, Henry. *A New Herball, or Historie of Plants.* London: Ninian Newton, 1586.

Mabberley, David. J. *The Plant-Book: A Portable Dictionary of the Vascular Plants.* Cambridge: Cambridge UP, 1997.

Mabey, Richard. *Flora Britannica.* London: Chatto & Windus, 1997.

MacCarthy, Fiona. *William Morris: A Life for Our Time.* London: Faber & Faber, 1994.

Mackail, J. W. *The Life of William Morris.* 2 vols. New York: Longmans, Green, 1899.

Mackenzie, Compton. *Sublime Tobacco.* New York: Macmillan, 1958.

MacLeod, Dawn. *The Gardener's London.* London: Gerald Duckworth, 1972.

Malthus, Thomas. *Essay on the Principles of Population.* London: J. Johnson, 1798.

———. *Essay on the Principles of Population.* London: Ward, Locke, 1890.

Marder, Michael. *Plant-Thinking: A Philosophy of Vegetal Life.* New York: Columbia UP, 2013.

Markham, Gervase. *The English Housewife* [1615]. Ed. Michael R. Best. Kingston and Montreal, Canada: McGill-Queens UP, 1986.

Marx, Karl. *The Eighteenth Brumaire of Louis Bonaparte.* New York: International Publishers, 1963.

Mann, John. *Murder, Magic, and Medicine.* Oxford: Oxford UP, 1992.

Matthews, Leslie G. *History of Pharmacy in Britain.* London: E. & S. Livingstone, 1962.

Maund, B. *The Botanic Garden.* Vol. 3. London: Simpkin & Marshall, 1829–30.

BIBLIOGRAPHY

Mayhew, Henry. *London Labour and the London Poor.* 4 vols. London: Griffin, Bohn, 1861.

McNeill, William. "How the Potato Changed the World's History." *Social Research,* vol. 66, issue 1 (Spring 1999): 67–83.

"Meaning of Bekan." *Notes and Queries.* 8th ser. Vols. 5 and 6. London: John Francis, 1894.

Medical and Philosophical Commentaries by a Society in Edinburgh. Vol. 3. London: J. Murray, 1775.

Meigs, Charles D. *Obstetrics: The Science and the Art.* 5th ed. Philadelphia: Henry C. Lea, 1867.

———. *Woman: Her Diseases and Remedies: A Series of Letters to His Class.* 4th ed. Philadelphia: Blanchard & Lea, 1859.

Meller, Henry James. *Nicotiana, or The Smoker's and Snuff-Taker's Companion.* London: Effingham Wilson, 1832.

Meredith, George. *The Ordeal of Richard Feverel* [1859]. Ed. John Halperin. New York: Oxford UP, 1984.

Michelet, Jules. *La sorcière (Satanism and Witchcraft)* [1862]. Trans. A. R. Allinson. New York: Citadel Press, 1992.

Middle English Dictionary. University of Michigan. https://quod.lib.umich.edu/m/middle-english-dictionary/dictionary.

Middleton, Thomas. *The Witch. The Works of Thomas Middleton.* Ed. A. H. Bullen. 8 vols. Boston: Houghton, Mifflin, 1885. Vol. 5: 351–453.

Mill, John Stuart. *England and Ireland.* London: Longmans, Green, Reader & Dyer, 1868.

Miller, Philip. *The Gardeners Dictionary.* 2 vols. London: C. Rivington, 1735.

———. *The Gardeners Dictionary.* 4th ed. 3 vols. London: John and Francis Rivington, 1754.

———. *The Gardeners Dictionary.* 8th ed. London: John and Francis Rivington, 1768.

———. *The Gardener's and Botanist's Dictionary.* Ed. Thomas Martyn. 2 vols. London: F. C. and J. Rivington, J. Johnson, G. and W. Nichol, et al., 1807.

Miller, Thomas. *The Poetical Language of Flowers, or The Pilgrimage of Love.* London: David Bogue, 1847.

Milton, John. *Lycidas. The Poetical Works.* Ed. L. Valentine. London: Frederick Warne, 1898: 97–102.

Minter, Sue. *The Apothecaries' Garden: A New History of the Chelsea Physic Garden.* Phoenix Mill, UK: Sutton, 2000.

Monardes, Nicholas. *Joyfull Newes out of the Newe Founde World.* Trans. John Frampton [1577]. London: Constable, 1925.

Moran, Maureen. "'Light No Smithfield Fires': Some Victorian Attitudes to Witchcraft." *The Journal of Popular Culture,* vol. 33, issue 4 (2000): 123–51.

Morris, William. *The Life and Death of Jason.* London: Bell & Daldy, 1867.

———. "The Story of an Unknown Church." *Oxford and Cambridge Magazine*, no. 1 (January 1856): 28–33. http://www.rossettiarchive.org/index.html.

Musgrave, Toby and Will. *An Empire of Plants: People and Plants That Changed the World.* London: Cassell, 2000.

Naj, Amal. *Peppers: A Story of Hot Pursuits.* New York: Alfred A. Knopf, 1992.

Neill, Patrick. *An Account of British Horticulture. Drawn Up for the "Edinburgh Encyclopedaeia."* Edinburgh: A. Balfour, 1817.

"Netley Abbey." *The Phytologist: A Botanical Journal*, vol. 2 (1857–58): 541–45.

New Grove Book of Operas. Ed. Stanley Sadie. Oxford: Oxford UP, 2006.

Nicholls, Philip A. *Homeopathy and the Medical Profession.* London: Croom Helm, 1988.

Nicholson, George, ed. *The Illustrated Dictionary of Gardening; A Practical and Scientific Encyclopaedia of Horticulture for Gardeners and Botanists.* 7 vols. London: L. Upcott Gill, 1887.

Official Descriptive and Illustrated Catalogue of the Great Exhibition of the Works of Art and Industry of All Nations, 1851. Ed. Robert Ellis. 3 vols. London: Spicer Brothers, W. Clowes & Sons, 1851.

The Old Bailey Proceedings Online, 1674–1913. Tim Hitchcock, Robert Shoemaker, Clive Emsley, Sharon Howard and Jamie McLaughlin, et al. www.oldbaileyonline .org, version 8.0.

"On the Legends and History of Certain Plants." *Hardwicke's Science-Gossip*, vol. 9 (1873): 149–53.

Ouida [Marie Louise Ramé]. *Under Two Flags* [1867]. Introduction and notes by Natalie Schroeder. Kansas City, MO: Valancourt Books, 2009.

Panayi, Panikos, *Fish and Chips.* London: Reaktion Books, 2014.

Paterson, Mrs. William. *Paterson's National Benefit: A Treatise on How to Propagate and Raise Potatoes, with Useful Hints on Farm and Domestic Management.* Dundee SCT: Printed at the Advertiser Office, 1872.

Paxton, Joseph. *What Is to Become of the Crystal Palace?* London: Bradbury & Evans, 1851.

Penn, W. A. *The Soverane Herbe: A History of Tobacco.* London: Grant Richards, 1902.

Pereira, Jonathan. *Elements of Materia Medica and Therapeutics.* Ed. Joseph Carson. 3rd American edition. 2 vols. Philadelphia: Blanchard & Lea, 1854.

———. *My Dear Mr. Bell: Letters from Dr. Jonathan Pereira to Mr. Jacob Bell, 1844 to 1853.* Ed. C. P. Cloughly, J G. L. Burnby, and M. P. Earles. Madison, WI: American Institute of the History of Pharmacy, 1987.

"Petunia Nyctaginiflora." *Curtis's Botanical Magazine*, vol. 52 (1825): plate 2552.

"Petunia nyctaginiflora violacea." *Paxton's Magazine of Botany*, vol. 2 (1836): 173.

"Petunia Violacea." *Paxton's Magazine of Botany*, vol. 1 (1834): 7.

Phillips, Henry. *Floral Emblems.* London: Saunders & Otley, 1825.

———. *History of Cultivated Vegetables.* 2 vols. London: H. Colburn, 1822.

BIBLIOGRAPHY

Piggott, J. R. *Palace of the People: The Crystal Palace at Sydenham, 1854–1936.* Madison: U of Wisconsin P, 2004.

Pliny the Elder. *The Natural History.* Trans. John Bostock and H. T. Riley. London: Taylor & Francis, 1855.

Plues, Margaret. *Rambles in Search of Wild Flowers and How to Distinguish Them.* London: Journal of Horticulture and Cottage Gardener Office, 1863.

Pollan, Michael. *The Botany of Desire.* New York: Random House, 2001.

Porter, Duncan M. "Charles Darwin's Vascular Plant Specimens from the Voyage of HMS *Beagle.*" *Botanical Journal of the Linnean Society,* vol. 93 (1986): 1–172.

Pratt, Anne. *The Flowering Plants and Ferns of Great Britain.* 4 vols. London: Society for Promoting Christian Knowledge, 1855.

———. *The Flowering Plants and Ferns of Great Britain.* Revised Edward Step. 4 vols. London: Frederick Warne, 1905.

———. *The Poisonous, Noxious, and Suspected Plants of Our Fields and Woods.* London: Society for Promoting Christian Knowledge, [n.d.].

———. *Wild Flowers.* 2 vols. London: Society for Promoting Christian Knowledge, 1857.

Pratt, Anne, and Thomas Miller. *The Language of Flowers; The Associations of Flowers; Popular Tales of Flowers.* London: Simpkin, Marshall, Hamilton, Kent, [18–].

The Private Life of the Queen. By One of Her Majesty's Servants. London: C. A. Pearson, 1898.

Promptorium parvulorum sive clericorum. Ed. Galfridis Anglicus (1440) and Albert Way (1805–74). London: Camden Society, 1865.

Purkiss, Diane. *The Witch in History: Early Modern and Twentieth-Century Representations.* London: Routledge, 1996.

Pulteney, Richard. *A General View of the Writings of Linnaeus.* 2nd ed. London: J. Mawman, 1805.

Radcliffe, Ann. *A Journey Made in the Summer of 1794: Through Holland and the Western Frontier of Germany, . . . to Which Are Added Observations of a Tour to the Lakes.* Dublin: William Porter Printer, 1795.

———. *The Mysteries of Udolpho* [1794]. 2 vols. London: J. M. Dent & Sons, 1959.

Ramsbotham, Francis. *The Principles and Practice of Obstetric Medicine and Surgery* [1841]. Philadelphia: Blanchard & Lea, 1861.

Rappaport, Helen. *Beautiful for Ever: Madame Rachel of Bond Street—Cosmetician, Con-Artist and Blackmailer.* Ebrington, Gloucesterhire, UK: Long Bam Books, 2010.

Reader, John. *Potato: A History of the Propitious Esculent.* New Haven, CT: Yale UP, 2009.

Rees, Abraham. *The Cyclopedia, or Universal Dictionary of Arts, Science, and Literature.* 39 vols. London: Longman, Hurst, Rees, Orme & Brown, 1819.

Reid, Clement. *The Origin of the British Flora.* London: Dulau, 1899.

Ristaino, Jean B., and Donald H. Pfister. "'What a Painfully Interesting Subject':

Charles Darwin's Studies of Potato Late Blight." *BioScience*, vol. 66 (December 2016): 1035–45.

Roberts, Margaret F., and Michael Wink, eds. *Alkaloids: Biochemistry, Ecology, and Medicinal Applications*. New York: Plenum Press, 1998.

Robinson, William. *The English Flower Garden*. 4th ed. London: John Murray, 1895.

———. *The Subtropical Garden, or Beauty of Form in the Flower Garden*. 2nd ed. London: John Murray, 1879.

Rossetti, Christina. *Sing-Song: A Nursery Rhyme Book* [1872]. New York: Dover Publications, 1968.

Rossetti, Dante Gabriel. *The Correspondence*. Ed. William E. Fredeman. Vol. 5: 1871–72. Cambridge: D. S. Brewer, 2005.

Rossetti, William Michael. *Dante Gabriel Rossetti: His Family Letters, with a Memoir*. Boston: Roberts Brothers, 1895.

Rousseau, Jean-Jacques. *Letters on the Elements of Botany, Addressed to a Lady*. Translated with Notes and Twenty-Four Additional Letters by Thomas Martyn. 2nd ed. London: B. White & Son, 1787.

Royle's Manual of Materia Medica and Therapeutics. 6th ed. Ed. John Harley. London: J. & A. Churchill, 1876.

Ruck, Carl A. P. "Gods and Plants in the Classical World." *Ethnobotany: Evolution of a Discipline*. Ed. Richard Evans Schultes and Siri von Reis. Portland, OR: Dioscorides Press, 1995: 131–43.

Ruddock, E. Harris. *The Stepping-Stone to Homeopathy and Health*. Chicago: C. S. Halsey, 1870.

Rundell, Maria ["By a Lady"]. *A New System of Domestic Cookery*. London: John Murray, 1833.

Ruskin, John. *The Works of John Ruskin*. 39 vols. Ed. E. T. Cook and Alexander Wedderburn. London: George Allen, 1903–12.

Sabine, Joseph. "On the *Love Apple* or *Tomato*, and an Account of Its Cultivation; with a Description of Several Varieties, and Some Observations on the Different Species of the Genus *Lycopersicum*." *Transactions of the Horticultural Society of London*, vol. 3 (1822): 342–54.

———. "On the Native Country of the Wild Potatoe, with an Account of Its Culture in the Garden of the Horticultural Society; and Observations on the Importance of Obtaining Improved Varieties of the Cultivated Plant." *Transactions of the Horticultural Society of London*, vol. 5 (1824): 249–59.

Salaman, Redcliffe N. *The History and Social Influence of the Potato* [1949]. Ed. J. G. Hawkes. Rpt. Cambridge: Cambridge UP, 1985.

Sanecki, Kay N. *History of the English Herb Garden*. London: Ward Lock, 1992.

The Saturday Review of Politics, Literature, Science and Art. London: Published at the Office.

Schama, Simon. *The Embarrassment of Riches: An Interpretation of Dutch Culture in the Golden Age*. Berkeley and Los Angeles: U of California P, 1988.

Schultes, Richard Evans, Albert Hofmann, and Christian Ratsch. *Plants of the Gods:*

BIBLIOGRAPHY

Their Sacred, Healing, and Hallucinogenic Powers. Rochester, VT: Healing Arts Press, 1992.

Schultes, Richard Evans, and Siri von Reis, eds. *Ethnobotany: Evolution of a Discipline.* Portland, OR: Dioscorides Press, 1995.

Scot, Reginald. *The Discoverie of Witchcraft.* Introduction by Montague Summers [1930]. Mineola, NY: Dover Publications, 1972.

Scourse, Nicolette. *The Victorians and Their Flowers.* Portland, OR: Timber Press, 1983.

Seaton, A. V. "Cope's and the Promotion of Tobacco in Victorian England." *Journal of Advertising History,* vol. 9, no. 2 (1986): 5–26.

Seaton, Beverly. *The Language of Flowers: A History.* Charlottesville: UP of Virginia, 1995.

Seeley, John G. "Liberty Hyde Bailey—Father of American Horticulture." *HortScience,* vol. 25, no. 10 (1990): 1204–10.

Shelley, Percy Bysshe. *Rosalind and Helen: A Modern Eclogue.* London: C. & J. Ollier, 1819.

———. *Shelley's Poetry and Prose.* Ed. Donald H. Reiman and Sharon B. Powers. Norton Critical Edition. New York: W. W. Norton, 1977.

Shoberl, Frederic. *The Language of Flowers.* London: Saunders & Otley, 1838.

Simpson, James Young. *Clinical Lectures on the Diseases of Women.* Philadelphia: Blanchard & Lea, 1863.

———. *Selected Obstetrical and Gynaecological Works Containing the Substance of His Lectures on Midwifery.* Ed. J. Watt Black. Edinburgh: Adam & Charles Black, 1871.

Skeat, Walter, ed. *An Etymological Dictionary of the English Language* [1879–82]. Oxford: Oxford UP, 1997.

Smith, Andrew F. *Potato: A Global History.* London: Reaktion Books, 2011.

———. *Pure Ketchup: A History of America's National Condiment.* Columbia: U of South Carolina P, 1996.

———. *The Tomato in America: Early History, Culture, and Cookery.* Columbia: U of South Carolina P, 1994.

Smith, Heywood. *Practical Gynaecology: A Handbook of the Diseases of Women.* London: J. & A. Churchill, 1877.

"Smokescreen: The Victorian Vogue for Tobacco." Exhibition arranged by Dr. Maureen Watry, Head of Special Collections and Archives University of Liverpool Library, May 2002. https://libguides.liverpool.ac.uk/library/sca/smokescreenexhib.

Soyer, Alexis. *The Gastronomic Regenerator.* London: Simpkin, Marshall, 1846.

———. *The Modern Housewife, or Ménagère.* London: Simpkin, Marshall, 1851.

Spratt, George. *Flora Medica.* 2 vols. London: Callow & Wilson, 1829.

Squire, Peter. *A Companion to the British Pharmacopoeia.* 3rd ed. London: John Churchill & Sons, 1866.

Shadwell, Thomas. *The Lancashire-Witches and Tegue O Divelly, the Irish-Priest.* Early English Books Online Text Creation Partnership, University of Mich-

330   BIBLIOGRAPHY

igan. https://quod.lib.umich.edu/cgi/t/text/text-idx?c=eebo;idno=A59429
.0001.001.

Shteir, Ann B. *Cultivating Women, Cultivating Science: Flora's Daughters and Botany in England, 1760 to 1860.* Baltimore, MD: Johns Hopkins UP, 1996.

Stafford Allen & Sons. *The Romance of Empire Drugs.* London: Stafford Allen & Sons, [n.d.].

Stearn, William T. *Botanical Latin.* 4th ed. Newton Abbot, Devon, UK: David & Charles, 1992.

———. *John Lindley, 1799–1865.* Woodbridge, Suffolk, UK: Antique Collectors' Club, 1999.

Stedman, Jane. *W. S. Gilbert: A Classic Victorian and His Theatre.* Oxford: Oxford UP, 1996.

Steel, Flora Annie. *English Fairy Tales.* Illustrated Arthur Rackham. New York: Macmillan, 1918.

Steinmetz, Andrew ["A Veteran of Smokedom"]. *The Smoker's Guide, Philosopher and Friend.* London: Hardwicke & Bogue, 1876.

———. *Tobacco: Its History, Cultivation, Manufacture, and Adulterations.* London: Richard Bentley, 1857.

Stephens, Walter. *Demon Lovers.* Chicago: U of Chicago P, 2002.

Stephenson, John, and James Morss Churchill. *Medical Botany: or Illustrations and Descriptions of the London, Edinburgh, and Dublin Pharmacopoeias; Comprising a Popular and Scientific Account of All the Poisonous Vegetables That Are Indigenous to Great Britain.* 2 vols. London: John Churchill, 1831.

Stirling, A. M. W. *Life's Little Day: Some Tales and Other Reminiscences.* London: T. Butterworth, 1925.

Stuart, David. *Dangerous Garden: The Quest for Plants to Change Our Lives.* Cambridge, MA: Harvard UP, 2004.

———. *The Garden Triumphant: A Victorian Legacy.* New York: Harper & Row, 1988.

———. *The Plants That Shaped Our Gardens.* Cambridge, MA: Harvard UP, 2002.

Sumner, Judith. *The Natural History of Medicinal Plants.* Portland, OR: Timber Press, 2000.

Sutton, Arthur. "Potatos." *Journal of the Royal Horticultural Society,* n.s., vol. 19 (1895–96): 387–430.

———. "Progress of Vegetable Cultivation during Queen Victoria's Reign." *Journal of the Royal Horticultural Society,* n.s., vol. 21 (1897–98): 363–93.

Swinburne, Algernon Charles. *The Complete Works.* Ed. Sir Edmund Gosse and Thomas James Wise. Vol. 5. London: William Heinemann Ltd., 1926.

Syme, John T. Boswell. *English Botany.* Illustrated James Sowerby. 12 vols. London: Robert Hardwicke, 1863–80.

Tallis's History and Description of the Crystal Palace. 3 vols. London: John Tallis, 1852.

Tanner, Heather, and Robin. *Woodland Plants.* London: Impact Books, 1981.

BIBLIOGRAPHY 331

Tarr, Clayton Carlyle. "Covent Goblin Market." *Victorian Poetry*, vol. 50, no. 3 (Fall 2012): 297–316.

Taylor, Alfred Swaine. *On Poisons in Relation to Medical Jurisprudence and Medicine* London: Lea & Blanchard, 1848.

Taylor, John Ellor. *The Sagacity and Morality of Plants: A Sketch of the Life and Conduct of the Vegetable Kingdom*. London: Chatto & Windus, 1884.

Tennyson, Alfred, Lord. *In Memoriam*. Ed. Robert H. Ross. Norton Critical Edition. New York: W. W. Norton, 1973.

Thackeray, William Makepeace. *The Irish Sketch Book, 1842*. London: J. M. Dent, 1903.

———. *Mrs. Perkins's Ball. Christmas Books*. London: Smith, Elder, 1866.

———. *Vanity Fair* [1848]. Ed. J. I. M. Stewart. Harmondsworth, Middlesex, UK: Penguin Books, 1986.

Thiselton-Dyer, T. H. *The Folk-Lore of Plants*. New York: D. Appleton, 1889.

Thompson, C. J. S. *The Mystic Mandrake* [1934]. Rpt. Detroit: Gale Research, 1975.

Thoms, Herbert. "Anesthésie à la reine." *American Journal of Obstetrics and Gynecology*, vol. 40, issue 2 (1940): 340–46.

Thomson, James. "Tobacco at the Opera." *Cope's Tobacco Plant*, vol. 2, no. 107 (February 1879): 293–94.

Thoreau, Henry David. *Wild Fruits*. Ed. Bradley P. Dean. New York: W. W. Norton, 2000.

Ticknor, Caroline. *Miss Belladonna; A Child of Today*. Boston: Little, Brown, 1897.

Tilt, Edward. *The Change of Life in Health and Disease*. London: John Churchill & Sons, 1870.

———. *A Handbook of Uterine Therapeutics and of Diseases of Women*. 4th ed. London: J. & A. Churchill, 1878.

Trevelyan, Charles E. *The Irish Crisis*. London: Longman, Brown, Green & Longmans, 1848.

Trollope, Anthony. *Framley Parsonage* [1861]. Ed. P. D. Williams. New York: Oxford UP, 1980.

———. *The Warden* [1855]. London: Penguin Books, 2004.

Turner, William. *A New Herball*. [1568]. Ed. George T. L. Chapman, Frank McCombie, and Anne U. Wesencraft. 2 vols. Cambridge: Cambridge UP, 1996.

Twamley, Louisa Anne. *The Romance of Nature, or The Flower-Seasons Illustrated*. London: Charles Tilt, 1836.

Tyas, Robert. *The Language of Flowers, or Floral Emblems*. London: George Routledge & Sons, 1869.

———. *The Sentiment of Flowers, or Language of Flora*. 9th ed. London: R. Tyas, 1842.

———. *The Sentiment of Flowers, or Language of Flora*. 10th ed. London: Houlston & Wright, 1869.

Umberger, Eugene. "In Praise of Lady Nicotine: A Bygone Era of Prose, Poetry . . .

and Presentation." *Smoke: A Global History of Smoking.* Ed. Sander L. Gilman and Zhou Xun. London: Reaktion Books, 2004: 236–47.

Velpeau, Alfred A. L. M. *An Elementary Treatise on Midwifery, or Principles of Tokology and Embryology* [1829]. Trans. Charles D. Meigs. Notes and Additions by William Harris. Philadelphia: Lindsay & Blakiston, 1845.

"Views of the Famine: Contemporary Newspaper Articles and Illustrations from the Great Hunger in Ireland, 1845–52." Compiled by Steve Taylor. WordPress.com. https://viewsofthefamine.wordpress.com.

Vickery, Roy. *A Dictionary of Plant-Lore.* Oxford: Oxford UP, 1995.

Wakefield, Priscilla. *An Introduction to Botany: In a Series of Familiar Letters* [1796]. 5th ed. London: Darton & Harvey, 1807.

Wallace, Eileen. *Children of the Labouring Poor: The Working Lives of Children in Nineteenth-Century Hertfordshire.* Hatfield, Herts.: U of Hertfordshire P, 2010.

Walpole, Horace. *Yale Edition of Horace Walpole's Correspondence.* 48 vols. New Haven, CT: Yale UP, 1937–83. Yale University Library Digital Collections.

Walton, John K. *Fish and Chips and the British Working Class, 1870–1940.* Leicester, UK: Leicester UP, 1992.

Warren, Eliza. *Cookery for Maids of All Work.* Groombridge & Sons, 1856.

Watson, Hewett Cottrell. *Cybele Britannica, or British Plants and Their Geographical Relations.* 4 vols. London: Longman, 1847–59.

Watts, Henry. *A Dictionary of Chemistry.* 5 vols. London: Longmans, Green, 1866–68.

West, Charles. *Lectures on the Diseases of Women.* 4th ed. London: J. & A. Churchill, 1879.

West, Thomas. *The Antiquities of Furness* [1774]. Ed. with additions by William Close. London: George Ashburner, 1805.

"Why Women Hate the Weed." *The Smoker's Garland.* Part 3. *Cope's Smoke Room Booklets,* No. 10. Liverpool, 1890: 157–59.

Wilde, Oscar. *The Picture of Dorian Gray* [1890]. Ed. Norman Page. Peterborough, Canada: Broadview Press, 2002.

Wilkinson, Anne. *The Victorian Gardener.* Brimscombe Port Shroud, Gloucestershire, UK: History Press, 2011.

Willes, Margaret. *The Gardens of the British Working Class.* New Haven, CT: Yale UP, 2014.

Williams, Leslie A. *Daniel O'Connell, The British Press and The Irish Famine: Killing Remarks.* Ed. William H. A. Williams. New York: Routledge, 2016.

Wiseman, Nicholas. Review of *Impediments to the Prosperity of Ireland* by Neilson Hancock. *The Dublin Review,* vol. 28 (June 1850): 399–420.

Wood, Jason. "Visualizations of Furness Abbey: From Romantic Ruin to Computer Model." *English Heritage Historical Review,* vol. 3, no. 1 (2008): 9–36.

Woodham-Smith, Cecil. *The Great Hunger, 1845–1849.* New York: Harper & Row, 1962.

Woodville, William. *Medical Botany.* 5 vols. London: John Bohn, 1832.

Woolsey, Sarah Chauncey. *The Letters of Jane Austen.* Boston: Little, Brown, 1908.

Wordsworth, William. *Selected Poems and Prefaces*. Ed. Jack Stillinger. Riverside Edition. Boston: Houghton Mifflin, 1965.

Wright, Thomas. *Narratives of Sorcery and Magic, from the Most Authentic Sources*. Clinton Hall, NY: Redfield, 1852.

Wygant, Amy. *Medea, Magic, and Modernity in France: 1553–1797*. London: Routledge, 2007.

Wyhe, John van, ed. *The Complete Work of Charles Darwin Online*. 2002–. http://darwin-online.org.uk.

Yonge, Charlotte. *The Herb of the Field*. 2nd ed. London: J. & C. Mozley, 1858.

Zuckerman, Larry. *The Potato: How the Humble Spud Rescued the Western World*. New York: North Point Press, 1998.

INDEX

Italicized page numbers refer to illustrations.

Abercrombie, John, 258–59, 264

aconite (*Aconitum napellus*): cultivation of, 104; homeopathy and, 66; poisonous nature of, 7, 73; ungendered identity of, 107; witchcraft and, 146, 147

Acton, Eliza, 182, 184, 267, 268

Actuarius, 256

addiction, 92–93, 195, 197, 204, 219–21, 236, 243, 279

affinity principle, 57, 256, 259

Ainsworth, Harrison, 139–40

Alcott, Louisa May, 95

Allen, David, 25, 62

Allen, Grant, 91–92

Allen, Stafford, 104–5, 278

Allingham, Helen, 254

allopathic treatments, 64–65, 95

Almeida, Hermione de, 39

Altick, Richard, 224, 229, 231, 243

Amherst, Alicia, 276

Anderson, Frank J., 24

Anderson, Katherine, 165

Andrews, James, 47, 48

anesthetics, 92, 98, 131–32, 135

Anne (queen of England), 212

aphrodisiacs, 7, 135, 163, 167, 262

Apperson, G. L., 195–96

arsenic, 66, 92

Atropa belladonna. See belladonna

atropia (atropine), 75, 88, 93–94, 98, 104, 129–30

Auber, Daniel, 67

aubergines. *See* eggplants

Austen, Jane, 32, 264–65

Babington, Charles Cardale, 90, 168

back-to-nature approach to landscaping, 254

Bacon, Francis, 136–40

Bagehot, Walter, 40

Bailey, John, 128–29

Bailey, Liberty Hyde, 255, 279

Baker, John Gilbert, 153, 158–59, 161

Balzac, Honoré de, 44, 219

Barrie, J. M., 200, 239–43

Bary, Anton de, 155

Baudelaire, Charles, 220

Bauhin, Gaspard, 153, 154, 163

Beaton, Donald, 245, 246, 252, 253, 255

Beaumont, Francis, 10, 108

Beckford, William, 34

bedding-out gardens, 244–55

Bede, 77

Beeton, Isabella, 13, 184, 267–68

Beeton, Samuel, 184, 268

belladonna (*Atropa belladonna*), 73–147; cosmetic applications, 10, 108–11, 125–28; cultivation of, 104–5, 278; dwale etymology and, 8, 75–77, 82–83; eradication efforts, 9, 75, 78, 86, 141, 147, 278; femme fatale and, 11, 106–25; *Gardeners' Chronicle* on, 80–84; habitat for, 82, 86–92, 142; homeopathy and, 66, 94–95, 106–7, 144; illustrations of, *74, 80–81;* in language of flowers, 110–12; in literature, 77, 95, 112–25, 123–24, 128, 289n13; lurid legacy of, 51, 54–56, 85; medicinal uses of, 37, 75, 77, 87, 93–104, 128–32, 276; misidentification of, 72, 73, 93–94; mydriatic properties of, 8, 72, 93, 99–100, 108, 110, 121, 125, 128, 147; native vs. denizen status of, 90–92; pain-relieving effects of, 75, 93, 128; poisonous nature of, 6–9, 73–75, 78–79, 82–88, 93–94, 99–101; psychoactive properties of, 7, 8, 51, 79, 96, 99–103; reputation of, 9–10, 75–79, 85–86, 93, 108–12, 115–17,

336 INDEX

belladonna (*continued*)
147; sedating effects of, 8, 39, 75–77, 93, 96, 128; seductive women and, 10, 108–9, 114–17, 125; smell of, 82, 85; whores and, 10, 11, 108, 110, 119, 147; witchcraft and, 8, 10–11, 88, 103, 110, 112, 115–17, 132–44, 147
Bennett, John, 267–68
Bentham, George, 22–23
Bentley, Robert, 103–5
Bessa, Pancrace, 44
Beverley, Robert, 101–2
Beyle, Marie-Henri (Stendhal), 44
Bitter, Georg, 161
bittersweet (*Solanum dulcamara*), 18–72; charlatanism and, 67–70; climbing ability of, 19–22, 33, 40, 47; cosmetic applications, 69, 72; femme fatale and, 39–40; at Furness Abbey, 33–37; habitat for, 19–20, 32; homeopathy and, 64, 66–67; illustrations of, 2, 22, 28; in language of flowers, 41, 44–48, 112; in literature, 18–19, 21, 26–36, 38–41; lurid legacy of, 52, 54, 56, 60; medicinal uses of, 13, 40, 60–64; metaphorical uses of term, 24; misidentification of, 23, 24, 60–61, 70, 72, 122, 147; pain-relieving effects of, 13; poisonous nature of, 2, 25–31, 47, 59; psychoactive properties of, 61; quackery and, 67, 69–70, 71; reputation of, 9, 26, 32, 34, 39; seductive women and, 46; semiannual leaf production of, 46; smell and taste of, 25–27, 47; witchcraft and, 61, 88, 143
Bizet, Georges, 219, 225
Blackmore, R. D., 18–19
black nightshade (*Solanum nigrum*): illustration of, 58; lurid legacy of, 52, 54, 55; medicinal uses of, 63, 64; poisonous nature of, 25, 59; reputation of, 9; witchcraft and, 135
black pepper (*Piper nigrum*), 256
blight (Potato Murrain), 12, 149, 154–58, 163, 168–71, 177–79, 182, 276
Bloomer, Amelia, 234
bloomers, 234–35, 235–37
Bodaeus, 100
Bordeaux Mixture, 156

Boswell, James, 217
botanical literacy, 50, 84, 276
botany: affinity principle in, 57, 256, 259; epistolary, 53–55; ethnobotany, 52, 85; geographic distribution studies in, 89–92; Linnaean classification system, 1, 5, 9–11, 14, 47–49, 107, 180; modernization and professionalization of, 1, 48–49; natural system of classification, 1, 13, 47, 48, 54–55, 57, 180–81, 246; paleobotany, 19, 90; sentimental, 10, 49, 112–17, 134, 141, 185; witchcraft and, 132–47; women's study of, 10, 48–50, 53–60, 109
Braddon, Mary Elizabeth, 128
Bragge, William, 223
Braithwait, Richard, 212
Brand, John, 133, 136–39
Bright, John, 69
Bromfeild, William, 63
Bromfield, William Arnold, 22, 86–89
Brontë, Charlotte, 230
Brown, Ford Madox, 120
Brown, Oliver Madox, 120–23
Browne, Janet, 157
bryony, 7, 26, 32, 52
Buchanan, George, 76, 77, 90
Bulwer-Lytton, Edward, 117, 119, 213
Burke, Edmund, 33–36, 38, 220
Burnand, Francis, 237
Burne-Jones, Edward, 23
Burnett, Gilbert, 99, 100
Burns, Robert, 30, 217
Burton, Robert, 14
Butler, Samuel, 136
Byron, Lord (George Gordon), 199–200, 213, 221

calceolarias, 245, 253, 254
Caldcleugh, Alexander, 150–53, 159
Campbell, Thomas, 30
Candolle, Augustin Pyramus de, 54
capsicums. *See* peppers
Carlyle, Thomas, 213, 224
carphologie, 101
Cathcart, Earl, 158, 160, 162
charlatans, 67–70, 97
Charlotte (queen of England), 216

INDEX   337

Chatto, William Andrew, 303n66
Chaucer, Geoffrey, 24, 77, 278
Chaussier, Francois, 131, 143
Cheer, Clarissa Lablache, 68
chewing tobacco, 15, 213, 228, 233
chloroform, 92, 93, 98, 131–32, 143
Christian, Fletcher, 199
Christian, Thomas, 93–94
Christianity, 27, 30, 139, 141
Churchill, James Morss, 27–28, 32
cigarettes, 15, 17, 218–21, 225–27, 227, 233–40, 242, 278
cigars, 15–17, 193–95, 199–201, 204–6, 209, 213–35, 228, 239
Clark, William, 27
Claubry, M. Gaultier de, 99–100, 292n99
Close, William, 37
Clusius, Carolus, 256, 257, 260, 262
Cobbett, William, 164–65
Cockerell, Hugh, 229
Coleridge, Samuel Taylor, 30–31
Coles, William, 61, 62, 77
Collins, Wilkie, 69, 71
Columbus, Christopher, 211, 256
Commerson, Philibert, 246
Conan Doyle, Arthur, 229, 230
Conquest, John Tricker, 131–32
cookbooks, 181–84, 190, 258, 266–68
Cooke, Mordecai, 16, 21, 73
Cope's *Smoke Room Booklets*, 223–29, 227–29, 241, 242–43
cosmetics, 10, 69, 72, 92, 108–11, 125–28
Crane, Walter, 123–24, 125
Crawford, Marion, 226
Cromwell, Oliver, 169
Crook, Nora, 40
Crowquill, Alfred (Alfred Henry Forrester), 206, 207
Cruikshank, George, 224
Crystal Palace, 3–5, 13, 16, 244–46, 252–54, 270, 273–74, 279
Cullen, William, 63
Culpeper, Nicholas, 61–62, 66, 76–78

dancing mania, 144
Darwin, Charles: *Beagle* voyage by, 149–50, 153; on Bentham's *Handbook*, 23; on bit-

tersweet, 20–21; evolutionary theory of, 89, 90, 148, 151, 181; *Gardeners' Chronicle* articles by, 81; Henslow as mentor to, 90, 150; *Journal of Researches*, 151, 153, 159; *The Movements and Habits of Climbing Plants*, 20; *On the Origin of Species*, 13, 20, 89, 151, 152, 157; potatoes and, 148–61, 181; *The Power of Movement in Plants*, 20; tobacco use by, 217; *The Variation of Animals and Plants under Domestication*, 156–58; *Voyage of the Beagle*, 151
Darwin, Francis, 20
Darwin potato (*Solanum maglia*), 150–53, 158–62, 160
Datura stramonium (thornapple), 7, 53, 55, 98, 101–3
Davis, Owen, 61
deadly nightshade. *See* belladonna
Delachénaye, B., 44
Democritus, 46
Dickens, Charles: *All the Year Round*, 117, 133; *American Notes for General Circulation*, 233; *Bleak House*, 164, 231, 233, 234; *A Christmas Carol*, 188; *David Copperfield*, 190, 231, 232; *Dombey and Son*, 232–34; *Great Expectations*, 233; *Hard Times*, 233; henbane utilized by, 102; *Household Words*, 69; *Little Dorrit*, 118, 190, 232–34; *Martin Chuzzlewit*, 231, 233; *The Mystery of Edwin Drood*, 233; *Nicholas Nickleby*, 187, 231; *The Old Curiosity Shop*, 188, 231–32; *Oliver Twist*, 187, 190, 232; *Our Mutual Friend*, 102, 118, 232; *Pickwick Papers*, 186–87, 230–31, 266; *Pictures from Italy*, 118; *Sketches by Boz*, 189; *A Tale of Two Cities*, 190; tobacco use by, 234
Dickens, Kate, 102, 187–88
Dioscorides, 61, 108
Doctrine of Signatures, 7, 11
Dodoens, Rembert, 256, 260, 263
Donizetti, Gaetano, 67–69
Douglas-Hamilton, Jill, 23
Drake, Francis, 196
Dudgeon, Robert Ellis, 94, 95
dulcamara. *See* bittersweet
Dunal, Michel Felix, 63, 159
dwale. *See* belladonna

eggplants (*Solanum melongena*): culinary uses, 56, 261, 262, 268; cultivation instructions, 273, 310n103; nightshade family redefined by, 3; origins of, 255–56, 259; popularity of, 12–13; reputation of, 259–61

Eiseley, Loren, 148

Eliot, George, 21, 22, 46, 191

Elliott, Brent, 245

Ellis, Henry, 133

Endersby, Jim, 3

Engels, Friedrich, 165–66

England. *See* Victorian England

environmental plasticity, 153

epistolary botany, 53–55

ethnobotany, 52, 85

Evelyn, John, 257–58

evil: blurring of boundaries between good and, 13; perceptions of evil in nature, 6, 37; poisonous plants as, 3, 9–11, 17, 38, 85, 89; potato and, 56, 170, 173–74, 176, 191

evolutionary theory, 3, 26, 89, 90, 148, 151, 181

Fairholt, F. W., 16, 201, 208–18, *210*, *213*, 222, 303n66

Fell, Thomas K., 36

females. *See* women

femmes fatales, 11, 39–40, 106–25, 130, 200

Fenn, Robert, 162–63, 176

fertility drugs, 7, 135

figworts (Scrophulariaceae family), 3, 6, 51

Fitch, John Nugent, 159

Fitzgerald, Percy, 117–18, 121

Fletcher, John, 10, 108

floriography. *See* language of flowers

flower books, 26–32, 48, 123, 125

Folkard, Richard, 124, 134–36

Forbes, Edward, 5–6, 9, 11–14, 16, 17, 84, 277

Forbes, T. R., 100, 108

Forrester, Alfred Henry (Alfred Crowquill), *206*, *207*

Forster, John, 234

Foster, Miles Birket, 254

Foster, Thomas Campbell, 171

Fox, W. D., 156–58

foxgloves, 6, 104

Francatelli, Charles, 182–84, 267

Fraser, John, 224

Frederick, Alan, 158

Frederick the Great (king of Prussia), 217

Friend, Hilderic, 10, 110, 124, 134, 136–39

Fuchs, Leonhart, 256

Furness Abbey, 33–37, 87

Galen, 264

Gardeners' Chronicle and Agricultural Gazette: on belladonna, 80–84; on blight (Potato Murrain), 155–56, 168; on cultivation of medicinal plants, 104, 105; greenhouses marketed in, 274; petunias' appearance in, 249, 251; on potato hybridization, 159–62; on tomatoes, 267, 269, 272

Gardeners' Magazine, 159–60, *160*, 261, 266, 273

garden nightshade. *See* black nightshade

Gataker, Thomas, 62–63

Gately, Iain, 200, 219, 229

Gauthier, Theophile, 219

Gentilcore, David, 259, 262

George III (king of England), 193, 216

George IV (king of England), 214

geraniums, 245, 253, 254

Gerard, John: on belladonna, 8–9, 11, 39, 75, 77, 108, 278; on bittersweet, 61; on eggplants, 260–61; on henbane, 15, 197; *The Herbal*, 8–9, 12, 14, 61, 108, 163, 196–97, 255; on peppers, 256–57; on potatoes, 11, 12, 163; on tobacco, 196–98, 209; on tomatoes, 262–64

Gilbert, William S., 42, 69

Gilfillan, George, 116

Gill, Gillian, 132

Gilpin, William, 34, 36, 38

Glasse, Hannah, 258

Glenny, George, 249–50

Glorious Revolution (1688), 199, 212

Goldoni, Carlo, 185

Goodman, Jordan, 15, 197, 199, 215, 218

Goody, Jack, 43

Gordon, George (Lord Byron), 199–200, 213, 221

Gothic literature, 26–29, 31–41, 88, 200

INDEX

Great Exhibition of 1851 (London): *Art-Journal* catalogue for, 5, 84; belladonna at, 11; Crystal Palace and, 3–5, 13, 16, 273; potatoes at, 12, 13; tobacco at, 16–17

Great Potato Boom (1903–4), 162–63

Greenaway, Kate, 43, 111–12, 186

greenhouses: affordability of, 272, 274, 275; Darwin's experimentation in, 20, 156; fumigation with tobacco smoke, 278; out-of-season growing in, 12; petunias grown in, 246, 249, 277. *See also* Crystal Palace

Grieve, Maud, 8

Griffiths, F. T., 101

Grimm, Jacob, 133

Guiton, Dere, 40

gynecological medicine, 129–32

Hahnemann, Samuel, 64–66, 77, 94–98, 100, 101

Hake, Thomas Gordon, 118, 119

Hall, Alaric, 60–61

Hallenburg, George, 62

Hancock, Neilson, 176–77

Hannen, James, 239

Hardy, Thomas, 123–24, 191

Harley, John, 95–100, 102–4, 129–30

Harrison, J. Bower, 129

Hart, Penny, 23

Hartley, L. P., 106, 123

Hatfield, Gabrielle, 25, 62

Hatton, Joseph, 226–29

Hawkes, John G., 161

hawthorn, 21, 31, 38

Heath, William, 173, 173–74

Hecker, Justus, 144

Heiser, Charles B., Jr., 261, 283n4

hemlock: cultivation of, 104; poisonous nature of, 7; sedating effects of, 96; ungendered identity of, 107; witchcraft and, 135, 136, 140

henbane (*Hyoscyamus*): cultivation of, 104, 105, 278; habitat for, 142; in literature, 30–31; lurid legacy of, 50–51, 54–56; medicinal uses of, 102–3, 130, 276; pain-relieving effects of, 51, 102; poisonous nature of, 7, 31, 54, 85; psychoactive properties of, 7, 51, 102–3; sedating effects of, 15, 51, 96, 102, 130, 197; smell of, 50, 51, 54; tobacco, similarities with, 14; ungendered identity of, 107; witchcraft and, 88, 103, 134–40, 143, 146, 147

Henderson, William, 136

Henry VIII (king of England), 33

Henslow, John Stevens, 90, 150–53, 155

Herbert, Sidney, 69

Hereman, Samuel, 274

Hernandez, Francisco, 211, 262

Hey, Rebecca, 26–32, 28, 48

Hibberd, Shirley, 159, 160, 244, 252–54, 273

Hibbert, Christopher, 127

Hillard, James, 79–80, 83, 86

Hilton, Matthew, 15, 195, 200, 218–19, 229

Hobbes, Thomas, 213

Hodgson, William, 25

Hogarth, William, 212

Hogg, Robert, 184

Holinshed, Raphael, 90, 289n13

Holland, James, 251–52

Holmes, E. M., 105

Holzman, Robert S., 98

homeopathy, 64–67, 94–95, 106–7, 140, 144, 288n87

honeysuckle, 21, 26

Hooker, Joseph Dalton, 23, 81, 159

Hooker, William Jackson, 90

hothouses. *See* greenhouses

housewives, 12, 118, 266, 277

Humboldt, Alexander von, 150, 152

humoral theory, 263, 308n63

hybridization experiments, 158–62

Hyman, Clarissa, 266

Hyman, Stanley Edgar, 20

Hyoscyamus. See henbane

Industrial Revolution, 148–49

Ingrouille, Martin, 88, 90

Ipomoea batatas (sweet potatoes), 163–64

Ireland: Home Rule movement in, 175; method of cooking potatoes in, 183, 184; peasant population in, 11–12, 163–72, 176, 177; Potato Famine in, 11–12, 149, 164, 168–71, 176, 186, 191; religion in, 69, 166–67, 171, 173–74

James I (king of England), 15, 198, 200, 203, 212, 224
Jekyll, Gertrude, 277
Jenner, Edward, 95
jimsonweed. *See* thornapple
Johns, Charles Alexander, 21, 22, 82–84, 86
Johnson, Samuel, 216–17
Johnson, Thomas, 8, 163, 257
Johnston, James F. W., 202, 204
Johnston, William, 116–17
Jonson, Ben, 135, 137, 139, 212
Jussieu, Antoine Laurent de, 13, 54, 246

Kane, Robert, 169
Kant, Immanuel, 220–21
Kaplan, Fred, 234
Keats, John, 23, 39
Kilmer, F. B., 105
King, Amy M., 107
King, William, 186
Kingsley, Charles, 196
kitchen gardens, 255–73; eggplants in, 12–13, 255–56, 259–62, 268, 273; peppers in, 12–13, 255–60, 268–69, 273; tomatoes in, 12–13, 255–56, 259, 262–73
Klein, Richard, 220–21

Lablache, Luigi, 68, *68*
Lachapelle, Marie, 131, 143
Lamarck, Jean-Baptiste, 246
Lamb, Charles, 30, 213
Lancet, 16, 93, 98, 132, 201, 202, 204, 206, 208, 267
Landow, George, 38
Lane, Ralph, 212
Lang, Andrew, 243
language of flowers: arbitrariness of, 43, 112, 125; belladonna in, 110–12; bittersweet in, 41, 44–48, 112; flower writing example, 45; origins and emergence of, 10, 42; petunias in, 248; potato in, 12, 185–87; romance and, 42–46, 113; woman-flower connection and, 49, 50, 57, 107
Lankester, Phebe, 125, 127
Latour, Charlotte de, 44–48, 110–11, 185–86
laudanum, 28, 92, 118

lazy-bed cultivation method, 166, 170
Leech, John, *175, 176, 178*, 234, 235, 236–37
Lees, Edwin, 87–88
Legrand, Brown, 132
Liebig, Justus, 202
Lindley, John: on affinity principle, 57; on belladonna, 10, *81*, 109–10, 115–16; on bittersweet, 2, *2*, 25; on black nightshade, 55, 58; *Flora Medica*, 25; *Gardeners' Chronicle* and, 81, 155; *An Introduction to the Natural System of Botany*, 55, 180–81; *Ladies' Botany*, 55–58, *58*, 181; *Medical and Oeconomical Botany*, 81, *81*, 209; on natural system of classification, 1, 13, 48, 180; on potatoes and potato blight, 155, 169, 180–81; on tobacco, 209–10; *The Vegetable Kingdom*, 1–2, *2*
Linnaeus, Carl: on belladonna, 101, 108, 136; classification system developed by, 1, 5, 9–11, 14, 47–49, 107, 180; eggplant's scientific name designated by, 261; Luridae as family name assigned to nightshades, 10, 50–51, 60, 142, 180, 277; tomato nomenclature and, 264
Linton, Eliza Lynn, 134
literature: belladonna in, 77, 95, 112–25, 123–24, 128, 289n13; bittersweet in, 18–19, 21, 26–36, 38–41; books of secrets, 138; flower books, 26–32, 48, 123, 125; Gothic, 26–29, 31–41, 88, 200; henbane in, 30–31; potatoes in, 164, 187–91; Romantic, 26, 29–32, 35, 39–40, 200; tobacco in, 16, 196, 199–201, 211–12, 222–43; transition from realism to modernism in Victorian fiction, 283n5; witches and witchcraft in, 139–45. *See also specific authors and works*
Lizars, John, 202–4
lobelias, 246, 254
Longfield, Mountifort, 177
Loudon, Jane, 57–59, 109, 247–49
Loudon, John Claudius: *Encyclopedia of Agriculture*, 184; *Encyclopedia of Gardening*, 180, 247, 259, 261–62, 266, 268; *Encyclopedia of Plants*, 10, 57, 109, 180, 181
love apples. *See* tomatoes
love potions, 67, 139

INDEX

Lucot, Alexis, 110–11, 113, 130, 146, 185–86, 265–66
Luridae, 10, 50–55, 60, 142, 180, 277
Lyte, Henry, 260–62

Mabberley, David, 62
Mabey, Richard, 26
Mackenzie, Compton, 194, 201, 215, 219, 229
mad apples. *See* eggplants
magic, 52, 61, 133–40, 145, 276
Malthus, Thomas, 168
mandrake (*Mandragora officinarum*): lore related to, 7, 29, 52, 55; medicinal uses of, 52; pain-relieving effects of, 51; poisonous nature of, 6–7; psychoactive properties of, 51, 103, 135; sedating effects of, 51; witchcraft and, 134–37, 139, 140
Mann, John, 137
Martyn, Thomas, 49–56, 78, 109, 111, 258
Marx, Karl, 166
mass hysteria, 144
mass poisonings, 99–102
Mattioli, Pietro Andrea, 108, 256, 259
Mayhew, Henry, 188, 190, 196, 215–16
McLean, Thomas, 173
McNeill, William, 149
Medici, Catherine de, 14, 211
Meigs, Charles, 131
Meller, Henry James, 222–23, 229, 232
Mendel, Gregor, 157
Meredith, George, 115
Mérimée, Prosper, 219, 225
Michelet, Jules, 140–46
Michiel, Pietro Antonio, 259
Middleton, Thomas, 139
midwives, 130–32, 140, 143
Mill, John Stuart, 171
Miller, Philip, 77–78, 258, 261, 263–64
Miller, Thomas, 112
Milton, John, 111
Molina, Ignatius, 150, 153
Monardes, Nicholas, 15, 196
Morley, Henry, 69
morphine, 64, 92, 103
Morris, William, 23–24, 31
Murphy, Patrick, 165

Napoleon Bonaparte, 217
natural system of classification, 1, 13, 47, 48, 54–55, 57, 180–81, 246
Neander, Johann, 211
Neill, Patrick, 179–80
Netley Abbey, 72, 87
nettles, 87, 168, 197
neuralgia facialis (*tic douloureux*), 128
neurotics, 96–103, 129
Newman, William, 172, 172
Newton, Isaac, 213
New Woman image, 15, 145
Nicholls, John, 185
Nicholls, Philip, 65
Nicot, Jean, 14, 211
Nicotiana. See tobacco
Nightingale, Florence, 132
nightshades. *See* Solanaceae family

obstetrics, 130–32, 143
O'Connell, Daniel, 171–76, 172–74
Oldham, John, 136
opium: belladonna with, 97, 102, 129–30; difficulty obtaining, 73; homeopathy and, 95; isolation of morphine from, 64; laudanum, 28, 92, 118; pain-relieving effects of, 93, 97; psychoactive properties of, 99; sedating effects of, 51, 64, 96, 102; witchcraft and, 136, 137
Ouida (Mary Louise Ramé), 235–38, 238
overdosing, 63, 92–96, 118
Oviedo, Gonzalo Fernández de, 211, 223

paleobotany, 19, 90
Panayi, Panikos, 190
Papaver somniferum (poppies), 7, 73, 104. *See also* opium
Paracelsus, 142, 144
Parmentier, Antoine-Augustin, 186
Parny (French love poet), 44–47
patent medicines, 70, 288n87
Paterson, William, 185
Paxton, Joseph, 4, 81, 244–45, 247, 248, 255, 273, 274
Peel, Robert, 169
Penn, W. A., 16, 218–21, 223, 279

342 INDEX

peppers (*Capsicum annum*): commercial availability of, 268–69, 309n92; culinary uses, 12, 255, 258, 268; cultivation instructions, 257–59, 273, 310n103; illustration of, *260*; nightshade family redefined by, 3, 17; origins of, 13, 255–56; popularity of, 12–13, 259, 280; therapeutic qualities of, 257

Pereira, Jonathan, 99–104

petunias, 245–55; critiques of, 249–51; hybridization of, 279; illustrations of, *2, 248*; in language of flowers, 248; nightshade family redefined by, 3, 17; in ornamental gardens, 1, 13–14, 245–47; *Petunia nyctaginiflora*, 247; *Petunia nyctaginiflora violacea*, 247, 248; *Petunia violacea (phoenicea)*, 1, 2, 247, 250–51, 253–54; popularity of, 248–49, 251–52, 255, 277

Pharmaceutical Journal, 79–81, *80*, 88, 92–96, 104–5

Pharmaceutical Society, 80, 93–95, 103

Pharmacy Act of 1868, 73, 93, 96

Phillips, Henry, 46–47, 111, 186

Physalis ixocarpa (tomatillos), 262

physic gardens, 77, 104, 258, 278

Piper nigrum (black pepper), 256

pipes, 17, 193–96, 199–201, 204, 206, 212–23, *215*, 231–34, 239–43

plague, 144, 203, 216

Plato, 96

Playfair, Lyon, 169

Plues, Margaret, 276

Plutarch, 99

poisonous nightshades: curse associated with, 3, 283n4; deaths caused by, 9, 28, 54, 73, 75, 78–79, 84, 86, 94; defense mechanisms of, 25–26; in language of flowers, 47; mass poisonings, 99–102; moral views of, 26, 27, 31; objective treatment of, 59; reputation for evil, 3, 9–11, 17, 38, 85, 89; solanine and, 25, 64, 154, 209; susceptibility to, 29–30. *See also specific species*

Pollan, Michael, 154

Pope, Alexander, 212, 216

poppies (*Papaver somniferum*), 7, 73, 104. *See also* opium

Porta, Giambattista della, 138–39

potatoes (*Solanum tuberosum*), 148–92; blight, 12, 149, 154–58, 163, 168–71, 177–79, 182, 276; cartoons involving, *172–73, 172–74, 177–78, 178–79*; "Cinderella of Nature" metaphor for, 162–63, 176; culinary uses, 56, 182–84, 187–90, *189*; Darwin and, 148–61, 181; domestication of, 154; environmental plasticity of, 153; evil and, 56, 170, 173–74, 176, 191; as food source, 11–13, 149, 164–67, 180; Great Exhibition displays, 12, 13; Great Potato Boom, 162–63; historical importance of, 149; hybridization experiments, 158–62; illustrations of, *152*; Irish Potato Famine, 11–12, 149, 164, 168–71, 176, 186, 191; in language of flowers, 12, 185–87; lazy-bed cultivation method for, 166, 170; in literature, 164, 187–91; nightshade family redefined by, 3, 149, 276; as providential food, 177, 185, 186, 191, 192; reputation of, 163, 182; social class and, 12, 164–67; witchcraft and, 191

potentizing method, 65–67

Potts, Thomas, 140

Pratt, Anne, 59, 59–60, 85–86, 112

Prendergast, M. B., 240, 242

Prescott, Henry, 209

prostitutes, 11, 40, 110, 118, 127

Punch (magazine): "Atropa Belladonna" letter to editor (1869), 71–73; "Bloomerism—an American Custom" (1851), 235, *237*; "Consolation for the Million.—The Loaf and the Potato" (1847), 177–78, *178*; "Dr. Dulcamara in Dublin" (1866), 69, *70*; "Dr. Dulcamara Up to Date; or Wanted, a Quack-Quelcher" (1893), 69–70, *71*; "False Fine Eyes" (1864), 126; "Fine Eyes for Foolish Girls" (1856), 126; "The Finest Balsams of Arabia" (1854), 288n87; "Grand Vegetable Banquet to the Potato on His Late Recovery" (1849), 178–79, *179*; "The Irish Cinderella and Her Haughty Sisters, Britannia and Caledonia" (1846), 175, 176; "Justice to Punch and Ireland" (1882), 176; "Punch's Fancy Portraits—No. 45. Ouida" (1881), 237, *238*; "The Real Potato Blight of Ire-

INDEX

land" (1845), 172, *172;* "Save Me from My Friends!" (1875), *174, 174*–76; "Something More Apropos of Bloomerism" (1851), 235, *236;* "The Song of the Passee Belle" (1869), 126–27, *127;* "Woman's Emancipation" (1851), 234–35, *235*

quackery, 61, 65, 67, 69–70, 71, 288n87
Quid (poet), 206, 208

Rackham, Arthur, 280–82, *281*
Radcliffe, Ann, 32–34, 36–37, 39
Rainier, Peter, 261
Raleigh, Walter, 15, 163, 178, 194, 196, 206, 212
Ramé, Mary Louise (Ouida), 235–38, *238*
Ramsbotham, Francis, 132
Ransom, William, 105, 278
Ray, John, 78
Reader, John, 149, 155, 162–63, 167
realism, 21, 44–45, 283n5
recipes. *See* cookbooks
Redwood, Theophilus, 103
Rees, Abraham, 148, 276
Reid, Clement, 19, 90
religion and spirituality: Christianity, 27, 30, 139, 141; hallucinogens for rituals, 7; in Ireland, 69, 166–67, 171, 173–74; science and, 30, 133. *See also* witches and witchcraft
return-to-nature principle, 254
Robinson, William, 254–55, 277
Rolfe, John, 198, 209
Romani, Felice, 67
Romantic literature, 26, 29–32, 35, 39–40, 49, 200
Ross, Georgina, 84
Rossetti, Christina, 119–20
Rossetti, Dante Gabriel, 118, 119, 145, 146
Rousseau, Jean-Jacques, 49, 50, 53–55, 78, 109
Royal Botanic Gardens at Kew, 153, 162, 245, 246
Royal Horticultural Society, 162, 270, 279
Ruck, Carl, 7
Ruddock, E. Harris, 66, 67
Rundell, Maria, 266–67

Ruskin, John: on belladonna, 107, 125; on bittersweet, 37–39, 57; *Modern Painters,* 37–38; on potatoes, 56, 191–92; *Proserpina,* 191; *The Queen of the Air,* 1, 38–39, 56, 191–92; *The Seven Lamps of Architecture,* 38; *The Stones of Venice,* 38; on tobacco, 16, 224
Russell, John, 169
Russell, Sarah Rachel, 69
Rymer, James Malcolm, 116

Sabine, Joseph, 150–52, 154, 159, 265
Salaman, Redcliffe, 11, 153, 161–64, 167, 174, 188
Sambourne, Linley, 237, *238*
Sand, George, 213
Sandys, Frederick, 72, *145,* 145–47
Scot, Reginald, 136, 138–39
Scott, Walter, 213
Scourse, Nicolette, 26–27
Scribe, Eugene, 67
Scrophulariaceae family (figworts), 3, 6, 51
Seaton, A. V., 223
Seaton, Beverley, 42, 45
sentimental botany, 10, 49, 112–17, 134, 141, 185
sentimental flower books, 26–32, 48
sex and sexuality: belladonna and, 8, 100, 101, 108–11, 116; botanical studies of, 9, 107; prostitutes, 11, 40, 110, 118, 127; sentimental botany and, 113; tobacco and, 16, 200, 219, 230, 238; whores, 10–12, 16, 40, 108, 110, 119, 147, 277
Shadwell, Thomas, 139
Shakespeare, William: *Antony and Cleopatra,* 51; belladonna and, 77, 82, 84, 289n13; *Hamlet,* 43; *Henry VI, Part 2,* 51; *Macbeth,* 51, 77, 82; mandrake and, 7, 29, 51; *Othello,* 51; *Romeo and Juliet,* 51, 91
Shelley, Percy Bysshe, 18, 39–40, 46
Shoberl, Frederic, 44–47, *45,* 111, 185–86
Shteir, Ann, 9–10, 48
Simmons, John, 23
Simpson, James Young, 132
Skeat, Walter, 11, 76–77, 83
sleepy nightshade. *See* belladonna
Smee, Alfred, 177

344 INDEX

Smith, Adam, 167–68
Smith, Andrew F., 149, 264, 265
snuff, 15, 193–95, 204, 212–14, 216–17, 231, 279
Socrates, 96
Solanaceae family (nightshades): conso-latory connotations of, 17, 142–43, 149, 180–81, 191, 277–78; etymology of family name, 13; Luridae as prior family name for, 10, 50–55, 60, 142, 180, 277; lurid leg-acy of, 7, 10–11, 13, 26, 48–60; prejudicial views of, 6, 56, 180; reputation of, 2–3, 9–11, 13, 16–17, 26, 57, 149, 194, 276–77. *See also* poisonous nightshades; *specific species*
solanine, 25, 64, 154, 209
Solanum crispum, 277
Solanum dulcamara. See bittersweet
Solanum jasminoides, 20, 277
Solanum lycopersicum. See tomatoes
Solanum maglia (Darwin potato), 150–53, 158–62, 160
Solanum melongena. See eggplants
Solanum nigrum. See black nightshade
Solanum tuberosum. See potatoes
Solly, Samuel, 201–4
Sowerby, James, 74, 74
Soyer, Alexis, 182–84, 267
Spenser, Edmund, 211
spirituality. *See* religion and spirituality
Sprague, Charles, 88
Spratt, George, 63–64, 259, 260
Squire, Peter, 75, 104–5, 278
Stearn, William, 180
Steel, Flora Annie, 280
Steinmetz, Andrew, 201–8, 207, 211, 221–23, 229, 238–39
Stendhal (Marie-Henri Beyle), 44
Stephenson, John, 27–28, 32
Stuart, David, 245
stump oratory, 176
sublime, in Gothic literary tradition, 32–41
Sumner, Judith, 25
Sussex, Duke of (Augustus Frederick), 193–95, 200, 213
Sutton, Arthur, 159, 161, 162, 185, 270–71, 279–80

sweet potatoes (*Ipomoea batatas*), 163–64
Swift, Jonathan, 212, 216
Swinburne, Algernon Charles, 146

Tanner, Heather, 74
Tarr, Clayton Carlyle, 120
Taylor, Alfred Swaine, 79
Taylor, John Ellor, 26
Tenniel, John, 69, 70, 174, 174–76, 234–35, 235
Tennyson, Alfred, Lord, 6, 213
Ternan, Ellen, 118
Thackeray, William Makepeace, 112–15, 117, 165, 213, 229, 230, 259
Theophrastus, 100
therapeutic window, 98
Thevet, André, 200
Thiselton-Dyer, T. F., 125, 134
Thomas, Owen, 270
Thomson, James, 225, 226
Thoreau, Henry David, 18, 19, 22, 26, 61
thornapple (*Datura stramonium*), 7, 53, 55, 98, 101–3
tic douloureux (neuralgia facialis), 128
Ticknor, Caroline, 106–7
Tilt, Edward, 129–31
Titian, 108
tobacco (*Nicotiana*), 193–243; accessories related to, 16–17, 193–95, 214–16, 215; addiction to, 92–93, 195, 197, 204, 219–21, 236, 243, 279; advertising for, 15, 196, 213, 213, 218, 223–26; anti-tobacco sentiment, 15, 16, 198, 201–4, 208, 212, 224; chewing, 15, 213, 228, 233; cigarettes, 15, 17, 218–21, 225–27, 227, 233–40, 242, 278; cigars, 15–17, 193–95, 199–201, 204–6, 209, 213–35, 228, 239; female employment at factories, 225–29, 227–29; femmes fatales and, 200; Great Exhibition displays, 16–17; Great Tobacco Controversy, 16, 201; greenhouse fumigation with smoke from, 278; health consequences of, 194, 200–202; henbane, similarities with, 14–15; in literature, 16, 196, 199–201, 211–12, 222–43; medicinal uses of, 14, 15, 196–97; nightshade family redefined by, 3, 16, 194; pipes, 17, 193–96, 199–201,

204, 206, 212–23, 215, 231–34, 239–43; seductive women and, 16, 200; sexuality and, 16, 200, 219, 230, 238; snuff, 15, 193–95, 204, 212–14, 216–17, 231, 279; social class and, 195–96, 212, 231, 232; types of, 14, 209–11, 210, 219; witchcraft and, 198; women's use of, 15–16, 196, 219, 234–39, 235–38, 243
tobaccomania, 194–96, 199, 202
tobacology, 208–22, 225, 229
tomatillos (*Physalis ixocarpa*), 262
tomatoes (*Solanum lycopersicum*): commercial availability of, 268–70, 309n92; culinary uses, 12, 56, 255, 263–68; cultivation instructions, 264, 266, 268, 273, 310n103; illustrations of, 271–72; nightshade family redefined by, 3, 17; origins of, 255–56, 259, 262; popularity of, 12–13, 259, 265, 268–73, 279–80; therapeutic qualities of, 267–68
Tournefort, Joseph de, 263–64
transmutation, 152
Tregear, G. S., 165
Trevelyan, Charles, 169–72, 174, 176–77
Trollope, Anthony, 191, 229, 232
tropane alkaloids, 7, 135, 137, 278
truth, in language of flowers, 41, 44–48, 112
Turner, J. M. W., 37, 38
Turner, William, 24–25
Twamley, Louisa Anne, 31–32, 48
Tweedie, James, 247, 250
Tyas, Robert, 47–48, 111, 186

vegetable neurotics, 96–103, 129
Velpeau, Alfred, 131
verbenas, 245–46, 252–55
Verlaine, Paul, 220
Vickery, Roy, 62
Victoria (queen of England), 68, 127, 131, 148–49, 183–85
Victorian England: addiction in, 92–93, 195, 197, 243, 279; Age of the Garden and, 4; citizenship rights in, 90; domestic economy in, 182; gender propriety in, 109; language of flowers in, 10, 42–43; popular culture in, 10, 15, 60, 69, 112, 125, 147,

200, 208; tobacco use in, 15–17, 194–96, 217, 243, 279; trade in flowers and vegetables, 4, 283n7; witches and witchcraft in, 133–34, 139–40
Victoria regia (water lily), 4, 273

Wakefield, Priscilla, 53–54, 78
Wallace, Alfred Russel, 91
Wallace, Eileen, 105
Wallace, John, 224, 227, 227–29, 241–42, 242–43
Walpole, Horace, 36
Walton, Isaak, 212
Wardian cases, 4, 274
Waring, George, 279
Warren, Eliza, 190
water lily (*Victoria regia*), 4, 273
Watson, Hewett Cottrell, 19, 89–91
Wellington, Duke of (Arthur Wellesley), 214, 215
West, Thomas, 36, 37
Weyer, Johann, 136
Whelan, J. H., 93, 98
whores, 10–12, 16, 40, 108, 110, 119, 147, 277
Wighton, J., 83–84
Wilde, Oscar, 220, 221, 278
Wildsmith, William, 255
Wilkinson, Anne, 250
Willes, Margaret, 250
William III (king of England), 199, 212
Williams, Leslie, 171
Wilmot, John, 265, 266
Wilson, Erasmus, 205
Winn, James, 131–32
Wiseman, Nicolas, 176
witches and witchcraft: aconite and, 146, 147; in art and theater, 145, 145–47, 296n135; belladonna and, 8, 10–11, 88, 103, 110, 112, 115–17, 132–44, 147; bittersweet and, 61, 88, 143; black nightshade and, 135; books of secrets on, 138; flying ointments and, 136–40; hemlock and, 135, 136, 140; henbane and, 88, 103, 134–40, 143, 146, 147; in literature, 139–45; mandrake and, 134–37, 139, 140; opium and, 136, 137; persecution of, 141, 142, 145; potatoes and, 191; tobacco and, 198

346 INDEX

women: bloomers worn by, 234–35, 235–37; botany studies by, 10, 48–50, 53–60, 109; connection with flowers, 49, 50, 57, 107; cosmetics and, 10, 69, 72, 92, 108–11, 125–28; employment at tobacco factories, 225–29, 227–29; femmes fatales, 11, 39–40, 106–25, 130, 200; housewives, 12, 118, 266, 277; midwives, 130–32, 140, 143; New Woman image, 15, 145; obstetrics and gynecological medicine for, 129–32, 143; prostitutes, 11, 40, 110, 118, 127; seductive, 10, 16, 46, 108–9, 114–17, 125, 200, 243; sentimental flower books for, 26, 48; tobacco use among, 15–16, 196, 219, 234–39, 235–38, 243; whores, 10–12, 16, 40, 108, 110, 119, 147, 277. *See also* sex and sexuality; witches and witchcraft

Woodham-Smith, Cecil, 164, 167, 168, 170–71

Woodville, William, 63, 64

woody nightshade. *See* bittersweet

Wordsworth, William, 35–36

Wright, David, 24

Wright, Thomas, 133–34

yew, 26, 39, 45

Yonge, Charlotte, 69, 71, 84–85

Zuckerman, Larry, 149, 167

RECENT BOOKS IN THE
Victorian Literature and Culture Series

Haunting Ecologies: Victorian Conceptions of Water
Ursula Kluwick

The Turn of Rhythm: How Victorian Poetry Shaped a New Concept
Ewan Jones

Narrative and Its Nonevents: The Unwritten Plots That Shaped Victorian Realism
Carra Glatt

Victorian Metafiction
Tabitha Sparks

Strangers in the Archive: Literary Evidence and London's East End
Heidi Kaufman

*Evangelical Gothic: The English Novel and the Religious
War on Virtue from Wesley to "Dracula"*
Christopher Herbert

Reading with the Senses in Victorian Literature and Science
David Sweeney Coombs

Parting Words: Victorian Poetry and Public Address
Justin A. Sider

The Physics of Possibility: Victorian Fiction, Science, and Gender
Michael Tondre

Willful Submission: Sado-Erotics and Heavenly Marriage in Victorian Poetry
Amanda Paxton

Pirating Fictions: Ownership and Creativity in Nineteenth-Century Popular Culture
Monica F. Cohen

Mathilde Blind: Late-Victorian Culture and the Woman of Letters
James Diedrick

Poetry and the Thought of Song in Nineteenth-Century Britain
Elizabeth K. Helsinger

The Antagonist Principle: John Henry Newman and the Paradox of Personality
Lawrence Poston

Personal Business: Character and Commerce in Victorian Literature and Culture
Aeron Hunt

Second Person Singular: Late Victorian Women Poets and the Bonds of Verse
Emily Harrington

The Ghost behind the Masks: The Victorian Poets and Shakespeare
W. David Shaw

Victorian Poets and the Changing Bible
Charles LaPorte

Liberal Epic: The Victorian Practice of History from Gibbon to Churchill
Edward Adams

Supposing "Bleak House"
John O. Jordan

Feeling for the Poor: Bourgeois Compassion, Social Action, and the Victorian Novel
Carolyn Betensky

The Science of Religion in Britain, 1860–1915
Marjorie Wheeler-Barclay

Reading for the Law: British Literary History and Gender Advocacy
Christine L. Krueger

The Dynamics of Genre: Journalism and the Practice
of Literature in Mid-Victorian Britain
Dallas Liddle

The Fowl and the Pussycat: Love Letters of Michael Field, 1876–1909
Edited by Sharon Bickle

Victorian Prism: Refractions of the Crystal Palace
Edited by James Buzard, Joseph W. Childers, and Eileen Gillooly

Nostalgia in Transition, 1780–1917
Linda M. Austin

The English Cult of Literature: Devoted Readers, 1774–1880
William R. McKelvy

Artist of Wonderland: The Life, Political Cartoons, and Illustrations of Tenniel
Frankie Morris

The Material Interests of the Victorian Novel
Daniel Hack

Behind Her Times: Transition England in the Novels of Mary Arnold Ward
Judith Wilt

The Circus and Victorian Society
Brenda Assael

www.ingramcontent.com/pod-product-compliance
Lightning Source LLC
Chambersburg PA
CBHW060309040325
22904CB00003B/113